Asymptotic Issues for
Some Partial Differential Equations

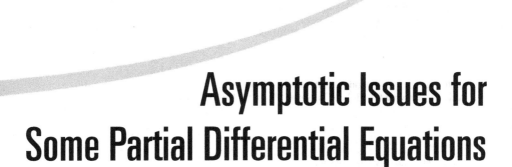

Asymptotic Issues for
Some Partial Differential Equations

Michel Marie Chipot
University of Zurich, Switzerland

Imperial College Press

Published by

Imperial College Press
57 Shelton Street
Covent Garden
London WC2H 9HE

Distributed by

World Scientific Publishing Co. Pte. Ltd.
5 Toh Tuck Link, Singapore 596224
USA office: 27 Warren Street, Suite 401-402, Hackensack, NJ 07601
UK office: 57 Shelton Street, Covent Garden, London WC2H 9HE

Library of Congress Cataloging-in-Publication Data
Names: Chipot, M. (Michel)
Title: Asymptotic issues for some partial differential equations / by Michel Marie Chipot
    (University of Zurich, Switzerland).
Description: New Jersey : Imperial College Press, 2016. |
    Includes bibliographical references and index.
Identifiers: LCCN 2016009613 | ISBN 9781783268917 (hardcover : alk. paper)
Subjects: LCSH: Differential equations, Partial--Asymptotic theory. |
    Differential equations--Asymptotic theory.
Classification: LCC QA372 .C5235 2016 | DDC 515/.353--dc23
LC record available at http://lccn.loc.gov/2016009613

British Library Cataloguing-in-Publication Data
A catalogue record for this book is available from the British Library.

Printed in Singapore

Foreword

Since the publication of the book "ℓ goes to plus infinity" a lot of new developments in this field took place. This is the goal of this new book to collect them in an affordable way for the potential students or researchers who find some interest in them.

As always in mathematics sometimes people are using results or recipes without having any proof of them but the feeling that they are correct when they are delivering reasonable outcomes. At the origin of the work developed here was some idea of numerical analysts. At the time when the computers were not so powerful some of them – I had the good luck to meet – and who were doing numerical computations on cylindrical domains came to the conclusion that, when a cylinder was long enough and the data of their problem depending only on the section of this cylinder, then the solution of this problem had a good chance to depend only on the section on a large part of the cylinder. Thus they were performing say two dimensional computations instead of three dimensional ones which has the advantage to save computer power and programming skills as well. For some reason I decided to check if these alchemists were right and during two years at least to my surprise nothing came out of this.

Then, some first results emerged and the problems started to spread in all directions. Let me stop here to explain the pattern of this unexpected developments.

As I said data depending only on the section of the cylinder force the solution of a problem to depend only on the section of the cylinder. Say if the longitudinal direction of the cylinder is the x_1 axis, data independent of x_1 force the solution of the problem at hand to be independent of x_1 at least when the cylinder length ℓ goes to plus infinity. Having proved this for different problems one comes to the idea that since to be independent of

x_1 means to be periodic of any period, it is very possible that periodic data at the limit force the solution to be periodic. This makes you enter a new world of new results. Then passing to the limit from bounded domains to unbounded ones makes you think that this could be a good opportunity to see if one could extend that way known existence results from bounded to unbounded domains. This is the case. In all the way you learn and profit from your mistakes. For instance one of our first ideas in dealing with a growing cylinder was a rescaling of it leading us to work on a fixed domain. This leads to nothing except that few years later this allowed us to deal with the problem of anisotropic singular perturbations.

Amazingly, a simple question lead to unexpected developments. We hope that the reader will enjoy the snow ball effect of this field and will get the energy to tackle new problems not necessarily related to "ℓ goes to plus infinity" but to his own believes with the certainty to discover a gold mine.

This book profited widely from the series of lectures I gave during a sabbatical semester at Imperial College and Oxford University. I am grateful to Professor José Antonio Carrillo and Professor John Ball for their kind hospitality at that time. I would like to thank also all my collaborators during the years we worked on these issues: Bernard Brighi, Arnaud Rougirel, Yitian Xie, Senoussi Guesmia, Sorin Mardare, Prosenjit Roy, Itai Shafrir. In addition I would like to thank Tomoyuki Nakatsuka, Karen Yeressian and Stephanie Zube for their careful reading of the manuscript. I am very grateful to Mrs Schacher for having helped me with the figures. Finally I would like to thank Ms Tan Rok Ting and Imperial college press for their work in publishing this book.

Contents

Chapter 1

Introduction

The goal of this chapter is to introduce in a simple setting some of the problems that we will investigate later on and to enlight the main ideas without technicalities.

1.1   The Dirichlet problem

It is not for every domain extending to infinity that the solution of the Dirichlet problem set in this domain possesses a limit. Consider for instance an interval $\Omega = (-1, 1)$ and set $\Omega_\ell = \ell\Omega = (-\ell, \ell)$. For $f \in L^1(\mathbb{R})$ let u_ℓ be the weak solution to the Dirichlet problem

$$- u_\ell'' = f \quad \text{in } \Omega_\ell, \quad u_\ell = 0 \quad \text{on } \partial\Omega_\ell = \{-\ell, \ell\}. \tag{1.1}$$

One has:

Theorem 1.1. *If*

$$\int_\mathbb{R} f \, dx \neq 0 \tag{1.2}$$

then u_ℓ solution to (1.1) cannot be bounded when $\ell \to \infty$.

Proof. Set $v_\ell(x) = u_\ell(\ell x)$ for $x \in \Omega$. One has

$$- v_\ell''(x) = -\ell^2 u_\ell''(\ell x) = \ell^2 f(\ell x) \quad \text{in } \Omega. \tag{1.3}$$

If u_ℓ is bounded so is v_ℓ and from (1.3) one deduces that

$$\ell f(\ell x) \to 0 \text{ in } \mathcal{D}'(\Omega).$$

(We suppose that the reader is familiar with the theory of distributions, see [71], [6], [49], [50], [67], [15]). Let $\varphi \in \mathcal{D}(\Omega)$ with $\varphi(0) \neq 0$. One has

1

then

$$\int_{\Omega} \ell f(\ell x) \varphi(x) \, dx = \int_{\Omega_{\ell}} f(y) \varphi(\frac{y}{\ell}) \, dy$$

$$= \int_{\mathbb{R}} \chi_{\Omega_{\ell}}(y) f(y) \varphi(\frac{y}{\ell}) \, dy \to \varphi(0) \int_{\mathbb{R}} f \, dx \neq 0$$

by the Lebesgue theorem ($\chi_{\Omega_{\ell}}$ denotes the characteristic function of Ω_{ℓ}). This is a contradiction if (1.2) holds and completes the proof. $\qquad\square$

We will consider later domains of the type $\Omega_{\ell} = \ell \omega_1 \times \omega_2$ where ω_1, ω_2 are bounded open sets of $\mathbb{R}^p$, $\mathbb{R}^{n-p}$ respectively so that Ω_{ℓ} is a "cylinder" in $\mathbb{R}^n$. The simplest such a domain is, in $\mathbb{R}^2$, the rectangle

$$\Omega_{\ell} = (-\ell, \ell) \times (-1, 1) \tag{1.4}$$

where $\omega_1 = \omega_2 = \omega = (-1, 1)$.

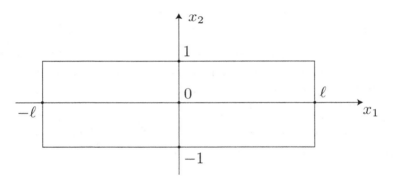

Fig. 1.1   The simplest cylinder.

Suppose that we are given a function $f = f(x_2)$ depending on x_2 only such that

$$f \in L^2(\omega).$$

Let u_{ℓ} be the weak solution of the Dirichlet problem

$$-\Delta u_{\ell} = f \text{ in } \Omega_{\ell}, \quad u_{\ell} = 0 \text{ on } \partial\Omega_{\ell}$$

i.e. u_{ℓ} satisfies

$$\begin{cases} u_{\ell} \in H_0^1(\Omega_{\ell}), \\ \int_{\Omega_{\ell}} \nabla u_{\ell} \cdot \nabla v \, dx = \int_{\Omega_{\ell}} f v \, dx \quad \forall v \in H_0^1(\Omega_{\ell}) \end{cases} \tag{1.5}$$

(we have denoted by $\partial\Omega_\ell$ the boundary of Ω_ℓ, we refer to [49], [50], [52], [16] for the definition of Sobolev spaces and the introduction to elliptic problems, see also the Appendix). Denote now by u_∞ the weak solution of the Dirichlet problem on the "section" of the rectangle Ω_ℓ – i.e. u_∞ satisfies

$$\begin{cases} u_\infty \in H_0^1(\omega), \\ \int_\omega \partial_{x_2} u_\infty \partial_{x_2} v \, dx_2 = \int_\omega f v \, dx_2 \quad \forall v \in H_0^1(\omega). \end{cases} \tag{1.6}$$

It is clear that if $f \neq 0$ then u_ℓ is depending on x_1. Indeed otherwise, since $u_\ell = 0$ for $x_1 = \pm\ell$, one would have $u_\ell \equiv 0$ and thus $f \equiv 0$. However when $\ell \to +\infty$, u_ℓ goes exponentially quickly toward u_∞ which is independent of x_1 - i.e. f independent of x_1 forces u_ℓ to be independent of x_1 at the limit. More precisely we have:

Theorem 1.2. *There exist positive constants C, α such that*

$$\left|\left|\nabla(u_\ell - u_\infty)\right|\right|_{2,\Omega_{\frac{\ell}{2}}} \leq Ce^{-\alpha\ell}. \tag{1.7}$$

$(|\quad|$ *denotes the usual Euclidean norm,* $|\quad|_{p,A}$ *the usual $L^p(A)$-norm).*

Proof. We give here a proof due to N. Bruyère (Cf. [11]). First we notice that if $v \in H_0^1(\Omega_\ell)$, $v(x_1, \cdot) \in H_0^1(\omega)$ for almost all $x_1 \in (-\ell, \ell)$ (see [13]). Thus from (1.6) we derive that

$$\int_\omega \partial_{x_2} u_\infty \partial_{x_2} v(x_1, \cdot) \, dx_2 = \int_\omega f v(x_1, \cdot) \, dx_2 \quad \forall v \in H_0^1(\Omega_\ell), \text{ a.e. } x_1 \in (-\ell, \ell).$$

Due to the fact that u_∞ is independent of x_1 this implies after an integration in x_1

$$\int_{\Omega_\ell} \nabla u_\infty \cdot \nabla v \, dx = \int_{\Omega_\ell} f v \, dx \quad \forall v \in H_0^1(\Omega_\ell).$$

By subtraction from (1.5) we get

$$\int_{\Omega_\ell} \nabla(u_\ell - u_\infty) \cdot \nabla v \, dx = 0 \quad \forall v \in H_0^1(\Omega_\ell). \tag{1.8}$$

(We have denoted by $\cdot$ the usual scalar product. Note that (1.8) expresses only the fact that $u_\ell - u_\infty$ is harmonic in Ω_ℓ.) We consider then the function ρ defined as

$$\rho = \rho(x_1) = e^{-\sigma|x_1|} - e^{-\sigma\ell}$$

where σ is a positive constant that we will choose later on. It is clear that

$$\rho(u_\ell - u_\infty) \in H_0^1(\Omega_\ell)$$

and thus from (1.8) we derive

$$\int_{\Omega_\ell} \nabla(u_\ell - u_\infty) \cdot \nabla\{\rho(u_\ell - u_\infty)\} \, dx = 0,$$

which is equivalent to

$$\int_{\Omega_\ell} \rho|\nabla(u_\ell - u_\infty)|^2 \, dx = -\int_{\Omega_\ell} \partial_{x_1}(u_\ell - u_\infty)\partial_{x_1}\rho \, (u_\ell - u_\infty) \, dx.$$

We have

$$\partial_{x_1}\rho = -\sigma e^{-\sigma|x_1|} \operatorname{sign}(x_1)$$

where sign denotes the sign-function and thus

$$\int_{\Omega_\ell} \rho|\nabla(u_\ell - u_\infty)|^2 \, dx \le \sigma \int_{\Omega_\ell} e^{-\sigma|x_1|}|\partial_{x_1}(u_\ell - u_\infty)||u_\ell - u_\infty| \, dx.$$

Using the Cauchy-Young inequality

$$ab \le \frac{1}{2}(a^2 + b^2)$$

we deduce

$$\int_{\Omega_\ell} \rho|\nabla(u_\ell - u_\infty)|^2 \, dx \le$$
$$\frac{\sigma}{2}\int_{\Omega_\ell} e^{-\sigma|x_1|}\{(\partial_{x_1}(u_\ell - u_\infty))^2 + (u_\ell - u_\infty)^2\} \, dx. \tag{1.9}$$

Applying the Poincaré inequality on the section ω (see [13] and the Appendix) one has

$$\int_\omega (u_\ell - u_\infty)^2(x_1, \cdot) \, dx_2 \le 2\int_\omega (\partial_{x_2}(u_\ell - u_\infty))^2(x_1, \cdot) \, dx_2.$$

Note that we did not choose here the best constant in this inequality. Thus

$$\int_{\Omega_\ell} e^{-\sigma|x_1|}(u_\ell - u_\infty)^2 \, dx = \int_{-\ell}^{\ell} e^{-\sigma|x_1|}\int_\omega (u_\ell - u_\infty)^2(x_1, x_2) \, dx_2 dx_1$$
$$\le 2\int_{\Omega_\ell} e^{-\sigma|x_1|}(\partial_{x_2}(u_\ell - u_\infty))^2 \, dx.$$

From (1.9) we then deduce

$$\int_{\Omega_\ell} (e^{-\sigma|x_1|} - e^{-\sigma\ell})|\nabla(u_\ell - u_\infty)|^2 \, dx \le \sigma \int_{\Omega_\ell} e^{-\sigma|x_1|}|\nabla(u_\ell - u_\infty)|^2 \, dx.$$

Taking $\sigma = \frac{1}{2}$ we obtain

$$\int_{\Omega_\ell} e^{-\frac{|x_1|}{2}}|\nabla(u_\ell - u_\infty)|^2 \, dx \le 2e^{-\frac{\ell}{2}}\int_{\Omega_\ell} |\nabla(u_\ell - u_\infty)|^2 \, dx$$

and thus

$$e^{-\frac{\ell}{4}} \int_{\Omega_{\frac{\ell}{2}}} |\nabla(u_\ell - u_\infty)|^2 \, dx \leq 2e^{-\frac{\ell}{2}} \int_{\Omega_\ell} |\nabla(u_\ell - u_\infty)|^2 \, dx$$

$$\Longleftrightarrow \int_{\Omega_{\frac{\ell}{2}}} |\nabla(u_\ell - u_\infty)|^2 \, dx \leq 2e^{-\frac{\ell}{4}} \int_{\Omega_\ell} |\nabla(u_\ell - u_\infty)|^2 \, dx. \qquad (1.10)$$

It remains to estimate this last integral. From (1.5), taking $v = u_\ell$ we derive

$$\int_{\Omega_\ell} |\nabla u_\ell|^2 \, dx = \int_{\Omega_\ell} f u_\ell \, dx \leq |f|_{2,\Omega_\ell} |u_\ell|_{2,\Omega_\ell}$$

$$\leq \sqrt{2} |f|_{2,\Omega_\ell} ||\nabla u_\ell||_{2,\Omega_\ell}$$

by the Poincaré inequality. Thus

$$\int_{\Omega_\ell} |\nabla u_\ell|^2 \, dx \leq 2|f|_{2,\Omega_\ell}^2 = 2 \int_{-\ell}^{\ell} \int_{-1}^{1} f^2(x_2) \, dx_2 dx_1 \qquad (1.11)$$

$$= 4\ell |f|_{2,\omega}^2.$$

Similarly taking $v = u_\infty$ in (1.6) we derive

$$\int_\omega (\partial_{x_2} u_\infty)^2 \, dx_2 = \int_\omega f u_\infty \, dx_2 \leq |f|_{2,\omega} |u_\infty|_{2,\omega}$$

$$\leq \sqrt{2} |f|_{2,\omega} |\partial_{x_2} u_\infty|_{2,\omega}.$$

Thus we have

$$\int_\omega (\partial_{x_2} u_\infty)^2 \, dx_2 \leq 2|f|_{2,\omega}^2$$

and also

$$\int_{\Omega_\ell} |\nabla u_\infty|^2 \, dx = \int_{-\ell}^{\ell} \int_{-1}^{1} (\partial_{x_2} u_\infty)^2 \, dx_2 dx_1 \leq 4\ell |f|_{2,\omega}^2$$

i.e. the same estimate as (1.11). Going back to (1.10) we have

$$|\nabla(u_\ell - u_\infty)|_{2,\Omega_{\frac{\ell}{2}}}^2 \leq 4e^{-\frac{\ell}{4}} \{ \int_{\Omega_\ell} |\nabla u_\ell|^2 \, dx + \int_{\Omega_\ell} |\nabla u_\infty|^2 \, dx \}$$

$$\leq 32\ell e^{-\frac{\ell}{4}} |f|_{2,\omega}^2 \leq Ce^{-\alpha\ell}$$

for any $\alpha < \frac{1}{4}$. This completes the proof of the theorem. $\qquad\square$

1.2   Periodic problems

We have seen in the previous section that a function independent of x_1 forces
- at the limit - u_ℓ to be independent of x_1. Another way to characterise
a function independent of x_1 is to say that it is periodic for any period.
Then the natural question coming in mind is to see if a periodic function
in x_1 forces u_ℓ the solution to (1.5) to be periodic at the limit. Let us then
consider a function f periodic in x_1 and of period P i.e. such that

$$f(x_1 + P, x_2) = f(x_1, x_2) \quad \text{a.e. } x \in S = \Omega_\infty = \mathbb{R} \times (-1,1). \qquad (1.12)$$

If C denotes the cell

$$C = (0, P) \times (-1, 1)$$

we will also assume that

$$f \in L^2(C), \qquad (1.13)$$

f being extended by periodicity to the whole domain S. By the Lax-
Milgram theorem there exists a unique u_ℓ solution to (1.5) i.e. such that

$$\begin{cases} u_\ell \in H_0^1(\Omega_\ell), \\ \int_{\Omega_\ell} \nabla u_\ell \cdot \nabla v \, dx = \int_{\Omega_\ell} fv \, dx \quad \forall v \in H_0^1(\Omega_\ell). \end{cases}$$

We denote by $H_{per}^1(C)$ the closure in $H^1(C)$ of

$$C_{per}^1(\overline{C}) = \{v \in C^1(\overline{C}) \mid v(0, x_2) = v(P, x_2)\}$$

and by $H_{0,per}^1(C)$ the closure in $H^1(C)$ of

$$C_{0,per}^1(\overline{C}) = \{v \in C_{per}^1(\overline{C}) \mid v(\cdot, \pm 1) = 0\}.$$

It is easy to show that $H_{0,per}^1(C)$ is a Hilbert space when equipped with
the Dirichlet norm and thus there exists a unique u_∞ solution to

$$\begin{cases} u_\infty \in H_{0,per}^1(C), \\ \int_C \nabla u_\infty \cdot \nabla v \, dx = \int_C fv \, dx \quad \forall v \in H_{0,per}^1(C). \end{cases} \qquad (1.14)$$

One will denote also by u_∞ the extension to S by periodicity of u_∞. Then
we have

Lemma 1.1. *The function u_∞ satisfies*

$$\int_S \nabla u_\infty \cdot \nabla \varphi \, dx = \int_S f\varphi \, dx \quad \forall \varphi \in \mathcal{D}(S). \qquad (1.15)$$

Proof. Let $\varphi \in \mathcal{D}(S)$. We denote by C_z, $z \in \mathbb{Z}$ the cell defined as

$$C_z = C + zPe_1$$

where $e_1 = (1,0)$ i.e. C_z is C shifted by zP in the x_1 direction. Clearly one has

$$\int_S \nabla u_\infty \cdot \nabla \varphi \, dx = \sum_{z \in \mathbb{Z}} \int_{C_z} \nabla u_\infty \cdot \nabla \varphi \, dx \qquad (1.16)$$

and since φ has compact support this sum has only a finite number of terms which are not equal to 0 (Cf. Figure 1.2 where Ω is a domain containing the support of φ).

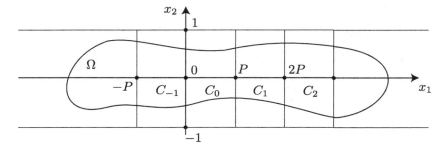

Fig. 1.2   The various C_z.

One has

$$\int_{C_z} \nabla u_\infty \cdot \nabla \varphi \, dx = \int_C \nabla u_\infty(x_1 + zP, x_2) \cdot \nabla \varphi(x_1 + zP, x_2) \, dx$$

$$= \int_C \nabla u_\infty(x_1, x_2) \cdot \nabla \varphi(x_1 + zP, x_2) \, dx$$

since, due to the periodicity of u_∞, ∇u_∞ is also periodic. Note that u_∞ has no jump on $x_1 = kP$, $k \in \mathbb{Z}$. From (1.16) we derive

$$\int_S \nabla u_\infty \cdot \nabla \varphi \, dx = \sum_{z \in \mathbb{Z}} \int_C \nabla u_\infty \cdot \nabla \varphi(x_1 + zP, x_2) \, dx$$

$$= \int_C \nabla u_\infty \cdot \{\sum_{z \in \mathbb{Z}} \nabla \varphi(x_1 + zP, x_2)\} \, dx$$

$$= \int_C \nabla u_\infty \cdot \nabla \{\sum_{z \in \mathbb{Z}} \varphi(x_1 + zP, x_2)\} \, dx$$

where the sum has only a finite number of non vanishing terms. Now clearly the function $\sum_{z \in \mathbb{Z}} \varphi(x_1 + zP, x_2)$ is periodic of period P in x_1. Thus from (1.14) we derive

$$\int_S \nabla u_\infty \cdot \nabla \varphi \, dx = \int_C f(x_1, x_2) \{ \sum_{z \in \mathbb{Z}} \varphi(x_1 + zP, x_2) \} \, dx$$

$$= \sum_{z \in \mathbb{Z}} \int_C f(x_1, x_2) \varphi(x_1 + zP, x_2) \, dx$$

$$= \sum_{z \in \mathbb{Z}} \int_{C_z} f(x_1 - zP, x_2) \varphi(x_1, x_2) \, dx$$

$$= \int_S f\varphi \, dx$$

since f is periodic of period P. This completes the proof of the lemma. $\square$

One can then prove:

Theorem 1.3. *Let f satisfy* (1.12), (1.13) *and let u_ℓ be the solution to* (1.5). *There exist positive constants C, α such that*

$$\big|\big|\nabla(u_\ell - u_\infty)\big|\big|_{2,\Omega_{\frac{\ell}{2}}} \le Ce^{-\alpha \ell}. \tag{1.17}$$

Proof. From (1.5), (1.15) we deduce

$$\int_{\Omega_\ell} \nabla(u_\ell - u_\infty) \cdot \nabla \varphi \, dx = 0 \quad \forall \varphi \in \mathcal{D}(\Omega_\ell).$$

Thus, by density of $\mathcal{D}(\Omega_\ell)$ in $H_0^1(\Omega_\ell)$, we get

$$\int_{\Omega_\ell} \nabla(u_\ell - u_\infty) \cdot \nabla v \, dx = 0 \quad \forall v \in H_0^1(\Omega_\ell) \tag{1.18}$$

which is exactly (1.8). Arguing as in the previous section one arrives to (1.10) -i.e. to

$$\int_{\Omega_{\frac{\ell}{2}}} |\nabla(u_\ell - u_\infty)|^2 \, dx \le 2e^{-\frac{\ell}{4}} \int_{\Omega_\ell} |\nabla(u_\ell - u_\infty)|^2 \, dx$$

$$\le 4e^{-\frac{\ell}{4}} \int_{\Omega_\ell} |\nabla u_\ell|^2 + |\nabla u_\infty|^2 \, dx.$$

We have now just to estimate u_ℓ and u_∞. Taking $v = u_\ell$ in (1.18) we get easily

$$\big|\big|\nabla u_\ell\big|\big|_{2,\Omega_\ell}^2 = \int_{\Omega_\ell} |\nabla u_\ell|^2 \, dx = \int_{\Omega_\ell} \nabla u_\ell \cdot \nabla u_\infty \, dx \le \big|\big|\nabla u_\ell\big|\big|_{2,\Omega_\ell} \big|\big|\nabla u_\infty\big|\big|_{2,\Omega_\ell}$$

i.e.

$$\int_{\Omega_\ell} |\nabla u_\ell|^2 \, dx \le \int_{\Omega_\ell} |\nabla u_\infty|^2 \, dx.$$

Due to the periodicity of $|\nabla u_\infty|$ and since Ω_ℓ can be covered by $2([\frac{\ell}{P}]+1)$ translated cells of C (we denote by [ ] the integer part of a number) one gets

$$\int_{\Omega_\ell} |\nabla u_\ell|^2 \, dx \le \int_{\Omega_\ell} |\nabla u_\infty|^2 \, dx \le 2(\frac{\ell}{P}+1) \int_C |\nabla u_\infty|^2 \, dx.$$

(1.17) follows then for any $\alpha < \frac{1}{4}$. $\qquad\qquad\qquad\qquad\qquad\square$

1.3 Problems in unbounded domains

When Ω is a bounded domain in $\mathbb{R}^n$ the Lax-Milgram theorem or even the Riesz representation theorem provides existence and uniqueness of a weak solution to the Dirichlet problem for very general right hand side f since the only condition required to get a solution in $H_0^1(\Omega)$ is that

$$f \in H^{-1}(\Omega)$$

where $H^{-1}(\Omega)$ denotes the dual space of $H_0^1(\Omega)$. Suppose now that we consider the strip

$$S = \Omega_\infty = \mathbb{R} \times (-1,1) = \mathbb{R} \times \omega.$$

Of course for $f \in H^{-1}(S)$ one still has existence and uniqueness of a solution to the Dirichlet problem in $H_0^1(S)$. There are some drawbacks:
- some very simple functions -like the constants- are not in $H^{-1}(S)$,
- one can lose uniqueness.

Indeed regarding this second point, for $f = 0$

$$v_k = e^{k\pi x_1} \sin k\pi x_2$$

is solution to

$$-\Delta v_k = 0 \text{ in } S, \quad v_k = 0 \text{ in } \partial S.$$

Of course the only solution in $H_0^1(S)$ is obtained for $k = 0$.

To see the first point above let us show

Lemma 1.2. *Let $f = f(x_2) \neq 0$, $f \in L^2(\omega)$ with $\omega = (-1,1)$. Then $f \notin H^{-1}(S)$.*

Proof. One can find an alternate proof of this result in [14] or in the next chapter. Consider $\rho_\epsilon = \rho_\epsilon(x_1)$ the function defined by the graph of Figure 1.3.

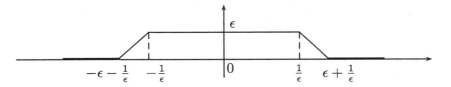

Fig. 1.3   The function ρ_ϵ.

Since $f \neq 0$ one can find $\psi \in \mathcal{D}(\omega)$ such that

$$\int_\omega f\psi \, dx_2 \neq 0.$$

We claim that

$$h_\epsilon = \rho_\epsilon(x_1)\psi(x_2) \to 0 \text{ in } H_0^1(S)$$

when $\epsilon \to 0$. Indeed one has

$$\partial_{x_1} h_\epsilon = \rho'_\epsilon(x_1)\psi(x_2), \quad \partial_{x_2} h_\epsilon = \rho_\epsilon(x_1)\psi'(x_2)$$

and thus

$$\int_S h_\epsilon^2 + |\nabla h_\epsilon|^2 \, dx = \int_S \rho_\epsilon^2(\psi^2 + (\psi')^2) + (\rho'_\epsilon\psi)^2 \, dx$$

$$\leq \int_{-\frac{1}{\epsilon}-\epsilon}^{\frac{1}{\epsilon}+\epsilon} \epsilon^2 \, dx_1 \int_\omega \psi^2 + (\psi')^2 \, dx_2$$

$$+ 2\int_{\frac{1}{\epsilon}}^{\frac{1}{\epsilon}+\epsilon} dx_1 \int_\omega \psi^2 \, dx_2$$

$$= 2\epsilon^2(\frac{1}{\epsilon}+\epsilon)\int_\omega \psi^2 + (\psi')^2 \, dx_2 + 2\epsilon \int_\omega \psi^2 \, dx_2 \to 0$$

when $\epsilon \to 0$. On the other hand

$$\int_S f h_\epsilon \, dx = \int_{-\frac{1}{\epsilon}-\epsilon}^{\frac{1}{\epsilon}+\epsilon} \rho_\epsilon dx_1 \int_\omega f\psi dx_2 = \epsilon(\frac{2}{\epsilon}+\epsilon)\int_\omega f\psi dx_2$$

$$\to 2\int_\omega f\psi dx_2 \neq 0$$

when $\epsilon \to 0$. This shows that the linear form defined by f is not continuous on $H_0^1(S)$ and completes the proof of the lemma. $\square$

So when $f \notin H^{-1}(S)$ an idea to construct a solution to the Dirichlet problem in the strip S is to use u_ℓ solution to (1.5). Indeed, writing (1.5) for Ω_ℓ and $\Omega_{\ell+r}$, with first $0 \leq r \leq 1$, one gets easily

$$\int_{\Omega_\ell} \nabla(u_\ell - u_{\ell+r}) \cdot \nabla v \, dx = 0 \quad \forall v \in H_0^1(\Omega_\ell)$$

i.e. something similar to (1.8). Arguing like in the first section we arrive to (1.10) where u_∞ is just replaced by $u_{\ell+r}$ that is to say

$$\int_{\Omega_{\frac{\ell}{2}}} |\nabla(u_\ell - u_{\ell+r})|^2 \, dx \leq 2e^{-\frac{\ell}{4}} \int_{\Omega_\ell} |\nabla(u_\ell - u_{\ell+r})|^2 \, dx$$

$$\leq 4e^{-\frac{\ell}{4}} \int_{\Omega_\ell} |\nabla u_\ell|^2 + |\nabla u_{\ell+r}|^2 \, dx.$$

As before if we can get estimates for the last integral and remove the condition $0 \leq r \leq 1$, we get that u_ℓ is a Cauchy sequence (for instance in any $H^1(\Omega_{\ell_0})$ such that $\Omega_{\ell_0} \subset \Omega_\ell$). Passing to the limit gives us then the solution that we are looking for. This is what we will implement later on.

1.4 Rescaling

One idea to attack the problem of finding the limit of u_ℓ solution to (1.5) would be to rescale it - i.e. to introduce v_ℓ defined as

$$v_\ell(x_1, x_2) = u_\ell(\ell x_1, x_2) \quad \forall x \in \Omega_1.$$

If one computes the different derivatives of v_ℓ one gets

$$\partial_{x_1} v_\ell(x_1, x_2) = \ell \partial_{x_1} u_\ell(\ell x_1, x_2), \quad \partial_{x_1}^2 v_\ell(x_1, x_2) = \ell^2 \partial_{x_1}^2 u_\ell(\ell x_1, x_2),$$

$$\partial_{x_2} v_\ell(x_1, x_2) = \partial_{x_2} u_\ell(\ell x_1, x_2), \quad \partial_{x_2}^2 v_\ell(x_1, x_2) = \partial_{x_2}^2 u_\ell(\ell x_1, x_2).$$

Thus if u_ℓ is the weak solution to

$$-\Delta u_\ell = f \text{ in } \Omega_\ell, \quad u_\ell = 0 \text{ on } \partial\Omega_\ell$$

one sees that v_ℓ is the weak solution to

$$-\frac{1}{\ell^2}\partial_{x_1}^2 v_\ell - \partial_{x_2}^2 v_\ell = f \text{ in } \Omega_1, \quad v_\ell = 0 \text{ on } \partial\Omega_1.$$

This does not allow any progress in finding the limit of u_ℓ. However the information that we collected on u_ℓ will allow us to attack anisotropic singular perturbation problems whose archetype could be (Cf. Chapter 5)

$$-\epsilon^2 \partial_{x_1}^2 v_\epsilon - \partial_{x_2}^2 v_\epsilon = f \text{ in } \Omega, \quad v_\epsilon = 0 \text{ on } \partial\Omega.$$

Ω is here any bounded open subset of $\mathbb{R}^2$.

1.5   Eigenvalue problems

Let us then determine the eigenvalues and eigenfunctions of the Laplace operator in the rectangle Ω_ℓ defined by (1.4) with Dirichlet boundary conditions. To achieve that, let us consider first the following one dimensional eigenvalue problem:

$$\begin{cases} -u'' = \lambda u & \text{in } (-a, a), \\ u(-a) = u(a) = 0. \end{cases} \tag{1.19}$$

It is clear that the general solution of the first equation of (1.19) is given by

$$u = A \sin \sqrt{\lambda}(x + a) + B \cos \sqrt{\lambda}(x + a)$$

$(\lambda > 0, A, B \in \mathbb{R})$. In order to match the boundary conditions one has to have

$$B = 0, \qquad \sin 2a\sqrt{\lambda} = 0.$$

($A = 0$ would give the trivial solution). Thus the eigenvalues of the problem (1.19) are given by the λ_k's such that

$$2a\sqrt{\lambda_k} = k\pi \qquad \Longleftrightarrow \qquad \lambda_k = \left(\frac{k\pi}{2a}\right)^2 \quad k = 1, 2, \ldots \tag{1.20}$$

and the corresponding eigenfunctions by

$$u_k = A_k \sin \frac{k\pi}{2a}(x + a).$$

If one wants to normalize u_k in such a way that its L^2-norm is equal to one, one is lead to choose A_k such that

$$A_k^2 \int_{-a}^{a} \sin^2 \frac{k\pi}{2a}(x + a) \, dx = 1$$

which leads to

$$u_k = \frac{1}{\sqrt{a}} \sin \frac{k\pi}{2a}(x + a).$$

Suppose now that we want to find the eigenvalues and eigenfunctions of the problem

$$\begin{cases} -\Delta u = \lambda_\ell u & \text{in } \Omega_\ell, \\ u = 0 & \text{on } \partial\Omega_\ell, \end{cases} \tag{1.21}$$

(Ω_ℓ defined by (1.4)). It is clear from above that

$$u = \frac{1}{\sqrt{\ell}} \sin \frac{k\pi}{2\ell}(x_1 + \ell) \sin \frac{m\pi}{2}(x_2 + 1) \qquad (1.22)$$

is an eigenfunction for (1.21) corresponding to the eigenvalue

$$\left(\frac{k\pi}{2\ell}\right)^2 + \left(\frac{m\pi}{2}\right)^2 \quad k, m = 1, 2, \dots. \qquad (1.23)$$

There are no other eigenvalues. Indeed, suppose that (v, λ_v) is another pair of an eigenfunction with the corresponding eigenvalue for (1.21). We have

$$-\Delta v = \lambda_v v.$$

Multiplying by u given by (1.22) and integrating by parts we obtain

$$\lambda_u(v, u) = \lambda_v(v, u)$$

where λ_u is given by (1.23) and $(\cdot, \cdot)$ denotes the $L^2(\Omega_\ell)$-scalar product. Since we have assumed $\lambda_v \neq \lambda_u$ we get

$$(v, u) = 0.$$

Thus we have $(v, u) = 0$ for any u given by (1.22). By density of the span of the functions $\sin \frac{k\pi}{2a}(x + a)$ in $L^2(-a, a)$ we obtain

$$\int_{\Omega_\ell} v\varphi(x_1)\psi(x_2)\, dx_1\, dx_2 = 0 \quad \forall\, \varphi \in L^2(-\ell, \ell), \ \forall\, \psi \in L^2(-1, 1).$$

It follows easily that $v = 0$ and λ_v cannot be an eigenvalue. This proves that (1.22), (1.23) give a complete set of eigenfunctions and eigenvalues of (1.21). Let us examine what happens when ℓ goes to plus infinity. The first eigenvalue of (1.21) is given by (1.23) for $k = 1$, $m = 1$ – i.e. is given by

$$\lambda_\ell^1 = \left(\frac{\pi}{2\ell}\right)^2 + \left(\frac{\pi}{2}\right)^2.$$

Clearly, it converges when $\ell \to +\infty$ towards the first eigenvalue

$$\mu^1 = \left(\frac{\pi}{2}\right)^2$$

of the problem

$$\begin{cases} -u'' = \lambda u & \text{in } \omega, \\ u = 0 & \text{on } \partial\omega, \end{cases} \qquad (1.24)$$

(see (1.19), (1.20)). Moreover, the rate of convergence is in $\frac{1}{\ell^2}$. One could expect next that the k-th eigenvalue to (1.21) converges also to the k-eigenvalue to (1.24). This is not what happens. Indeed, going back to (1.23), if λ_ℓ^k denotes the k-th eigenvalue to (1.21) one sees that

$$\lambda_\ell^k = \left(\frac{k\pi}{2\ell}\right)^2 + \left(\frac{\pi}{2}\right)^2$$

for

$$\left(\frac{k\pi}{2\ell}\right)^2 + \left(\frac{\pi}{2}\right)^2 < \pi^2 \quad \text{i.e. for } \sqrt{3}\ell > k.$$

(π^2 is the second eigenvalue of (1.24)). Thus, when $\ell \to +\infty$, one has

$$\lambda_\ell^k \longrightarrow \mu^1$$

and all the eigenvalues of (1.21) are piling up on μ^1 when $\ell \to +\infty$. We will see later that this phenomenon happens also in the general case. Of course if one allows k to depend on ℓ the behaviour of λ_ℓ^k could be different. For instance if $\sqrt{3}\ell$ is an integer

$$\lambda_\ell^{\sqrt{3}\ell} = \mu^2$$

where $\mu^2(=\pi^2)$ is the second eigenvalue of (1.24).

The behaviour, when $\ell \to +\infty$, of the eigenfunctions is touchy as well. Indeed, consider the first eigenfunction of (1.21). It is given – see (1.22) – by

$$u = \frac{1}{\sqrt{\ell}} \sin \frac{\pi}{2\ell}(x_1 + \ell) \sin \frac{\pi}{2}(x_2 + 1).$$

When $\ell \to +\infty$, it converges toward 0. To obtain something converging toward the first eigenfunction to (1.24) one has to rescale it properly. Indeed

$$\bar{u} = \sin \frac{\pi}{2\ell}(x_1 + \ell) \sin \frac{\pi}{2}(x_2 + 1) \to \sin \frac{\pi}{2}(x_2 + 1)$$

when $\ell \to +\infty$, i.e. $\bar{u}$ is such that

$$\int_{\Omega_\ell} \bar{u}^2 \, dx_1 \, dx_2 = \ell$$

and thus to expect convergence of the eigenfunctions from (1.21) toward the eigenfunctions to (1.24) some proper scaling is in order.

The convergence rate in the H^1 norm of the eigenfunctions in any subdomain is then exactly $\frac{1}{\ell^2}$. This contrasts with the non resonance case where the convergence in any subdomain is at least exponential as we saw in the section 1.1 above.

These are the kinds of issues that we would like to analyse in the chapter devoted to eigenvalues. We will see in particular that the situation for Dirichlet or Neumann at the extremities of the rectangle are completely different.

1.6 Calculus of variations problems

Suppose that F is a function from $\mathbb{R}^2$ to $\mathbb{R}$ such that

$$\nabla F \text{ is uniformly Lipschitz continuous} \qquad (1.25)$$

and

$$\{\nabla F(X) - \nabla F(Y)\} \cdot (X - Y) \geq \mu|X - Y|^2 \quad \forall X, Y \in \mathbb{R}^2 \qquad (1.26)$$

where μ is a positive number. Let Ω_ℓ be the rectangle defined by (1.4) i.e. $\Omega_\ell = \ell\omega \times \omega$ with $\omega = (-1, 1)$ and f be a function in $L^2(\omega)$. Let u_ℓ, u_∞ be the minimizers of

$$\int_{\Omega_\ell} F(\nabla v) - f(x_2)v \, dx, \qquad \int_\omega F(0, \partial_{x_2} v) - f(x_2)v \, dx_2$$

on $H_0^1(\Omega_\ell)$ and $H_0^1(\omega)$ respectively i.e. $u_\ell \in H_0^1(\Omega_\ell)$, $u_\infty \in H_0^1(\omega)$ satisfy

$$\int_{\Omega_\ell} F(\nabla u_\ell) - f(x_2)u_\ell \, dx$$
$$\leq \int_{\Omega_\ell} F(\nabla v) - f(x_2)v \, dx \quad \forall v \in H_0^1(\Omega_\ell), \qquad (1.27)$$

$$\int_\omega F(0, \partial_{x_2} u_\infty) - f(x_2)u_\infty \, dx_2$$
$$\leq \int_\omega F(0, \partial_{x_2} v) - f(x_2)v \, dx_2 \quad \forall v \in H_0^1(\omega). \qquad (1.28)$$

Note that thanks to our assumptions one can show easily that the solutions to (1.27), (1.28) respectively exist and are unique (see [48]). Then we have

Theorem 1.4. *There exist positive constants C, α such that*

$$\big|\big|\nabla(u_\ell - u_\infty)\big|\big|_{2, \Omega_{\frac{\ell}{2}}} \leq Ce^{-\alpha\ell}.$$

Proof. For $\ell_1 \leq \ell - 1$ we denote by $\rho = \rho_{\ell_1}(x_1)$ the function whose graph is given by Figure 1.4.

For any $\lambda \in (0, 1)$ one has a.e. $x_1 \in \ell\omega$

$$u_\ell - \lambda\rho(u_\ell - u_\infty) \in H_0^1(\Omega_\ell), \quad u_\infty + \lambda\rho(u_\ell(x_1, .) - u_\infty) \in H_0^1(\omega).$$

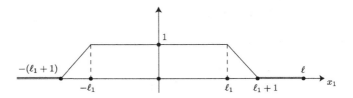

Fig. 1.4   The function ρ.

Using these two functions as test functions in (1.27), (1.28) we get

$$\int_{\Omega_\ell} F(\nabla u_\ell) - f(x_2)u_\ell \, dx$$
$$\leq \int_{\Omega_\ell} F(\nabla\{u_\ell - \lambda\rho(u_\ell - u_\infty)\}) - f(x_2)\{u_\ell - \lambda\rho(u_\ell - u_\infty)\} \, dx, \quad (1.29)$$

$$\int_\omega F(0, \partial_{x_2} u_\infty) - f(x_2)u_\infty \, dx_2$$
$$\leq \int_\omega F(0, \partial_{x_2}\{u_\infty + \lambda\rho(u_\ell - u_\infty)\}) - f(x_2)\{u_\infty + \lambda\rho(u_\ell - u_\infty)\} \, dx_2.$$

Integrating this last inequality in x_1 we obtain

$$\int_{\Omega_\ell} F(0, \partial_{x_2} u_\infty) - f(x_2)u_\infty \, dx$$
$$\leq \int_{\Omega_\ell} F(0, \partial_{x_2}\{u_\infty + \lambda\rho(u_\ell - u_\infty)\}) - f(x_2)\{u_\infty + \lambda\rho(u_\ell - u_\infty)\} dx.$$

Summing up this inequality with (1.29) we obtain

$$\int_{\Omega_\ell} F(\nabla\{u_\ell - \lambda\rho(u_\ell - u_\infty)\}) - F(\nabla u_\ell)$$
$$+ F(0, \partial_{x_2}\{u_\infty + \lambda\rho(u_\ell - u_\infty)\}) - F(0, \partial_{x_2} u_\infty) \, dx \geq 0.$$

Since the above integrand vanishes on the set where $\rho = 0$ it is enough to integrate on Ω_{ℓ_1+1}. Using the formula

$$F(Y) - F(X) = \int_0^1 \frac{d}{dt} F(tY + (1-t)X) dt$$
$$= \int_0^1 \nabla F(tY + (1-t)X) \cdot (Y - X) dt \quad (1.30)$$

we obtain

$$\int_{\Omega_{\ell_1+1}} \int_0^1 \nabla F(\nabla\{u_\ell - t\lambda\rho(u_\ell - u_\infty)\}) \cdot \lambda\nabla\{-\rho(u_\ell - u_\infty)\}$$
$$+ \partial_2 F(0, \partial_{x_2}\{u_\infty + t\lambda\rho(u_\ell - u_\infty)\})\lambda\,\partial_{x_2}\{\rho(u_\ell - u_\infty)\} \, dtdx \geq 0.$$

Dividing by λ and letting $\lambda \to 0$ we get

$$\int_{\Omega_{\ell_1+1}} \nabla F(\nabla u_\ell) \cdot \nabla\{-\rho(u_\ell - u_\infty)\} \tag{1.31}$$
$$- \partial_2 F(0, \partial_{x_2} u_\infty) \partial_{x_2}\{-\rho(u_\ell - u_\infty)\}\, dx \geq 0.$$

Since u_∞ is independent of x_1 one has $(0, \partial_{x_2} u_\infty) = \nabla u_\infty$ and

$$\int_{\Omega_{\ell_1+1}} \partial_1 F(0, \partial_{x_2} u_\infty) \partial_{x_1}\{-\rho(u_\ell - u_\infty)\}\, dx = 0.$$

This allows us to write (1.31) as

$$\int_{\Omega_{\ell_1+1}} \nabla F(\nabla u_\ell) \cdot \nabla\{-\rho(u_\ell - u_\infty)\} - \nabla F(\nabla u_\infty) \cdot \nabla\{-\rho(u_\ell - u_\infty)\}\, dx \geq 0.$$

This leads to

$$\int_{\Omega_{\ell_1+1}} \rho\{\nabla F(\nabla u_\ell) - \nabla F(\nabla u_\infty)\} \cdot \nabla(u_\ell - u_\infty) dx$$
$$\leq -\int_{\Omega_{\ell_1+1}} \{\nabla F(\nabla u_\ell) - \nabla F(\nabla u_\infty)\} \cdot \nabla\rho(u_\ell - u_\infty)\, dx.$$

Using (1.25), (1.26) and the fact that $\rho = 1$ on Ω_{ℓ_1} we deduce

$$\mu \int_{\Omega_{\ell_1}} |\nabla(u_\ell - u_\infty)|^2 dx$$
$$\leq \int_{\Omega_{\ell_1+1} \setminus \Omega_{\ell_1}} |\nabla F(\nabla u_\ell) - \nabla F(\nabla u_\infty)||u_\ell - u_\infty|\, dx$$
$$\leq C \int_{\Omega_{\ell_1+1} \setminus \Omega_{\ell_1}} |\nabla(u_\ell - u_\infty)||u_\ell - u_\infty|\, dx$$

where C is the Lipschitz constant of ∇F. Using now the Cauchy-Schwarz inequality and the Poincaré inequality on the section, (see the Appendix), we get setting $D_{\ell_1} = \Omega_{\ell_1+1} \setminus \Omega_{\ell_1}$

$$\mu \int_{\Omega_{\ell_1}} |\nabla(u_\ell - u_\infty)|^2 dx$$
$$\leq C\{\int_{D_{\ell_1}} |\nabla(u_\ell - u_\infty)|^2 dx\}^{\frac{1}{2}} \{\int_{D_{\ell_1}} |(u_\ell - u_\infty)|^2 dx\}^{\frac{1}{2}}$$
$$\leq C' \int_{\Omega_{\ell_1+1} \setminus \Omega_{\ell_1}} |\nabla(u_\ell - u_\infty)|^2\, dx$$

for some constant C' independent of ℓ. Thus, setting $\gamma = \frac{C'}{\mu}$, we have

$$\int_{\Omega_{\ell_1}} |\nabla(u_\ell - u_\infty)|^2 dx$$

$$\leq \gamma \{ \int_{\Omega_{\ell_1+1}} |\nabla(u_\ell - u_\infty)|^2 \, dx - \int_{\Omega_{\ell_1}} |\nabla(u_\ell - u_\infty)|^2 \, dx \}$$

i.e.

$$\int_{\Omega_{\ell_1}} |\nabla(u_\ell - u_\infty)|^2 dx \leq \delta \int_{\Omega_{\ell_1+1}} |\nabla(u_\ell - u_\infty)|^2 \, dx$$

where $\delta = \frac{\gamma}{\gamma+1} < 1$. We then iterate the above inequality $[\frac{\ell}{2}]$ times starting from $\ell_1 = \frac{\ell}{2}$. ([ ] denotes the integer part of a number. Note that in the following we will use several times this technique. We will repeat it to make each chapter easier to read). Thus iterating we get

$$\int_{\Omega_{\frac{\ell}{2}}} |\nabla(u_\ell - u_\infty)|^2 dx \leq \delta^{[\frac{\ell}{2}]} \int_{\Omega_{\frac{\ell}{2}+[\frac{\ell}{2}]}} |\nabla(u_\ell - u_\infty)|^2 \, dx.$$

Since

$$\frac{\ell}{2} - 1 \leq [\frac{\ell}{2}] \leq \frac{\ell}{2}$$

this leads to

$$\int_{\Omega_{\frac{\ell}{2}}} |\nabla(u_\ell - u_\infty)|^2 dx \leq \frac{1}{\delta}\delta^{\frac{\ell}{2}} \int_{\Omega_\ell} |\nabla(u_\ell - u_\infty)|^2 \, dx$$

i.e. to

$$\int_{\Omega_{\frac{\ell}{2}}} |\nabla(u_\ell - u_\infty)|^2 dx \leq Ce^{-\beta\ell} \int_{\Omega_\ell} |\nabla(u_\ell - u_\infty)|^2 \, dx \qquad (1.32)$$

with $C = \frac{1}{\delta}$ and $\beta = \frac{1}{2}\ln\frac{1}{\delta}$.

It remains to evaluate this last integral. For this one notices that since u_ℓ is the minimizer of our functional one has

$$\int_{\Omega_\ell} F(\nabla u_\ell) - f(x_2)u_\ell \, dx \leq \int_{\Omega_\ell} F(0) - f(x_2)0 \, dx = \int_{\Omega_\ell} F(0) \, dx.$$

This implies that

$$\int_{\Omega_\ell} F(\nabla u_\ell) - F(0) - \nabla F(0) \cdot \nabla u_\ell \, dx \leq \int_{\Omega_\ell} f(x_2)u_\ell \, dx$$

($\int_{\Omega_\ell} \nabla F(0) \cdot \nabla u_\ell \, dx = 0$ since u_ℓ vanishes on $\partial\Omega_\ell$). Using (1.30) we deduce

$$\int_{\Omega_\ell} \int_0^1 \{\nabla F(t\nabla u_\ell) - \nabla F(0)\} \cdot \nabla u_\ell \, dt dx \leq \int_{\Omega_\ell} f(x_2) u_\ell \, dx.$$

By (1.26) it follows that

$$\int_{\Omega_\ell} \int_0^1 \mu t |\nabla u_\ell|^2 \, dt dx \leq \int_{\Omega_\ell} f(x_2) u_\ell \, dx.$$

Thus

$$\frac{\mu}{2} \int_{\Omega_\ell} |\nabla u_\ell|^2 \, dx \leq \int_{\Omega_\ell} f(x_2) u_\ell \, dx.$$

Using the Cauchy-Schwarz inequality and the Poincaré inequality it comes

$$\frac{\mu}{2} ||\nabla u_\ell||^2_{2,\Omega_\ell} \leq |f|_{2,\Omega_\ell} |u_\ell|_{2,\Omega_\ell} \leq C |f|_{2,\Omega_\ell} ||\nabla u_\ell||_{2,\Omega_\ell}.$$

Thus

$$||\nabla u_\ell||_{2,\Omega_\ell} \leq \frac{2C}{\mu} |f|_{2,\Omega_\ell} = \sqrt{2\ell} \frac{2C}{\mu} |f|_{2,\omega} = \kappa \sqrt{\ell}.$$

Similarly one has

$$||\nabla u_\infty||_{2,\Omega_\ell} \leq \kappa \sqrt{\ell}.$$

Going back to (1.32) one deduces easily

$$\int_{\Omega_{\frac{\ell}{2}}} |\nabla(u_\ell - u_\infty)|^2 dx \leq C\ell e^{-\beta\ell}$$

and the result follows for any $\alpha < \beta$. This completes the proof of the theorem. $\qquad\square$

Remark 1.1. One can replace the minimization on $H_0^1(\Omega_\ell)$ and $H_0^1(\omega)$ by a minimization on some convex subsets of these spaces provided that the test functions that we used above are in it.

1.7 Parabolic problems

We will restrict ourselves to the case of the section 1.1 where

$$\Omega_\ell = (-\ell, \ell) \times (-1, 1) = \ell\omega \times \omega$$

with $\omega = (-1, 1)$. We would like to show that when the data of the parabolic problem set in $(0, T) \times \Omega_\ell$, $T > 0$ are depending only on the section ω of Ω_ℓ, then when $\ell \to \infty$, the solution of this problem converges with an exponential rate of convergence toward the solution of a parabolic problem set on the section of Ω_ℓ i.e. the behaviour of the parabolic problems follows the same path as the elliptic ones when $\ell \to \infty$. Let us be more precise. Our data will be two functions depending only on the section of Ω_ℓ namely

$$u_0 = u_0(x_2) \in L^2(\omega), \quad f = f(t, x_2) \in L^2(0, T; L^2(\omega)).$$

(We refer to [49], [62], [13] for the definition and the notation of the spaces introduced to solve parabolic problems. The reader can admit in a first reading that all the functions at hand are smooth).

We denote by $u_\ell = u_\ell(t, x)$ the weak solution to

$$\begin{cases} \partial_t u_\ell - \Delta u_\ell = f & \text{in } (0, T) \times \Omega_\ell, \\ u_\ell = 0 & \text{on } (0, T) \times \partial\Omega_\ell, \\ u_\ell(0, \cdot) = u_0 & \text{in } \Omega_\ell. \end{cases} \tag{1.33}$$

It is well known (see [49], [62], [13]) that (1.33) admits a weak solution in the sense that

$$\begin{cases} u_\ell \in L^2(0, T; H_0^1(\Omega_\ell)) \cap C([0, T]; L^2(\Omega_\ell)), \\ \qquad\qquad \partial_t u_\ell \in L^2(0, T; H^{-1}(\Omega_\ell)), \\ \langle \partial_t u_\ell, v \rangle + \int_{\Omega_\ell} \nabla u_\ell \cdot \nabla v dx = \int_{\Omega_\ell} f v dx \\ \qquad\qquad \text{in } \mathcal{D}'(0, T), \quad \forall v \in H_0^1(\Omega_\ell), \\ u_\ell(0, \cdot) = u_0. \end{cases} \tag{1.34}$$

Similarly for the section of Ω_ℓ there exists a unique u_∞ solution to

$$\begin{cases} u_\infty \in L^2(0, T; H_0^1(\omega)) \cap C([0, T]; L^2(\omega)), \\ \qquad\qquad \partial_t u_\infty \in L^2(0, T; H^{-1}(\omega)), \\ \langle \partial_t u_\infty, v \rangle + \int_\omega \partial_{x_2} u_\infty \partial_{x_2} v dx_2 = \int_\omega f v dx_2 \\ \qquad\qquad \text{in } \mathcal{D}'(0, T), \quad \forall v \in H_0^1(\omega), \\ u_\infty(0, \cdot) = u_0, \end{cases} \tag{1.35}$$

i.e. u_∞ is the weak solution to

$$\begin{cases} \partial_t u_\infty - \partial_{x_2}^2 u_\infty = f & \text{in } (0, T) \times \omega, \\ u_\infty = 0 & \text{on } (0, T) \times \partial\omega, \\ u_\infty(0, \cdot) = u_0 & \text{in } \omega. \end{cases}$$

(In the equations above $\langle, \rangle$ denotes the H^{-1}, H_0^1 duality bracket, ∂ stands for the boundary of a domain). Then we can show

Theorem 1.5. *There exist positive constants C, α such that*

$$|u_\ell - u_\infty|_{L^\infty(0, T; L^2(\Omega_{\frac{\ell}{2}}))} + \int_0^T \int_{\Omega_{\frac{\ell}{2}}} |\nabla(u_\ell - u_\infty)|^2 dx dt \leq C e^{-\alpha\ell} \tag{1.36}$$

i.e. in particular $u_\ell \to u_\infty$ with an exponential rate of convergence for a certain norm on every domain of the type $(0, T) \times \Omega_{\ell_0}$, ℓ_0 being a fixed number.

Proof. We are going first to use the fact that in a weak sense one has

$$\partial_t(u_\ell - u_\infty) - \Delta(u_\ell - u_\infty) = 0 \quad \text{in } (0, T) \times \Omega_\ell. \tag{1.37}$$

More precisely let $\ell_1 \leq \ell - 1$ and consider $\rho = \rho(x_1)$ the function introduced in the proof of Theorem 1.4. One has

$$\rho(u_\ell - u_\infty) \in L^2(0, T; H_0^1(\Omega_\ell)).$$

Thus from (1.34), one deduces

$$\langle \partial_t u_\ell, \rho(u_\ell - u_\infty) \rangle + \int_{\Omega_\ell} \nabla u_\ell \cdot \nabla(\rho(u_\ell - u_\infty)) dx$$
$$= \int_{\Omega_\ell} f\rho(u_\ell - u_\infty) dx \quad \text{a.e. } t \in (0, T).$$

Similarly

$$\rho(x_1)(u_\ell - u_\infty)(\cdot, x_1, \cdot) \in L^2(0, T; H_0^1(\omega)) \quad \text{a.e. } x_1 \in \ell\omega$$

and from (1.35) one deduces for a.e. $x_1 \in \ell\omega$

$$\langle \partial_t u_\infty, \rho(u_\ell - u_\infty) \rangle + \int_\omega \partial_{x_2} u_\infty \partial_{x_2}(\rho(u_\ell - u_\infty)) dx_2$$
$$= \int_\omega f\rho(u_\ell - u_\infty) dx_2 \quad \text{a.e. } t \in (0, T).$$

Integrating in x_1 one gets (recall that u_∞ is independent of x_1)

$$\int_{-\ell}^\ell \langle \partial_t u_\infty, \rho(u_\ell - u_\infty) \rangle + \int_{\Omega_\ell} \nabla u_\infty \cdot \nabla(\rho(u_\ell - u_\infty)) dx$$
$$= \int_{\Omega_\ell} f\rho(u_\ell - u_\infty) dx \quad \text{a.e. } t \in (0, T).$$

Assuming –to avoid technicalities– that u_ℓ, u_∞ are smooth, by subtraction we get

$$\int_{\Omega_\ell} \partial_t(u_\ell - u_\infty)\rho(u_\ell - u_\infty) dx$$
$$+ \int_{\Omega_\ell} \nabla(u_\ell - u_\infty) \cdot \nabla(\rho(u_\ell - u_\infty)) dx = 0 \quad \text{a.e. } t \in (0, T).$$

(Compare to (1.37)). This can also be written as

$$\frac{1}{2}\frac{d}{dt}\int_{\Omega_\ell}(u_\ell-u_\infty)^2\rho dx$$

$$+\int_{\Omega_\ell}\nabla(u_\ell-u_\infty)\cdot\nabla(\rho(u_\ell-u_\infty))dx=0 \quad \text{a.e. } t\in(0,T).$$

Integrating on $(0,t)$ and using the fact that $(u_\ell-u_\infty)(0,\cdot)=0$ we get

$$\frac{1}{2}\int_{\Omega_\ell}(u_\ell-u_\infty)^2(t,\cdot)\rho dx+\int_0^t\int_{\Omega_\ell}\nabla(u_\ell-u_\infty)\cdot\nabla(\rho(u_\ell-u_\infty))dxdt=0.$$

Since ρ vanishes outside of Ω_{ℓ_1+1} it is enough to integrate in the integrals above on Ω_{ℓ_1+1} and we get

$$\frac{1}{2}\int_{\Omega_{\ell_1+1}}(u_\ell-u_\infty)^2(t,\cdot)\rho dx+\int_0^t\int_{\Omega_{\ell_1+1}}|\nabla(u_\ell-u_\infty)|^2\rho dxdt$$

$$=-\int_0^t\int_{D_{\ell_1}}\partial_{x_1}(u_\ell-u_\infty)(\partial_{x_1}\rho)(u_\ell-u_\infty)dxdt$$

($D_{\ell_1}=\Omega_{\ell_1+1}\backslash\Omega_{\ell_1}$, we used the fact that $\partial_{x_1}\rho=0$ on Ω_{ℓ_1}). Taking into account that $\rho=1$ on Ω_{ℓ_1}, $|\partial_{x_1}\rho|\leq 1$ and the Young inequality we obtain

$$\frac{1}{2}\int_{\Omega_{\ell_1}}(u_\ell-u_\infty)^2(t,\cdot)dx+\int_0^t\int_{\Omega_{\ell_1}}|\nabla(u_\ell-u_\infty)|^2dxdt$$

$$\leq\frac{1}{2}\int_0^t\int_{D_{\ell_1}}(\partial_{x_1}(u_\ell-u_\infty))^2+(u_\ell-u_\infty)^2dxdt.$$

Using now the Poincaré inequality

$$\int_\omega(u_\ell-u_\infty)^2dx_2\leq 2\int_\omega(\partial_{x_2}(u_\ell-u_\infty))^2dx_2$$

in the last integral we obtain easily

$$\frac{1}{2}\int_{\Omega_{\ell_1}}(u_\ell-u_\infty)^2(t,\cdot)dx+\int_0^t\int_{\Omega_{\ell_1}}|\nabla(u_\ell-u_\infty)|^2dxdt$$

$$\leq\int_0^t\int_{D_{\ell_1}}|\nabla(u_\ell-u_\infty)|^2dxdt$$

$$\leq\frac{1}{2}\int_{D_{\ell_1}}(u_\ell-u_\infty)^2(t,\cdot)dx+\int_0^t\int_{D_{\ell_1}}|\nabla(u_\ell-u_\infty)|^2dxdt.$$

Let us set –dropping the dependence on t

$$F(\ell_1)=\frac{1}{2}\int_{\Omega_{\ell_1}}(u_\ell-u_\infty)^2(t,\cdot)dx+\int_0^t\int_{\Omega_{\ell_1}}|\nabla(u_\ell-u_\infty)|^2dxdt.$$

The inequality above reads

$$F(\ell_1) \le F(\ell_1 + 1) - F(\ell_1) \quad \Leftrightarrow \quad F(\ell_1) \le \frac{1}{2}F(\ell_1 + 1).$$

Starting from $\ell_1 = \frac{\ell}{2}$ (see the previous section) and iterating this inequality $[\frac{\ell}{2}]$ times we obtain

$$F(\frac{\ell}{2}) \le (\frac{1}{2})^{[\frac{\ell}{2}]}F(\frac{\ell}{2} + [\frac{\ell}{2}]) \le (\frac{1}{2})^{\frac{\ell}{2}-1}F(\ell) = 2e^{-\frac{\ell}{2}\ln 2}F(\ell). \tag{1.38}$$

It remains to estimate $F(\ell)$. It is easy to see that

$$\begin{aligned} F(\ell) \le 2\{\frac{1}{2}\int_{\Omega_\ell} u_\ell^2(t,\cdot)dx &+ \int_0^t\int_{\Omega_\ell} |\nabla u_\ell|^2 dxdt \\ &+ \frac{1}{2}\int_{\Omega_\ell} u_\infty^2(t,\cdot)dx + \int_0^t\int_{\Omega_\ell} |\nabla u_\infty|^2 dxdt\}. \end{aligned} \tag{1.39}$$

Taking $v = u_\ell$ in (1.34) one has

$$\langle \partial_t u_\ell, u_\ell \rangle + \int_{\Omega_\ell} |\nabla u_\ell|^2 dx = \int_{\Omega_\ell} f u_\ell dx.$$

Integrating in t it comes

$$\frac{1}{2}\int_{\Omega_\ell} u_\ell^2(t,\cdot)dx + \int_0^t\int_{\Omega_\ell} |\nabla u_\ell|^2 dxdt = \int_0^t\int_{\Omega_\ell} f u_\ell dxdt + \frac{1}{2}\int_{\Omega_\ell} u_0^2 dx.$$

Using the Young inequality

$$ab \le a^2 + \frac{1}{4}b^2$$

we derive

$$\begin{aligned} \frac{1}{2}\int_{\Omega_\ell} u_\ell^2(t,\cdot)dx &+ \int_0^t\int_{\Omega_\ell} |\nabla u_\ell|^2 dxdt \\ &\le \int_0^t\int_{\Omega_\ell} f^2 dxdt + \frac{1}{4}\int_0^t\int_{\Omega_\ell} u_\ell^2 dxdt + \frac{1}{2}\int_{\Omega_\ell} u_0^2 dx \\ &\le \int_0^t\int_{\Omega_\ell} f^2 dxdt + \frac{1}{2}\int_0^t\int_{\Omega_\ell} |\nabla u_\ell|^2 dxdt + \frac{1}{2}\int_{\Omega_\ell} u_0^2 dx, \end{aligned}$$

(we have used the Poincaré inequality again in the middle integral). Thus

$$\begin{aligned} \frac{1}{4}\int_{\Omega_\ell} u_\ell^2(t,\cdot)dx &+ \frac{1}{2}\int_0^t\int_{\Omega_\ell} |\nabla u_\ell|^2 dxdt \\ &\le \int_0^t\int_{\Omega_\ell} f^2 dxdt + \frac{1}{2}\int_{\Omega_\ell} u_0^2 dx \\ &\le \int_0^t\int_{-\ell}^{\ell}\int_\omega f^2 dxdt + \frac{1}{2}\int_{-\ell}^{\ell}\int_\omega u_0^2 dx \\ &= 2\ell\int_0^t\int_\omega f^2 dx_2 dt + \ell\int_\omega u_0^2 dx_2. \end{aligned} \tag{1.40}$$

This leads to

$$\frac{1}{2}\int_{\Omega_\ell} u_\ell^2(t,\cdot)dx + \int_0^t \int_{\Omega_\ell} |\nabla u_\ell|^2 dx dt \leq 4\ell\{|f|_{2,Q_T}^2 + |u_0|_{2,\omega}^2\}$$

where $Q_T = (0,T) \times \omega$, recall that $|\ |_{2,A}$ denotes the usual $L^2(A)$-norm. Using now $v = u_\infty$ in (1.35) and proceeding the same way will lead to (see (1.40))

$$\frac{1}{4}\int_\omega u_\infty^2(t,\cdot)dx_2 + \frac{1}{2}\int_0^t \int_\omega |\partial_{x_2}u_\infty|^2 dx_2 dt \leq \int_0^t \int_\omega f^2 dx_2 dt + \frac{1}{2}\int_\omega u_0^2 dx_2$$

i.e.

$$\frac{1}{2}\int_\omega u_\infty^2(t,\cdot)dx_2 + \int_0^t \int_\omega |\partial_{x_2}u_\infty|^2 dx_2 dt \leq 2\{|f|_{2,Q_T}^2 + |u_0|_{2,\omega}^2\}.$$

Integrating in x_1 we arrive to

$$\frac{1}{2}\int_{\Omega_\ell} u_\infty^2(t,\cdot)dx + \int_0^t \int_{\Omega_\ell} |\partial_{x_2}u_\infty|^2 dx dt \leq 4\ell\{|f|_{2,Q_T}^2 + |u_0|_{2,\omega}^2\}.$$

Thus going back to (1.38), (1.39) we obtain

$$F(\frac{\ell}{2}) \leq 32\ell e^{-\frac{\ell}{2}\ln 2}\{|f|_{2,Q_T}^2 + |u_0|_{2,\omega}^2\}.$$

Choosing $\alpha < \frac{1}{2}\ln 2$ we obtain then for some constant C

$$\frac{1}{2}\int_{\Omega_{\frac{\ell}{2}}}(u_\ell - u_\infty)^2(t,\cdot)dx + \int_0^t \int_{\Omega_{\frac{\ell}{2}}} |\nabla(u_\ell - u_\infty)|^2 dx dt \leq Ce^{-\alpha\ell}$$

a.e. $t \in (0.T)$. This completes the proof of the theorem.   □

Remark 1.2. This estimate is uniform in t. One can get different estimates when α is allowed to depend on t (see [78]).

1.8   Strongly nonlinear problems

We take advantage of this new class of problems to introduce a more general framework. We denote by ω_1, ω_2 two bounded open subsets of $\mathbb{R}^p$ and $\mathbb{R}^{n-p}$ respectively and we suppose that

$$0 \in \omega_1 \text{ and } \omega_1 \text{ is star shaped with respect to 0.} \qquad (1.41)$$

For $\ell > 0$ we set

$$\Omega_\ell = \ell\omega_1 \times \omega_2.$$

The points $x = (x_1, x_2, ..., x_n)$ in $\mathbb{R}^n$ are split into two components X_1, X_2 where

$$X_1 = x_1, ..., x_p, \quad X_2 = x_{p+1}, ..., x_n$$

i.e. $x = (X_1, X_2)$.

Let us denote by f a function independent of X_1, i.e.

$$f = f(X_2)$$

and consider u_ℓ the weak solution to

$$\begin{cases} -\partial_{x_i}(|\partial_{x_i} u_\ell|^{q-2} \partial_{x_i} u_\ell) = f & \text{in } \Omega_\ell, \\ \\ u_\ell = 0 & \text{on } \partial\Omega_\ell, \end{cases}$$

(with the summation convention on $i = 1, ..., n$). When $q = 2$ the above operator is the Laplace operator. We will assume in what follows that

$$q \geq 2.$$

Then, if for instance

$$f \in L^{q'}(\omega_2), \quad \frac{1}{q'} + \frac{1}{q} = 1,$$

the problem above admits a unique weak solution in $W_0^{1,q}(\Omega_\ell)$ -i.e. there exists a unique u_ℓ satisfying (see [16] or the Theorem A.1 in the Appendix)

$$\begin{cases} u_\ell \in W_0^{1,q}(\Omega_\ell), \\ \\ \int_{\Omega_\ell} |\partial_{x_i} u_\ell|^{q-2} \partial_{x_i} u_\ell \partial_{x_i} v \, dx = \int_{\Omega_\ell} fv \, dx \quad \forall v \in W_0^{1,q}(\Omega_\ell). \end{cases} \tag{1.42}$$

We would like to find out the limit behaviour of u_ℓ when $\ell \to +\infty$. For that we will assume that ω_1 -in addition to (1.41)- possesses the following property. For every $a > 0$ there exists a function $\rho_a = \rho_a(X_1)$ such that

$$0 \leq \rho_a \leq 1, \ \rho_a = 1 \text{ on } a\omega_1, \ \rho_a = 0 \text{ outside } (a+1)\omega_1,$$
$$|\nabla_{X_1} \rho_a| \leq C \tag{1.43}$$

for some constant C, $\nabla_{X_1} = (\partial_{x_1}, ..., \partial_{x_p})$ is the gradient in X_1.

Remark 1.3. If ω_1 is convex the property above holds true and for simplicity we will often assume this in the rest of the book.

Note also that one could assume ω_2 bounded in one direction (see the Poincaré inequality in the Appendix). We will stick to ω_2 bounded to remain close to the usual cylindrical domains.

Then, if u_∞ denotes the solution to

$$\begin{cases} u_\infty \in W_0^{1,q}(\omega_2), \\ \int_{\omega_2} |\partial_{x_i} u_\infty|^{q-2} \partial_{x_i} u_\infty \partial_{x_i} v \, dX_2 = \int_{\omega_2} fv \, dX_2 \quad \forall v \in W_0^{1,q}(\omega_2), \end{cases} \tag{1.44}$$

(note that in the integral above the summation convention is made for $i = p+1, ..., n$, $dX_2 = dx_{p+1}...dx_n$), we have:

Theorem 1.6. *There exist positive constants C, α such that*

$$\left\| \nabla (u_\ell - u_\infty) \right\|_{q, \Omega_{\frac{\ell}{2}}} \leq C e^{-\alpha \ell}. \tag{1.45}$$

Proof. Let us denote by ℓ_1 a real number such that $0 < \ell_1 \leq \ell - 1$. It is clear that

$$\rho_{\ell_1}(X_1)(u_\ell - u_\infty) \in W_0^{1,q}(\Omega_\ell)$$

and thus it is a suitable test function for (1.42). Moreover for a.e. X_1 (Cf. [13])

$$\rho_{\ell_1}(X_1)(u_\ell - u_\infty)(X_1, .) \in W_0^{1,q}(\omega_2)$$

and thus it is a suitable test function for (1.44). Thus, for a.e. X_1 it holds

$$\int_{\omega_2} |\partial_{x_i} u_\infty|^{q-2} \partial_{x_i} u_\infty \partial_{x_i} \{\rho_{\ell_1}(u_\ell - u_\infty)\} dX_2$$

$$= \int_{\omega_2} f \rho_{\ell_1}(u_\ell - u_\infty) dX_2. \tag{1.46}$$

(Note that in the first integral above i can run from 1 to n since u_∞ is independent of X_1). Integrating (1.46) in X_1 we get

$$\int_{\Omega_\ell} |\partial_{x_i} u_\infty|^{q-2} \partial_{x_i} u_\infty \partial_{x_i} \{\rho_{\ell_1}(u_\ell - u_\infty)\} dx = \int_{\Omega_\ell} f \rho_{\ell_1}(u_\ell - u_\infty) dx.$$

Taking $v = \rho_{\ell_1}(u_\ell - u_\infty)$ in (1.42) and subtracting the equality above we obtain

$$\int_{\Omega_\ell} \{|\partial_{x_i} u_\ell|^{q-2} \partial_{x_i} u_\ell - |\partial_{x_i} u_\infty|^{q-2} \partial_{x_i} u_\infty\} \partial_{x_i} \{\rho_{\ell_1}(u_\ell - u_\infty)\} dx = 0.$$

Since ρ_{ℓ_1} vanishes outside Ω_{ℓ_1+1} it is enough in the integral above to integrate on Ω_{ℓ_1+1} and we obtain easily

$$\int_{\Omega_{\ell_1+1}} \{|\partial_{x_i} u_\ell|^{q-2} \partial_{x_i} u_\ell - |\partial_{x_i} u_\infty|^{q-2} \partial_{x_i} u_\infty\} \partial_{x_i}(u_\ell - u_\infty) \rho_{\ell_1} dx$$

$$= -\int_{\Omega_{\ell_1+1} \backslash \Omega_{\ell_1}} |\partial_{x_i} u_\ell|^{q-2} \partial_{x_i} u_\ell \partial_{x_i} \rho_{\ell_1}(u_\ell - u_\infty) dx. \tag{1.47}$$

(We used the fact that $\rho_{\ell_1} = 1$ on Ω_{ℓ_1}. In the last integral i runs from 1 to p since ρ_{ℓ_1} is independent of X_2). It is well known that for some positive constant C_q one has (see for instance [16])

$$C_q |a - b|^q \leq (|a|^{q-2}a - |b|^{q-2}b)(a - b) \quad \forall a, b \in \mathbb{R}$$

and thus from (1.47) we derive easily, (with a summation in i from 1 to n in the first integral)

$$C_q \int_{\Omega_{\ell_1+1}} |\partial_{x_i}(u_\ell - u_\infty)|^q \rho_{\ell_1} dx \leq \int_{\Omega_{\ell_1+1} \setminus \Omega_{\ell_1}} |\partial_{x_i} u_\ell|^{q-1} |\partial_{x_i} \rho_{\ell_1}| |u_\ell - u_\infty| dx.$$

Since $\rho_{\ell_1} = 1$ on Ω_{ℓ_1} and u_∞ is independent of X_1 this implies that

$$C_q \int_{\Omega_{\ell_1}} |\partial_{x_i}(u_\ell - u_\infty)|^q dx$$
$$\leq \int_{\Omega_{\ell_1+1} \setminus \Omega_{\ell_1}} |\partial_{x_i}(u_\ell - u_\infty)|^{q-1} |\partial_{x_i} \rho_{\ell_1}| |u_\ell - u_\infty| dx. \tag{1.48}$$

(Note that in the last integral i is running from 1 to p). Using the equivalence of norms in $\mathbb{R}^n$ and (1.43) we derive easily

$$\int_{\Omega_{\ell_1}} |\nabla(u_\ell - u_\infty)|^q dx \leq C \int_{\Omega_{\ell_1+1} \setminus \Omega_{\ell_1}} |\nabla_{X_1}(u_\ell - u_\infty)|^{q-1} |u_\ell - u_\infty| dx$$

where $C = C(p, q, n)$ is a constant depending on p, q and n. Using then the Young inequality

$$ab \leq \frac{a^q}{q} + \frac{b^{q'}}{q'}$$

with $q' = \frac{q}{q-1}$ we arrive to

$$\int_{\Omega_{\ell_1}} |\nabla(u_\ell - u_\infty)|^q dx$$
$$\leq C \int_{\Omega_{\ell_1+1} \setminus \Omega_{\ell_1}} |\nabla_{X_1}(u_\ell - u_\infty)|^q + |u_\ell - u_\infty|^q dx. \tag{1.49}$$

By the Poincaré inequality we have for a.e. $X_1 \in \ell \omega_1$

$$\int_{\omega_2} |u_\ell - u_\infty|^q (X_1, X_2) dX_2 \leq C \int_{\omega_2} |\nabla_{X_2}(u_\ell - u_\infty)|^q (X_1, X_2) dX_2$$

for some constant C independent of ℓ or ℓ_1, $\nabla_{X_2} = (\partial_{x_{p+1}}, ..., \partial_{x_n})$ is the gradient in X_2. Integrating in X_1 we derive

$$\int_{\Omega_{\ell_1+1} \setminus \Omega_{\ell_1}} |u_\ell - u_\infty|^q dx \leq C(\omega_2) \int_{\Omega_{\ell_1+1} \setminus \Omega_{\ell_1}} |\nabla_{X_2}(u_\ell - u_\infty)|^q dx.$$

Thus going back to (1.49) we get for some constant C

$$\int_{\Omega_{\ell_1}} |\nabla(u_\ell - u_\infty)|^q dx \leq C \int_{\Omega_{\ell_1+1} \backslash \Omega_{\ell_1}} |\nabla(u_\ell - u_\infty)|^q dx$$

which leads to

$$\int_{\Omega_{\ell_1}} |\nabla(u_\ell - u_\infty)|^q dx \leq \frac{C}{C+1} \int_{\Omega_{\ell_1+1}} |\nabla(u_\ell - u_\infty)|^q dx.$$

Starting from $\ell_1 = \frac{\ell}{2}$ and iterating this inequality $[\frac{\ell}{2}]$-times where $[\ ]$ denotes the integer part of a real number, we get

$$\int_{\Omega_{\frac{\ell}{2}}} |\nabla(u_\ell - u_\infty)|^q dx \leq (\frac{C}{C+1})^{[\frac{\ell}{2}]} \int_{\Omega_{\frac{\ell}{2}+[\frac{\ell}{2}]}} |\nabla(u_\ell - u_\infty)|^q dx.$$

Setting $\gamma = \frac{C}{C+1} < 1$ and noticing that

$$\frac{\ell}{2} - 1 < [\frac{\ell}{2}] \leq \frac{\ell}{2}$$

we obtain

$$\int_{\Omega_{\frac{\ell}{2}}} |\nabla(u_\ell - u_\infty)|^q dx \leq \gamma^{\frac{\ell}{2}-1} \int_{\Omega_\ell} |\nabla(u_\ell - u_\infty)|^q dx. \qquad (1.50)$$

We try now to estimate the integral in the right hand side above. Taking $v = u_\ell$ in (1.42) we get easily using in particular the Hölder and the Poincaré inequality:

$$\int_{\Omega_\ell} |\partial_{x_i} u_\ell|^q dx = \int_{\Omega_\ell} f u_\ell dx$$

$$\leq |f|_{q',\Omega_\ell} |u_\ell|_{q,\Omega_\ell}$$

$$\leq |f|_{q',\Omega_\ell} C \big\| \nabla_{X_2} u_\ell \big\|_{q,\Omega_\ell}$$

$$\leq C |f|_{q',\Omega_\ell} \big\| \nabla u_\ell \big\|_{q,\Omega_\ell}$$

and thus, by equivalence of the norms in $\mathbb{R}^n$ (note that we have a summation in i in the first integral)

$$\big\| \nabla u_\ell \big\|_{q,\Omega_\ell}^q \leq C |f|_{q',\Omega_\ell} \big\| \nabla u_\ell \big\|_{q,\Omega_\ell}$$

$$\Rightarrow \big\| \nabla u_\ell \big\|_{q,\Omega_\ell}^q \leq C |f|_{q',\Omega_\ell}^{\frac{q}{q-1}} = C \int_{\ell\omega_1} \int_{\omega_2} |f(X_2)|^{q'} dx$$

$$\leq C \ell^p |f|_{q',\omega_2}^{\frac{q}{q-1}}$$

where $C = C(\omega_1, \omega_2, p, q, n)$. Similarly taking $v = u_\infty$ in (1.44) we derive

$$\big\| \nabla u_\infty \big\|_{q,\omega_2}^q \leq C |f|_{q',\omega_2}^{\frac{q}{q-1}}$$

and thus

$$\big|\big|\nabla u_\infty\big|\big|_{q,\Omega_\ell}^q \leq C\ell^p |f|_{q',\omega_2}^{\frac{q}{q-1}}.$$

Going back to (1.50), we get

$$\{\int_{\Omega_{\frac{\ell}{2}}} |\nabla(u_\ell - u_\infty)|^q dx\}^{\frac{1}{q}} \leq (\frac{e^{\frac{\ell}{2}\ln\gamma}}{\gamma})^{\frac{1}{q}}\{\int_{\Omega_\ell} |\nabla(u_\ell - u_\infty)|^q dx\}^{\frac{1}{q}}$$

$$\leq C|f|_{q',\omega_2}^{\frac{1}{q-1}}\ell^{\frac{p}{q}}e^{\frac{\ell}{2q}\ln\gamma}$$

$$\leq Ce^{-\alpha\ell},$$

for any $\alpha < -\frac{\ln\gamma}{2q}$. This completes the proof of the theorem.

$\square$

Remark 1.4. In the case $q = 2$ we have obtained at the same time an alternate proof of Theorem 1.2. Note that in the case of the q-Laplace operator i.e. when

$$-\Delta_q u = -\partial_{x_i}\{|\nabla u|^{q-2}\partial_{x_i}\}$$

the technique above does not work since all the derivatives are involved in the operator (see (1.48)).

1.9 Higher order operators

We consider a thin clamped plate occupying the domain

$$\Omega_\ell \times (-\varepsilon, \varepsilon)$$

where $\Omega_\ell = (-\ell, \ell) \times (-1, 1)$. The displacement of this plate under the action of a force density f is given by u_ℓ solution to

$$\begin{cases} \Delta^2 u_\ell = f \text{ in } \Omega_\ell, \\ u_\ell = \frac{\partial u_\ell}{\partial \nu} = 0 \text{ on } \partial\Omega_\ell. \end{cases} \quad (1.51)$$

(ν denotes the outward unit normal to $\partial\Omega_\ell$ the boundary of Ω_ℓ, see [45], [46] for the derivation of this system via a particular scaling). Let us assume that

$$f = f(x_2) \in L^2(\omega)$$

with $\omega = (-1, 1)$. When $\ell \to \infty$ one expects that the displacement of the plate becomes at the limit independent of x_1. To be able to work with the weak formulation of the problem we introduce some new spaces. If O is an open subset of $\mathbb{R}^k$, we set

$$H^2(O) = \{v \in L^2(O) \mid \partial^\alpha v \in L^2(O) \ \forall |\alpha| \leq 2\}$$

($\alpha = (\alpha_1, \alpha_2, ..., \alpha_k) \in \mathbb{N}^k$, $|\alpha| = \sum_{i=1}^k \alpha_i$, $\partial^\alpha = \partial_{x_1}^{\alpha_1} \cdots \partial_{x_k}^{\alpha_k}$ the α-derivative in the distributional sense – see [49], [50], [16]). Note that we will use this space here only with $k = 1$ or 2.

If $\mathcal{D}(O)$ is the usual space of $C^\infty(O)$-functions with compact support in O, $H_0^2(O)$ denotes the closure of $\mathcal{D}(O)$ in $H^2(O)$ for the norm

$$||v||_{H^2(O)} = \{\sum_{|\alpha| \leq 2} |\partial^\alpha v|_{2,O}^2\}^{\frac{1}{2}}.$$

It is well known or easy to show that $H^2(O)$, $H_0^2(O)$ are Hilbert spaces for the above norm. Then the weak formulation to (1.51) is simply to find u_ℓ such that

$$\begin{cases} u_\ell \in H_0^2(\Omega_\ell), \\ \int_{\Omega_\ell} \Delta u_\ell \Delta v \, dx = \int_{\Omega_\ell} fv \, dx \quad \forall v \in H_0^2(\Omega_\ell). \end{cases} \tag{1.52}$$

The existence and uniqueness of a solution to (1.52) follow from the Lax-Milgram theorem combined with the following lemma.

Lemma 1.3. *There exist constants c, C independent of ℓ such that*

$$c||v||_{H^2(\Omega_\ell)} \leq |\Delta v|_{2,\Omega_\ell} \leq C||v||_{H^2(\Omega_\ell)} \quad \forall v \in H_0^2(\Omega_\ell).$$

Proof. By density of $\mathcal{D}(\Omega_\ell)$ in $H_0^2(\Omega_\ell)$ it is enough to prove these inequalities for $v \in \mathcal{D}(\Omega_\ell)$. Since

$$(\Delta v)^2 = (\partial_{x_1}^2 v + \partial_{x_2}^2 v)^2 \leq 2\{(\partial_{x_1}^2 v)^2 + (\partial_{x_2}^2 v)^2\}$$

one has clearly after integration on Ω_ℓ

$$|\Delta v|_{2,\Omega_\ell} \leq \sqrt{2}||v||_{H^2(\Omega_\ell)}$$

which proves the right hand side inequality.

Integrating by parts for $v \in \mathcal{D}(\Omega_\ell)$, one has

$$\int_{\Omega_\ell} \partial_{x_1}^2 v \partial_{x_2}^2 v \, dx = \int_{\Omega_\ell} (\partial_{x_1 x_2}^2 v)^2 \, dx.$$

Thus it comes

$$|\Delta v|_{2,\Omega_\ell}^2 = \int_{\Omega_\ell} (\Delta v)^2 dx \geq \sum_{|\alpha|=2} |\partial^\alpha v|_{2,\Omega_\ell}^2. \tag{1.53}$$

Moreover, applying the Poincaré inequality in ω for the functions

$$v(x_1, \cdot), \partial_{x_i} v(x_1, \cdot)$$

we obtain easily

$$|v(x_1, \cdot)|_{2,\omega}^2 \le 2|\partial_{x_2} v(x_1, \cdot)|_{2,\omega}^2,$$

$$|\partial_{x_i} v(x_1, \cdot)|_{2,\omega}^2 \le 2|\partial_{x_2 x_i}^2 v(x_1, \cdot)|_{2,\omega}^2.$$

Integrating these inequalities in x_1 we get for some constant C independent of ℓ

$$|v|_{2,\Omega_\ell}^2 + ||\nabla v||_{2,\Omega_\ell}^2 \le C \sum_{|\alpha|=2} |\partial^\alpha v|_{2,\Omega_\ell}^2. \tag{1.54}$$

The result follows then combining (1.53) and (1.54). This completes the proof of the lemma. $\qquad\square$

A natural candidate for the limit of u_ℓ when $\ell \to \infty$ is u_∞ the weak solution to

$$\begin{cases} u_\infty \in H_0^2(\omega), \\ \int_\omega \partial_{x_2}^2 u_\infty \, \partial_{x_2}^2 v \, dx_2 = \int_\omega fv dx_2 \quad \forall v \in H_0^2(\omega). \end{cases} \tag{1.55}$$

Note that the existence and uniqueness of a solution to this problem follows easily by the Lax-Milgram theorem since arguing as in Lemma 1.3 one can show easily that $|\partial_{x_2}^2 v|_{2,\omega}$ is equivalent on $H_0^2(\omega)$ to the $H_0^2(\omega)$-norm. Then we have:

Theorem 1.7. *There exist positive constants C and α such that*

$$||u_\ell - u_\infty||_{H^2(\Omega_{\ell/2})} \le Ce^{-\alpha\ell}. \tag{1.56}$$

Proof. Let $\ell_1 \le \ell - 1$. Consider a smooth function $\rho = \rho_{\ell_1}$ such that

$$0 \le \rho \le 1, \quad \rho = 1 \text{ on } (-\ell_1, \ell_1), \quad \rho = 0 \text{ outside } (-\ell_1 - 1, \ell_1 + 1),$$
$$|\rho'|, |\rho''| \le C, \tag{1.57}$$

where C is independent of ℓ and ℓ_1. One has then

$$(u_\ell - u_\infty)\rho \in H_0^2(\Omega_\ell).$$

Approximating possibly this function by a function in $\mathcal{D}(\Omega_\ell)$ one derives from (1.52), (1.55)

$$\int_{\Omega_\ell} \Delta(u_\ell - u_\infty)\Delta((u_\ell - u_\infty)\rho)dx = 0.$$

(Note that $\partial_{x_2}^2 u_\infty = \Delta u_\infty$ and integrate (1.55) in x_1). Since ρ vanishes outside Ω_{ℓ_1+1} it is enough in the above integral to integrate on Ω_{ℓ_1+1} and one gets

$$\int_{\Omega_{\ell_1+1}} |\Delta(u_\ell - u_\infty)|^2 \rho dx$$

$$= -\int_{\Omega_{\ell_1+1}} \Delta(u_\ell - u_\infty)\{(u_\ell - u_\infty)\Delta\rho + 2\nabla(u_\ell - u_\infty) \cdot \nabla\rho\}dx.$$

(We used the formula $\Delta(uv) = \Delta u v + 2\nabla u \cdot \nabla v + u\Delta v$). Using the properties of ρ given in (1.57) it is easy to derive that

$$\int_{\Omega_{\ell_1}} |\Delta(u_\ell - u_\infty)|^2 dx$$

$$\leq C\int_{D_{\ell_1}} |\Delta(u_\ell - u_\infty)|\{|u_\ell - u_\infty| + |\partial_{x_1}(u_\ell - u_\infty)|\}dx$$

where $D_{\ell_1} = \Omega_{\ell_1+1}\backslash\Omega_{\ell_1}$. Using the Cauchy-Young inequality $ab \leq \frac{1}{2}(a^2 + b^2)$ we derive easily for some constant C independent of ℓ and ℓ_1

$$\int_{\Omega_{\ell_1}} |\Delta(u_\ell - u_\infty)|^2 dx$$

$$\leq C\int_{D_{\ell_1}} \Delta(u_\ell - u_\infty)^2 + (u_\ell - u_\infty)^2 + \partial_{x_1}(u_\ell - u_\infty)^2 dx.$$

From Lemma 1.3 it follows then that, for some constant C independent of ℓ and ℓ_1, one has

$$\|u_\ell - u_\infty\|_{H^2(\Omega_{\ell_1})} \leq C\|u_\ell - u_\infty\|_{H^2(D_{\ell_1})}$$

$$\leq C\|u_\ell - u_\infty\|_{H^2(\Omega_{\ell_1+1})} - C\|u_\ell - u_\infty\|_{H^2(\Omega_{\ell_1})}$$

and thus

$$\|u_\ell - u_\infty\|_{H^2(\Omega_{\ell_1})} \leq a\|u_\ell - u_\infty\|_{H^2(\Omega_{\ell_1+1})}$$

where $a = \frac{C}{C+1} < 1$.

Iterating this formula $[\frac{\ell}{2}]$ times starting from $\ell_1 = \frac{\ell}{2}$ and recalling that $\frac{\ell}{2} - 1 \leq [\frac{\ell}{2}] \leq \frac{\ell}{2}$ we arrive easily to

$$\|u_\ell - u_\infty\|_{H^2(\Omega_{\ell/2})} \leq \frac{1}{a}e^{-\frac{\ell}{2}\ln\frac{1}{a}}\|u_\ell - u_\infty\|_{H^2(\Omega_\ell)}. \tag{1.58}$$

Thus it remains now to estimate u_ℓ and u_∞. Taking $v = u_\ell$ in (1.52) we get

$$\int_{\Omega_\ell} (\Delta u_\ell)^2 dx = \int_{\Omega_\ell} f u_\ell dx \leq |f|_{2,\Omega_\ell}|u_\ell|_{2,\Omega_\ell}$$

$$\leq C\ell|f|_{2,\omega}|\Delta u_\ell|_{2,\Omega_\ell}$$

(by Lemma 1.3). It follows that

$$||u_\ell||_{H^2(\Omega_\ell)} \leq C\ell|f|_{2,\omega}.$$

Similarly taking $v = u_\infty$ in (1.55) and integrating in x_1 lead to

$$||u_\infty||_{H^2(\Omega_\ell)} \leq C\ell|f|_{2,\omega}.$$

Going back to (1.58) we obtain

$$||u_\ell - u_\infty||_{H^2(\Omega_{\ell/2})} \leq C\ell e^{-\frac{\ell}{2}\ln\frac{1}{a}}$$

and the result follows by taking $\alpha < \frac{1}{2}\ln\frac{1}{a}$. This completes the proof of the theorem. $\qquad\square$

Remark 1.5. We refer the reader to [9] [56] for more results on higher order elliptic problems.

1.10 Higher order estimates

We suppose here that we are under the assumptions of Section 1.1 with in particular $\Omega_\ell = (-\ell, \ell) \times \omega$. First notice that even if f is not very smooth $u_\ell - u_\infty$ is since by (1.8) one has

$$-\Delta(u_\ell - u_\infty) = 0 \quad \text{in } \Omega_\ell. \tag{1.59}$$

Note also that (1.7) reads –since u_∞ is independent of x_1

$$\int_{\Omega_{\frac{\ell}{2}}} (\partial_{x_1} u_\ell)^2 + (\partial_{x_2}(u_\ell - u_\infty))^2 dx \leq C^2 e^{-2\alpha\ell}. \tag{1.60}$$

Suppose one is interested in finding estimates for the second derivatives of $u_\ell - u_\infty$. Taking the derivative in x_1 of (1.59) one gets

$$-\Delta(\partial_{x_1} u_\ell) = 0 \quad \text{in } \Omega_\ell. \tag{1.61}$$

Consider the function $\rho = \rho_{\ell_1}$ defined in Figure 1.4. Since $\partial_{x_1} u_\ell$ is smooth it is clear that

$$(\partial_{x_1} u_\ell)\rho^2 \in H_0^1(\Omega_\ell).$$

(Note that $u_\ell = 0$ on $(-\ell, \ell) \times \partial\omega$ implies $\partial_{x_1} u_\ell = 0$ on $(-\ell, \ell) \times \partial\omega$). Thus we derive from (1.61)

$$\int_{\Omega_\ell} \nabla(\partial_{x_1} u_\ell) \cdot \nabla\{(\partial_{x_1} u_\ell)\rho^2\} dx = 0. \tag{1.62}$$

By simple computations one has for any function v

$$\nabla v \cdot \nabla(v\rho^2) = \nabla v \cdot \{2\rho v \nabla\rho + \rho^2 \nabla v\} = \rho^2 |\nabla v|^2 + 2\rho v \nabla v \cdot \nabla\rho,$$

$$|\nabla(\rho v)|^2 = |\rho\nabla v + v\nabla\rho|^2 = \rho^2|\nabla v|^2 + 2\rho v\nabla v \cdot \nabla\rho + v^2|\nabla\rho|^2,$$

and thus

$$\nabla v \cdot \nabla(v\rho^2) = |\nabla(\rho v)|^2 - v^2|\nabla\rho|^2.$$

From (1.62) we derive then

$$\int_{\Omega_\ell} |\nabla(\rho \, \partial_{x_1} u_\ell)|^2 dx = \int_{\Omega_\ell} (\partial_{x_1} u_\ell)^2 |\nabla\rho|^2 dx.$$

In particular taking into account the properties of $\rho = \rho_{\ell_1}$, we obtain

$$\int_{\Omega_{\ell_1}} |\nabla(\partial_{x_1} u_\ell)|^2 dx \le \int_{\Omega_{\ell_1+1}} (\partial_{x_1} u_\ell)^2 dx.$$

Choosing $\ell_1 = \frac{\ell}{2} - 1$ and taking into account (1.60), we get

$$\int_{\Omega_{\frac{\ell}{2}-1}} |\nabla(\partial_{x_1} u_\ell)|^2 dx \le C^2 e^{-2\alpha\ell}$$

i.e.

$$\int_{\Omega_{\frac{\ell}{2}-1}} (\partial_{x_1}^2 u_\ell)^2 + (\partial_{x_1 x_2}^2 u_\ell)^2 dx \le C^2 e^{-2\alpha\ell}.$$

From the equation (1.59), one has

$$\partial_{x_2}^2(u_\ell - u_\infty) = -\partial_{x_1}^2(u_\ell - u_\infty) = -\partial_{x_1}^2 u_\ell$$

and thus we have obtained (Cf the previous section)

$$\|u_\ell - u_\infty\|_{H^2(\Omega_{\frac{\ell}{2}-1})} \le Ce^{-\alpha\ell}.$$

This implies for any $\ell_0 \le \frac{\ell}{2} - 1$

$$\|u_\ell - u_\infty\|_{H^2(\Omega_{\ell_0})} \le Ce^{-\alpha\ell}$$

and an exponential rate of convergence for the second derivatives of $u_\ell - u_\infty$. Since

$$H^2(\Omega_{\ell_0}) \subset W^{1,p}(\Omega_{\ell_0}) \subset C^{1,\gamma}(\Omega_{\ell_0}) \quad \forall p, \quad \forall \gamma \in (0,1)$$

(we refer the reader to [49], [50] for details) this gives us in this case pointwise convergence (compare to Section 2.3). It is possible then to take more derivatives of the equation above and to obtain an exponential rate of convergence for any derivatives.

1.11 Global convergence

An important issue is to know if the estimate (1.7) is somehow sharp i.e. if the convergence could be extended to Ω_ℓ or not. It is easy to see if $a \in (0,1)$ that one can obtain for some constants C and α'

$$\left\| \nabla(u_\ell - u_\infty) \right\|_{2,\Omega_{a\ell}} \leq Ce^{-\alpha'\ell}.$$

However one has

Theorem 1.8. *Under the assumptions of Theorem 1.2, one has*

$$\int_{\Omega_\ell} |\nabla(u_\ell - u_\infty)|^2 dx \leq 2|u_\infty|^2_{H^1(\omega)}, \qquad (1.63)$$

$$\int_{\Omega_\ell} |\nabla(u_\ell - u_\infty)|^2 dx \not\to 0 \quad when \quad \ell \to \infty. \qquad (1.64)$$

$(|u_\infty|^2_{H^1(\omega)} = \int_\omega (u_\infty)^2 + (\partial_{x_2} u_\infty)^2 dx_2).$

Proof. One considers the following function $\eta = \eta_\ell$ whose graph is depicted on Figure 1.5.

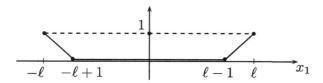

Fig. 1.5   The graph of the function η.

Clearly

$$u_\ell - (1 - \eta)u_\infty \in H_0^1(\Omega_\ell)$$

and by (1.8) we get

$$\int_{\Omega_\ell} \nabla(u_\ell - u_\infty) \cdot \nabla(u_\ell - u_\infty + \eta u_\infty)dx = 0$$

and thus by the Cauchy-Schwarz inequality

$$\int_{\Omega_\ell} |\nabla(u_\ell - u_\infty)|^2 dx = -\int_{\Omega_\ell} \nabla(u_\ell - u_\infty) \cdot \nabla(\eta u_\infty)dx$$

$$\leq \left(\int_{\Omega_\ell} |\nabla(u_\ell - u_\infty)|^2 dx \right)^{\frac{1}{2}} \left(\int_{\Omega_\ell} |\nabla(\eta u_\infty)|^2 dx \right)^{\frac{1}{2}}.$$

It follows that

$$\int_{\Omega_\ell} |\nabla(u_\ell - u_\infty)|^2 dx \le \int_{\Omega_\ell} |\nabla(\eta u_\infty)|^2 dx$$

$$= \int_{\Omega_\ell \setminus \Omega_{\ell-1}} |(u_\infty \partial_{x_1} \eta, \eta \partial_{x_2} u_\infty)|^2 dx$$

$$= \int_{\Omega_\ell \setminus \Omega_{\ell-1}} (u_\infty \partial_{x_1} \eta)^2 + (\eta \partial_{x_2} u_\infty)^2 dx$$

$$\le 2 \int_\omega (u_\infty)^2 + (\partial_{x_2} u_\infty)^2 dx_2.$$

This shows (1.63).

Let us denote by $u_\infty = \cos(\frac{\pi}{2} x_2)$ the first eigenfunction of the eigenvalue problem

$$\begin{cases} -\partial_{x_2}^2 u_\infty = \lambda_1 u_\infty & \text{in } \omega, \\ u_\infty = 0 & \text{at } \pm 1, \end{cases} \tag{1.65}$$

(see Section 1.5, $\lambda_1 = \frac{\pi^2}{4}$). The solution to

$$\begin{cases} -\Delta u_\ell = \lambda_1 u_\infty & \text{in } \Omega_\ell, \\ u_\ell = 0 & \text{on } \partial\Omega_\ell, \end{cases}$$

is given by

$$u_\ell = \Big(1 - \frac{\cosh\sqrt{\lambda_1} x_1}{\cosh\sqrt{\lambda_1} \ell}\Big) u_\infty.$$

This follows immediately from (1.65) and the fact that

$$\partial_{x_1}^2 u_\ell = -\lambda_1 \frac{\cosh\sqrt{\lambda_1} x_1}{\cosh\sqrt{\lambda_1}\ell} u_\infty, \quad \partial_{x_2}^2 u_\ell = \Big(1 - \frac{\cosh\sqrt{\lambda_1} x_1}{\cosh\sqrt{\lambda_1}\ell}\Big)\partial_{x_2}^2 u_\infty.$$

Now one has

$$\int_{\Omega_\ell} |\nabla(u_\ell - u_\infty)|^2 dx = \int_{\Omega_\ell} \Big|\nabla\Big\{\Big(-\frac{\cosh\sqrt{\lambda_1} x_1}{\cosh\sqrt{\lambda_1}\ell}\Big) u_\infty\Big\}\Big|^2 dx$$

$$\ge \int_{\Omega_\ell} \Big(\frac{\cosh\sqrt{\lambda_1} x_1}{\cosh\sqrt{\lambda_1}\ell}\Big)^2 (\partial_{x_2} u_\infty)^2 dx$$

$$= \int_{-\ell}^{\ell} \Big(\frac{\cosh\sqrt{\lambda_1} x_1}{\cosh\sqrt{\lambda_1}\ell}\Big)^2 dx_1 \int_\omega (\partial_{x_2} u_\infty)^2 dx_2.$$

By a simple calculation one has

$$\int_{-\ell}^{\ell}\Big(\frac{\cosh\sqrt{\lambda_1}x_1}{\cosh\sqrt{\lambda_1}\ell}\Big)^2 dx_1 = \int_{-\ell}^{\ell}\Big(\frac{e^{\sqrt{\lambda_1}x_1}+e^{-\sqrt{\lambda_1}x_1}}{e^{\sqrt{\lambda_1}\ell}+e^{-\sqrt{\lambda_1}\ell}}\Big)^2 dx_1$$

$$= \int_{-\ell}^{\ell}\Big(\frac{e^{2\sqrt{\lambda_1}x_1}+2+e^{-2\sqrt{\lambda_1}x_1}}{e^{2\sqrt{\lambda_1}\ell}+2+e^{-2\sqrt{\lambda_1}\ell}}\Big)dx_1$$

$$= \frac{1}{2\sqrt{\lambda_1}}\Big[\frac{e^{2\sqrt{\lambda_1}x_1}+4\sqrt{\lambda_1}x_1-e^{-2\sqrt{\lambda_1}x_1}}{e^{2\sqrt{\lambda_1}\ell}+2+e^{-2\sqrt{\lambda_1}\ell}}\Big]_{-\ell}^{\ell}$$

$$= \frac{1}{\sqrt{\lambda_1}}\frac{e^{2\sqrt{\lambda_1}\ell}+4\sqrt{\lambda_1}\ell-e^{-2\sqrt{\lambda_1}\ell}}{e^{2\sqrt{\lambda_1}\ell}+2+e^{-2\sqrt{\lambda_1}\ell}}.$$

Thus one has clearly

$$\int_{-\ell}^{\ell}\Big(\frac{\cosh\sqrt{\lambda_1}x_1}{\cosh\sqrt{\lambda_1}\ell}\Big)^2 dx_1 \int_{\omega}(\partial_{x_2}u_\infty)^2 dx_2 \to \frac{1}{\sqrt{\lambda_1}}\int_{\omega}(\partial_{x_2}u_\infty)^2 dx_2 \neq 0$$

when $\ell \to \infty$. This shows (1.64) and completes the proof of the theorem.

$\square$

1.12 Correctors for the Dirichlet problem

If, as we saw in the previous section, the convergence fails near the extremities of the rectangle $\Omega_\ell = (-\ell, \ell) \times (-1, 1)$, one can somehow have some estimates of the discrepancy of the global convergence and introduce correctors which give more information on the behaviour of u_ℓ near the extremities of the domain. We are going to explain this in the case where

$$\Omega_\ell = (-\ell, \ell) \times \omega'.$$

ω' is here a bounded open set of $\mathbb{R}^{n-1}$. We will denote by

$$x = (x_1, x'), \quad x' = x_2, \dots, x_n$$

the points in Ω_ℓ. For

$$f = f(x') \in L^2(\omega')$$

we denote by u_ℓ the solution to

$$\begin{cases} u_\ell \in H_0^1(\Omega_\ell), \\ \int_{\Omega_\ell} \nabla u_\ell \cdot \nabla v\, dx = \int_{\Omega_\ell} fv\, dx, \quad \forall v \in H_0^1(\Omega_\ell), \end{cases} \tag{1.66}$$

and u_∞ the solution to

$$\begin{cases} u_\infty \in H_0^1(\omega'), \\ \int_{\omega'} \nabla' u_\infty \cdot \nabla' v\, dx' = \int_{\omega'} fv\, dx', \quad \forall v \in H_0^1(\omega') \end{cases} \tag{1.67}$$

$(\nabla' = (\partial_{x_2}, \ldots, \partial_{x_n}))$ and with an obvious notation for dx'). Let us denote by Ω_ℓ^+ and Ω_ℓ^- the sets defined by

$$\Omega_\ell^+ = (0, \ell) \times \omega', \quad \Omega_\ell^- = (-\ell, 0) \times \omega'$$

and set

$$V_\ell = \{v \in H^1(\Omega_\ell^+) \mid v = 0 \text{ on } \partial\Omega_\ell^+ \backslash (\{0\} \times \omega')\}.$$

We will need a series of lemmas. First we have

Lemma 1.4. *If u_ℓ is solution to* (1.66) *one has*

$$u_\ell(-x_1, x') = u_\ell(x_1, x'). \tag{1.68}$$

Proof. For $v \in H_0^1(\Omega_\ell)$ we denote by $\tilde{v}$ the function defined as

$$\tilde{v}(x_1, x') = v(-x_1, x').$$

One has

$$\int_{\Omega_\ell} \nabla\tilde{u}_\ell \cdot \nabla v \, dx = \int_{\Omega_\ell} \nabla\tilde{u}_\ell \cdot \nabla\tilde{\tilde{v}} \, dx = \int_{\Omega_\ell} \nabla u_\ell \cdot \nabla\tilde{v} \, dx.$$

(We made the change of variable $-x_1 \to x_1$ in the middle integral). Thus, by (1.66)

$$\int_{\Omega_\ell} \nabla\tilde{u}_\ell \cdot \nabla v \, dx = \int_{\Omega_\ell} f\tilde{v} \, dx = \int_{\Omega_\ell} f v \, dx$$

by making again the change of variable $-x_1 \to x_1$. Thus by uniqueness of the solution to (1.66) we have

$$\tilde{u}_\ell = u_\ell$$

which is (1.68). This completes the proof of the lemma. $\qquad\square$

We have also

Lemma 1.5. *We keep denoting by u_ℓ the restriction of u_ℓ to Ω_ℓ^+. Then u_ℓ is the unique solution to*

$$\begin{cases} u_\ell \in V_\ell, \\ \int_{\Omega_\ell^+} \nabla u_\ell \cdot \nabla v \, dx = \int_{\Omega_\ell^+} f v \, dx, \quad \forall v \in V_\ell. \end{cases} \tag{1.69}$$

Proof. Consider $v \in V_\ell$ and define

$$\hat{v}(x_1, x') = \begin{cases} v(x_1, x') & \text{for } x_1 \geq 0, \\ v(-x_1, x') & \text{for } x_1 \leq 0. \end{cases}$$

Clearly $\hat{v} \in H_0^1(\Omega_\ell)$. Moreover

$$
\int_{\Omega_\ell} \nabla u_\ell \cdot \nabla \hat{v} dx = \int_{\Omega_\ell^+} \nabla u_\ell \cdot \nabla v dx
$$
$$
+ \int_{\Omega_\ell^-} \nabla u_\ell(x_1, x') \cdot \nabla \{v(-x_1, x')\} dx. \tag{1.70}
$$

The second integral I above can be written as

$$
I = \int_{\Omega_\ell^-} -\partial_{x_1} u_\ell(x_1, x')(\partial_{x_1} v)(-x_1, x') + \nabla' u_\ell(x_1, x') \cdot \nabla' v(-x_1, x') dx
$$
$$
= \int_{\Omega_\ell^-} \partial_{x_1} u_\ell(-x_1, x')(\partial_{x_1} v)(-x_1, x') + \nabla' u_\ell(-x_1, x') \cdot \nabla' v(-x_1, x') dx
$$

since by Lemma 1.4

$$
\partial_{x_1} u_\ell(-x_1, x') = -\partial_{x_1} u_\ell(x_1, x'), \ \nabla' u_\ell(-x_1, x') = \nabla' u_\ell(x_1, x').
$$

By the change of variable $-x_1 \to x_1$ we get then

$$
I = \int_{\Omega_\ell^-} \nabla u_\ell(-x_1, x') \cdot \nabla v(-x_1, x') dx = \int_{\Omega_\ell^+} \nabla u_\ell \cdot \nabla v dx
$$

and from (1.70) we deduce

$$
\int_{\Omega_\ell} \nabla u_\ell \cdot \nabla \hat{v} dx = 2 \int_{\Omega_\ell^+} \nabla u_\ell \cdot \nabla v dx \ \ \forall v \in V_\ell.
$$

By (1.66) one has

$$
\int_{\Omega_\ell} \nabla u_\ell \cdot \nabla \hat{v} dx = \int_{\Omega_\ell} f \hat{v} dx = 2 \int_{\Omega_\ell^+} f v dx \ \ \forall v \in V_\ell.
$$

Thus one has

$$
\int_{\Omega_\ell^+} \nabla u_\ell \cdot \nabla v dx = \int_{\Omega_\ell^+} f v dx \ \ \forall v \in V_\ell.
$$

This completes the proof of the lemma. $\qquad \square$

We can then prove

Lemma 1.6. *Let u_ℓ, u_∞ the solutions to (1.66), (1.67) respectively. Then one has*

$$
\int_{\Omega_\ell^+} \nabla(u_\ell - u_\infty) \cdot \nabla v dx = 0, \ \ \forall v \in V_\ell. \tag{1.71}
$$

Proof. For $v \in V_\ell$, $v(x_1, \cdot) \in H_0^1(\omega')$ for a.e. $x_1 \in (0, \ell)$ and thus by (1.67)

$$\int_{\omega'} \nabla' u_\infty \cdot \nabla' v(x_1, \cdot) dx' = \int_{\omega'} f v(x_1, \cdot) dx'.$$

Integrating this equality between 0 and ℓ and taking into account the independence of u_∞ in x_1, we get

$$\int_{\Omega_\ell^+} \nabla u_\infty \cdot \nabla v dx = \int_{\Omega_\ell^+} f v dx.$$

It follows then from Lemma 1.5 that

$$\int_{\Omega_\ell^+} \nabla(u_\ell - u_\infty) \cdot \nabla v dx = 0, \quad \forall v \in V_\ell.$$

This completes the proof of the lemma. □

Let us set

$$\Omega_\infty^+ = (0, \infty) \times \omega'$$

and introduce w the solution to

$$\begin{cases} w \in H^1(\Omega_\infty^+), \ w = -u_\infty \text{ on } \{0\} \times \omega', w = 0 \text{ on } (0, \infty) \times \partial \omega', \\ \int_{\Omega_\infty^+} \nabla w \cdot \nabla v dx = 0, \quad \forall v \in H_0^1(\Omega_\infty^+). \end{cases} \quad (1.72)$$

We have

Lemma 1.7. *There exists a unique weak solution to* (1.72).

Proof. Let $\rho = \rho(x_1)$ be a smooth function such that

$$\rho(0) = 1, \quad \rho(x_1) = 0 \text{ for } x_1 \geq 1.$$

By the Lax-Milgram theorem there exists a unique u solution to

$$\begin{cases} u \in H_0^1(\Omega_\infty^+), \\ \int_{\Omega_\infty^+} \nabla u \cdot \nabla v dx = \int_{\Omega_\infty^+} \nabla(\rho u_\infty) \cdot \nabla v dx, \quad \forall v \in H_0^1(\Omega_\infty^+). \end{cases}$$

Then $w = u - \rho u_\infty$ is solution to (1.72). This completes the proof of the lemma. □

We denote by w_ℓ the function defined as

$$w_\ell(x_1, x') = w(\ell - x_1, x') \text{ a.e. } x_1 < \ell.$$

One can then show

Lemma 1.8. *One has*

$$\int_{\Omega_\ell^+} \nabla w_\ell \cdot \nabla v dx = -\int_{\Omega_\ell^-} \nabla w_\ell \cdot \nabla \hat{v} dx, \quad \forall v \in V_\ell.$$

($\hat{v}$ is the function defined in the proof of Lemma 1.5).

Proof. If $v \in V_\ell$ then $\hat{v} \in H_0^1(\Omega_\ell)$ and $\hat{v}(\ell - x_1, x') \in H_0^1(\Omega_{2\ell}^+)$. Thus by (1.72)

$$\int_{\Omega_{2\ell}^+} \nabla w \cdot \nabla\{\hat{v}(\ell - x_1, x')\}dx = 0. \tag{1.73}$$

One has first

$$\int_{\Omega_\ell^+} \nabla w \cdot \nabla\{\hat{v}(\ell - x_1, x')\}dx = \int_{\Omega_\ell^+} \nabla w \cdot \nabla\{v(\ell - x_1, x')\}dx$$

$$= \int_{\Omega_\ell^+} -\partial_{x_1} w(x_1, x')\partial_{x_1} v(\ell - x_1, x') + \nabla' w(x_1, x') \cdot \nabla' v(\ell - x_1, x')dx.$$

Making the change of variable $\ell - x_1 \to x_1$, this can be written as

$$\int_{\Omega_\ell^+} \nabla w \cdot \nabla\{\hat{v}(\ell - x_1, x')\}dx$$

$$= \int_{\Omega_\ell^+} \partial_{x_1}\{w(\ell - x_1, x')\}\partial_{x_1} v(x_1, x') + \nabla' w(\ell - x_1, x') \cdot \nabla' v(x_1, x')dx$$

$$= \int_{\Omega_\ell^+} \nabla w_\ell \cdot \nabla v dx.$$

One has also

$$\int_{\Omega_{2\ell}^+ \setminus \Omega_\ell^+} \nabla w \cdot \nabla\{\hat{v}(\ell - x_1, x')\}dx$$

$$= \int_{\Omega_{2\ell}^+ \setminus \Omega_\ell^+} \partial_{x_1} w(x_1, x')\partial_{x_1} v(x_1 - \ell, x') + \nabla' w \cdot \nabla' v(x_1 - \ell, x')dx.$$

Making the change of variable $\ell - x_1 \to x_1$ it comes

$$\int_{\Omega_{2\ell}^+ \setminus \Omega_\ell^+} \nabla w \cdot \nabla\{\hat{v}(\ell - x_1, x')\}dx$$

$$= \int_{\Omega_\ell^-} -\partial_{x_1}\{w(\ell - x_1, x')\}\partial_{x_1} v(-x_1, x') + \nabla' w(\ell - x_1, x') \cdot \nabla' v(-x_1, x')dx$$

$$= \int_{\Omega_\ell^-} \partial_{x_1}\{w(\ell - x_1, x')\}\partial_{x_1}\hat{v}(x_1, x') + \nabla' w(\ell - x_1, x') \cdot \nabla'\hat{v}(x_1, x')dx$$

$$= \int_{\Omega_\ell^-} \nabla w_\ell \cdot \nabla\hat{v}dx.$$

Then the result follows from (1.73). This completes the proof of the lemma.

$\square$

Let us show now

Lemma 1.9. *One has*

$$\int_{\Omega_\ell^+} |\nabla(u_\ell - u_\infty - w_\ell)|^2 dx \leq \int_{\Omega_{2\ell}^+ \setminus \Omega_\ell^+} |\nabla w|^2 dx. \qquad (1.74)$$

Proof. We notice first that $u_\ell - u_\infty - w_\ell \in V_\ell$. Thus by Lemma 1.6 one has

$$\int_{\Omega_\ell^+} \nabla(u_\ell - u_\infty) \cdot \nabla(u_\ell - u_\infty - w_\ell) dx = 0.$$

This can be written taking into account Lemma 1.8

$$\begin{aligned}
\int_{\Omega_\ell^+} |\nabla(u_\ell - u_\infty - w_\ell)|^2 dx &= -\int_{\Omega_\ell^+} \nabla w_\ell \cdot \nabla(u_\ell - u_\infty - w_\ell) dx \\
&= \int_{\Omega_\ell^-} \nabla w_\ell \cdot \nabla(\widehat{u_\ell - u_\infty} - w_\ell) dx \\
&\leq \{\int_{\Omega_\ell^-} |\nabla w_\ell|^2 dx\}^{\frac{1}{2}} \{\int_{\Omega_\ell^-} |\nabla(\widehat{u_\ell - u_\infty} - w_\ell)|^2 dx\}^{\frac{1}{2}} \\
&= \{\int_{\Omega_\ell^-} |\nabla w_\ell|^2 dx\}^{\frac{1}{2}} \{\int_{\Omega_\ell^+} |\nabla(u_\ell - u_\infty - w_\ell)|^2 dx\}^{\frac{1}{2}}
\end{aligned}$$

after a change of variable in the last integral. It follows that

$$\begin{aligned}
\int_{\Omega_\ell^+} |\nabla(u_\ell - u_\infty - w_\ell)|^2 dx &\leq \int_{\Omega_\ell^-} |\nabla w_\ell|^2 dx \\
&= \int_{\Omega_\ell^-} |\nabla w(\ell - x_1, x')|^2 dx \\
&= \int_{\Omega_{2\ell}^+ \setminus \Omega_\ell^+} |\nabla w|^2 dx.
\end{aligned}$$

This completes the proof of the lemma. $\qquad \square$

Denote by S_ℓ the "strip" defined as

$$S_\ell = (\ell, +\infty) \times \omega', \quad \ell \geq 0.$$

We have

Lemma 1.10. *There exist positive constants C and α such that*

$$||\nabla w||_{2,S_\ell}^2 \leq C e^{-\alpha \ell} ||\nabla w||_{2,S_0}^2. \qquad (1.75)$$

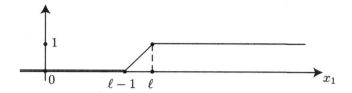

Fig. 1.6   Graph of the function ρ.

Proof. For $\ell \geq 1$ let ρ be the function whose graph is depicted on Figure 1.6.

Since $\rho w \in H_0^1(\Omega_\infty^+) = H_0^1(S_0)$ we have (see (1.72))

$$\int_{S_0} \nabla w \cdot \nabla(\rho w)dx = 0.$$

This implies

$$\int_{S_0} |\nabla w|^2 \rho\, dx = -\int_{S_0} \nabla w \cdot \nabla \rho\, w\, dx \leq \int_{S_{\ell-1}\setminus S_\ell} |\nabla w||w| dx$$

$$\leq \frac{1}{2}\int_{S_{\ell-1}\setminus S_\ell} |\nabla w|^2 dx + \frac{1}{2}\int_{S_{\ell-1}\setminus S_\ell} |w|^2 dx$$

by the Young inequality. Using the Poincaré inequality we deduce for some constant C independent of ℓ (note that $\rho = 1$ on S_ℓ)

$$\int_{S_\ell} |\nabla w|^2 dx \leq C \int_{S_{\ell-1}\setminus S_\ell} |\nabla w|^2 dx$$

$$= C \int_{S_{\ell-1}} |\nabla w|^2 dx - C \int_{S_\ell} |\nabla w|^2 dx$$

i.e.

$$\int_{S_\ell} |\nabla w|^2 dx \leq \frac{C}{1+C} \int_{S_{\ell-1}} |\nabla w|^2 dx.$$

Note that $\frac{C}{1+C} < 1$. Iterating $[\ell]$ times this formula we get

$$\int_{S_\ell} |\nabla w|^2 dx \leq (\frac{C}{1+C})^{[\ell]} \int_{S_{\ell-[\ell]}} |\nabla w|^2 dx.$$

Since $\ell - 1 \leq [\ell] \leq \ell$ this leads to

$$\int_{S_\ell} |\nabla w|^2 dx \leq e^{-\alpha(\ell-1)} \int_{S_0} |\nabla w|^2 dx$$

with $\alpha = \ln \frac{1+C}{C}$. This completes the proof of the lemma. $\qquad \square$

Then we can show

Theorem 1.9. *We have*

$$\int_{\Omega_\ell} |\nabla(u_\ell - u_\infty - \hat{w}_\ell)|^2 dx \leq 2Ce^{-\alpha\ell} ||\nabla w||^2_{2,S_0}. \tag{1.76}$$

For some positive constants C', α' we have

$$\int_{\Omega_{\frac{\ell}{2}}} |\nabla(u_\ell - u_\infty)|^2 dx \leq C'e^{-\alpha'\ell} \tag{1.77}$$

but for any $a > 0$

$$\int_{\Omega_{\ell-a}} |\nabla(u_\ell - u_\infty)|^2 dx \not\to 0. \tag{1.78}$$

Proof. $\hat{w}_\ell$ is a corrector allowing global convergence on Ω_ℓ. To prove (1.76) note that by (1.68), (1.74), (1.75)

$$\int_{\Omega_\ell} |\nabla(u_\ell - u_\infty - \hat{w}_\ell)|^2 dx = 2\int_{\Omega_\ell^+} |\nabla(u_\ell - u_\infty - w_\ell)|^2 dx$$

$$\leq 2\int_{S_\ell} |\nabla w|^2 dx \leq 2Ce^{-\alpha\ell} ||\nabla w||^2_{2,S_0}.$$

To show (1.77) using (1.76) note that

$$\int_{\Omega_{\frac{\ell}{2}}} |\nabla(u_\ell - u_\infty)|^2 dx = \int_{\Omega_{\frac{\ell}{2}}} |\nabla(u_\ell - u_\infty - \hat{w}_\ell + \hat{w}_\ell)|^2 dx$$

$$\leq 2\int_{\Omega_{\frac{\ell}{2}}} |\nabla(u_\ell - u_\infty - \hat{w}_\ell)|^2 dx + 2\int_{\Omega_{\frac{\ell}{2}}} |\nabla\hat{w}_\ell|^2 dx$$

$$\leq 4Ce^{-\alpha\ell} ||\nabla w||^2_{2,S_0} + 4\int_{\Omega_{\frac{\ell}{2}}^+} |\nabla w_\ell|^2 dx$$

$$\leq 4Ce^{-\alpha\ell} ||\nabla w||^2_{2,S_0} + 4\int_{\Omega_{\frac{\ell}{2}}^+} |\nabla w(\ell - x_1, x')|^2 dx$$

$$\leq 4Ce^{-\alpha\ell} ||\nabla w||^2_{2,S_0} + 4||\nabla w||^2_{2,S_{\frac{\ell}{2}}}$$

and (1.77) follows from (1.75).

To show (1.78) note that

$$\int_{\Omega_{\ell-a}} |\nabla(u_\ell - u_\infty)|^2 dx$$

$$= \int_{\Omega_{\ell-a}} |\nabla(u_\ell - u_\infty - \hat{w}_\ell + \hat{w}_\ell)|^2 dx$$

$$\geq \frac{1}{2} \int_{\Omega_{\ell-a}} |\nabla\hat{w}_\ell|^2 dx - \int_{\Omega_{\ell-a}} |\nabla(u_\ell - u_\infty - \hat{w}_\ell)|^2 dx$$

$$\geq \frac{1}{2} \int_{\Omega_{\ell-a}} |\nabla\hat{w}_\ell|^2 dx - \int_{\Omega_\ell} |\nabla(u_\ell - u_\infty - \hat{w}_\ell)|^2 dx$$

$$= \int_{\Omega_{\ell-a}^+} |\nabla w_\ell|^2 dx - \int_{\Omega_\ell} |\nabla(u_\ell - u_\infty - \hat{w}_\ell)|^2 dx$$

$$= \int_{\Omega_{\ell-a}^+} |\nabla w(\ell - x_1, x')|^2 dx - \int_{\Omega_\ell} |\nabla(u_\ell - u_\infty - \hat{w}_\ell)|^2 dx$$

$$= \int_{\Omega_\ell^+ \backslash \Omega_a^+} |\nabla w|^2 dx - \int_{\Omega_\ell} |\nabla(u_\ell - u_\infty - \hat{w}_\ell)|^2 dx.$$

When $\ell \to +\infty$, the right hand side of the above inequality converges toward

$$\int_{S_a} |\nabla w|^2 dx.$$

This last integral cannot vanish. Indeed, it would imply $w = 0$ on S_a. Since w is harmonic one would have $w \equiv 0$ which is impossible in the case where $u_\infty \neq 0$ i.e. when $f \not\equiv 0$. To see this if w^+, respectively w^- denote the solutions to (1.72) with respectively $w^+ = (-u_\infty)^+$, $w^- = (-u_\infty)^-$ on $\{0\} \times \omega'$ one has

$$w = w^+ - w^-,$$

$(-u_\infty)^\pm$ denote respectively the positive and negative part of $-u_\infty$.

Now if

$$\int_{S_a} |\nabla w|^2 dx = 0$$

then $\nabla w^+, \nabla w^-, w^+, w^-$ vanish on S_a. But w^+, w^- are non negative and harmonic and by the maximum principle they would have to vanish identically on Ω_∞^+ i.e. one would have $w = 0$ on Ω_∞^+ which is impossible. This completes the proof of the theorem. $\qquad\square$

Chapter 2

The Dirichlet problem in some unbounded domains

2.1 The linear case

We consider a domain Ω such that

$$\Omega \subset \mathbb{R}^p \times \omega_2$$

where ω_2 is a bounded domain in $\mathbb{R}^{n-p}$. The figure below describes the situation when $n = 2$, $p = 1$, $\omega_2 = (-a, a)$.

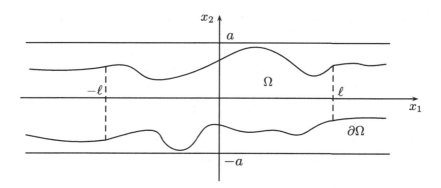

Fig. 2.1   The two dimensional situation.

A point x in Ω will be split into two parts

$$x = (X_1, X_2)$$

where $X_1 \in \mathbb{R}^p$ and $X_2 \in \mathbb{R}^{n-p}$. We denote by $A = A(x)$ a $n \times n$-matrix defined on Ω and such that for some positive constants λ, Λ one has

$$\lambda |\xi|^2 \leq A(x)\xi \cdot \xi, \quad |A(x)\xi| \leq \Lambda |\xi| \quad \forall \xi \in \mathbb{R}^n, \quad \text{a.e. } x \in \Omega. \tag{2.1}$$

(We denote by a dot the usual scalar product in $\mathbb{R}^n$, $A(x)\xi$ denotes the vector obtained when one multiplies $A(x)$ and ξ).

For $f \in H^{-1}(\Omega) = (H_0^1(\Omega))'$ the dual of $H_0^1(\Omega)$ by the Lax-Milgram theorem there exists a unique weak solution to

$$-\nabla \cdot (A(x)\nabla u) = f \text{ in } \Omega, \quad u = 0 \text{ on } \partial\Omega,$$

i.e. a unique solution to

$$\begin{cases} u \in H_0^1(\Omega), \\ \int_\Omega A(x)\nabla u \cdot \nabla v \, dx = \langle f, v \rangle \quad \forall v \in H_0^1(\Omega). \end{cases} \tag{2.2}$$

(∂A denotes the boundary of A, $\langle,\rangle$ the duality bracket between $H^{-1}(\Omega)$ and $H_0^1(\Omega)$). However as we saw in Chapter 1 there are very simple functions which are not in $H^{-1}(\Omega)$. We give here another argument. Suppose for instance $n = 2$, $\Omega = \mathbb{R} \times (-1,1)$, $f = f(x_2)$, $A(x) = Id$. If $f \in H^{-1}(\Omega)$ there exists a unique solution to

$$u \in H_0^1(\Omega), \quad \int_\Omega \nabla u \cdot \nabla v dx = \langle f, v \rangle \quad \forall v \in H_0^1(\Omega).$$

By uniqueness it is easy to show that u is translation invariant in the x_1 direction. This implies that $u = u(x_2)$. But this function cannot be in $L^2(\Omega)$ unless $u = f = 0$. So in order to establish existence of a solution to (2.2) some new arguments have to be developed.

We introduce

$$\Omega_\ell = (\ell\omega_1 \times \omega_2) \cap \Omega$$

where ω_1 is a bounded convex subset of $\mathbb{R}^p$ containing 0,

$$\ell\omega_1 = \{\ell X_1 \mid X_1 \in \omega_1\}.$$

One can think for instance for ω_1 to the unit ball in $\mathbb{R}^p$ or to the set $(-1,1)^p$. Set

$$V_\ell = H_0^1(\Omega_\ell) = \{v \in H^1(\Omega_\ell) : v = 0 \text{ on } \partial\Omega_\ell\}.$$

We suppose V_ℓ normed with the Dirichlet norm defined as

$$|u|_{V_\ell}^2 = \int_{\Omega_\ell} |\nabla u|^2 dx = ||\nabla u||_{2,\Omega_\ell}^2.$$

Let us denote by V'_ℓ the dual of V_ℓ and suppose that $f \in V'_\ell$ for every $\ell > 0$. Then there exists a unique u_ℓ solution to

$$u_\ell \in V_\ell, \quad \int_{\Omega_\ell} A(x)\nabla u_\ell \cdot \nabla v dx = \langle f, v \rangle \quad \forall v \in V_\ell. \tag{2.3}$$

($\langle \, , \, \rangle$ denotes the V'_ℓ, V_ℓ duality). Then we have:

Theorem 2.1. *Suppose that for some $\gamma > 0$*

$$|f|_{V'_\ell} = O(\ell^\gamma), \tag{2.4}$$

where $|\;|_{V'_\ell}$ is the strong dual norm in V'_ℓ. Then u_ℓ converges toward the unique weak solution to

$$\begin{cases} u_\infty \in H^1_{loc}(\overline{\Omega}), \quad u_\infty = 0 \quad on \quad \partial\Omega, \\[2mm] -\nabla \cdot (A(x)\nabla u_\infty) = f \quad in \quad \Omega, \quad \left\|\nabla u_\infty\right\|_{2,\Omega_\ell} = O(\ell^\gamma). \end{cases} \tag{2.5}$$

Moreover

$$\left\|\nabla(u_\ell - u_\infty)\right\|_{2,\Omega_{\frac{\ell}{2}}} \leq Ce^{-\beta\ell} \tag{2.6}$$

for some positive constants C and β.

Remark 2.1. We denoted by $H^1_{loc}(\overline{\Omega})$ the space defined as

$$H^1_{loc}(\overline{\Omega}) = \{v \in H^1(K), \forall K \subset \Omega, K \text{ bounded }\}.$$

Note that in order to have an existence and uniqueness result one has to impose some growth at infinity for u_∞. Indeed considering $n = 2$, $\Omega = \mathbb{R} \times (-1, 1)$, $f = 0$, $A(x) = Id$ it is clear as seen in Chapter 1 that for every constant C

$$u_\infty = Ce^{\frac{\pi}{2}x_1} \cos(\frac{\pi}{2}x_2)$$

is solution to

$$-\Delta u_\infty = 0 \text{ in } \Omega, \ u_\infty = 0 \text{ on } \partial\Omega$$

but the only solution for which the growth is bounded is obtained for $C = 0$.

Proof. For $\ell_1 \leq \ell - 1$ consider a function $\rho = \rho_{\ell_1} = \rho_{\ell_1}(X_1)$ such that

$$0 \leq \rho_{\ell_1} \leq 1, \ \rho_{\ell_1} = 1 \text{ on } \ell_1\omega_1, \ \rho_{\ell_1} = 0 \text{ outside } (\ell_1 + 1)\omega_1, \\ |\nabla_{X_1}\rho_{\ell_1}| \leq C \tag{2.7}$$

where C is some constant independent of ℓ_1 and $\nabla_{X_1} = (\partial_{x_1}, \cdots, \partial_{x_p})$ is the gradient in $\mathbb{R}^p$. It is clear from our assumptions that such a function exists. Then we proceed in several steps in order to show that u_ℓ is a Cauchy sequence.

a) Estimate of $u_\ell - u_{\ell+r}$ for $0 \leq r \leq 1$

One has clearly

$$(u_\ell - u_{\ell+r})\rho(X_1) \in V_\ell \cap V_{\ell+r}.$$

Thus from (2.3) one deduces (note that $(u_\ell - u_{\ell+r})\rho$ vanishes outside Ω_{ℓ_1+1})

$$\int_{\Omega_{\ell_1+1}} A(x)\nabla(u_\ell - u_{\ell+r}) \cdot \nabla\{(u_\ell - u_{\ell+r})\rho\}dx = 0.$$

This implies easily by (2.1)

$$\lambda \int_{\Omega_{\ell_1}} |\nabla(u_\ell - u_{\ell+r})|^2 \, dx$$

$$\leq \int_{\Omega_{\ell_1+1}} A(x)\nabla(u_\ell - u_{\ell+r}) \cdot \nabla(u_\ell - u_{\ell+r})\rho dx$$

$$= -\int_{\Omega_{\ell_1+1}\backslash\Omega_{\ell_1}} A(x)\nabla(u_\ell - u_{\ell+r}) \cdot \nabla\rho \, (u_\ell - u_{\ell+r})dx$$

$$\leq C\Lambda \int_{D_{\ell_1}} |\nabla(u_\ell - u_{\ell+r})||u_\ell - u_{\ell+r}|dx$$

$$\leq C\Lambda\{\frac{\epsilon}{2} \int_{D_{\ell_1}} |\nabla(u_\ell - u_{\ell+r})|^2 dx + \frac{1}{2\epsilon} \int_{D_{\ell_1}} (u_\ell - u_{\ell+r})^2 dx\}$$

($D_{\ell_1} = \Omega_{\ell_1+1}\backslash\Omega_{\ell_1}$, the last inequality follows from Young's inequality). One has the Poincaré inequality (without choosing here the best constant, d being for instance the diameter of ω_2)

$$\int_{D_{\ell_1}} (u_\ell - u_{\ell+r})^2 dx \leq \frac{d^2}{2} \int_{D_{\ell_1}} |\nabla_{X_2}(u_\ell - u_{\ell+r})|^2 dx$$

and thus we derive

$$\int_{\Omega_{\ell_1}} |\nabla(u_\ell - u_{\ell+r})|^2 dx$$

$$\leq \frac{C\Lambda\epsilon}{2\lambda} \int_{D_{\ell_1}} |\nabla(u_\ell - u_{\ell+r})|^2 dx + \frac{C\Lambda d^2}{4\epsilon\lambda} \int_{D_{\ell_1}} |\nabla_{X_2}(u_\ell - u_{\ell+r})|^2 dx.$$

$$(2.8)$$

Choosing $\epsilon = d$ we obtain for $a = \frac{3\Lambda d}{4\lambda}$

$$\int_{\Omega_{\ell_1}} |\nabla(u_\ell - u_{\ell+r})|^2 dx \leq a \int_{\Omega_{\ell_1+1}\backslash\Omega_{\ell_1}} |\nabla(u_\ell - u_{\ell+r})|^2 dx$$

i.e.

$$\int_{\Omega_{\ell_1}} |\nabla(u_\ell - u_{\ell+r})|^2 dx \le \frac{a}{a+1} \int_{\Omega_{\ell_1+1}} |\nabla(u_\ell - u_{\ell+r})|^2 dx.$$

Starting from $\frac{\ell}{2}$ and iterating $[\frac{\ell}{2}]$ times this inequality we get

$$\int_{\Omega_{\frac{\ell}{2}}} |\nabla(u_\ell - u_{\ell+r})|^2 dx \le \frac{a+1}{a} \left(\frac{a}{a+1}\right)^{\frac{\ell}{2}} \int_{\Omega_\ell} |\nabla(u_\ell - u_{\ell+r})|^2 dx$$

$$\le Ce^{-\alpha\ell} \int_{\Omega_\ell} |\nabla(u_\ell - u_{\ell+r})|^2 dx \qquad (2.9)$$

with $C = \frac{a+1}{a}$, $\alpha = \frac{1}{2}\ln(\frac{a+1}{a})$.

b) Estimate of u_ℓ

We just take $v = u_\ell$ in (2.3) to get

$$\lambda \int_{\Omega_\ell} |\nabla u_\ell|^2 dx \le \int_{\Omega_\ell} A(x)\nabla u_\ell \cdot \nabla u_\ell dx = \langle f, u_\ell \rangle \le |f|_{V'_\ell} ||\nabla u_\ell||_{2,\Omega_\ell}.$$

It follows by (2.4) that

$$\int_{\Omega_\ell} |\nabla u_\ell|^2 dx \le \frac{|f|^2_{V'_\ell}}{\lambda^2} = O(\ell^{2\gamma}). \qquad (2.10)$$

c) u_ℓ is a Cauchy sequence

Combining (2.9) and (2.10) we have for different constants C and $2\beta < \alpha$

$$\int_{\Omega_{\frac{\ell}{2}}} |\nabla(u_\ell - u_{\ell+r})|^2 dx \le Ce^{-\alpha\ell} 2\{||\nabla u_\ell||^2_{2,\Omega_\ell} + ||\nabla u_{\ell+r}||^2_{2,\Omega_{\ell+r}}\}$$

$$\le Ce^{-\alpha\ell}\{\ell^{2\gamma} + (\ell+r)^{2\gamma}\}$$

$$\le C\ell^{2\gamma}e^{-\alpha\ell}\{1 + (1 + \frac{r}{\ell})^{2\gamma}\} \le C^2 e^{-2\beta\ell}$$

when $\ell > 1$ -recall that $0 \le r \le 1$. This can be written as

$$|u_\ell - u_{\ell+r}|_{V_{\frac{\ell}{2}}} \le Ce^{-\beta\ell}.$$

Thus for $t > 0$ arbitrary we derive then

$$|u_\ell - u_{\ell+t}|_{V_{\frac{\ell}{2}}} \le |u_\ell - u_{\ell+1}|_{V_{\frac{\ell}{2}}}$$

$$+ |u_{\ell+1} - u_{\ell+2}|_{V_{\frac{\ell+1}{2}}} + \cdots + |u_{\ell+[t]} - u_{\ell+t}|_{V_{\frac{\ell+[t]}{2}}}$$

$$\le Ce^{-\beta\ell} + Ce^{-\beta(\ell+1)} + \cdots + Ce^{-\beta(\ell+[t])} \qquad (2.11)$$

$$\le Ce^{-\beta\ell}\{1 + e^{-\beta} + \cdots + e^{-\beta[t]}\}$$

$$\le C\frac{1}{1 - e^{-\beta}}e^{-\beta\ell} = C'e^{-\beta\ell}$$

independently of t.

Suppose that we choose $\ell_0 \leq \frac{\ell}{2}$ then it follows from above that u_ℓ is a Cauchy sequence in $H^1(\Omega_{\ell_0})$. It converges toward u_∞ such that $u_\infty = 0$ on $\partial\Omega_{\ell_0} \cap \partial\Omega$. Moreover, passing to the limit in t in (2.11) provides (2.6).

d) Limit problem

From above we have for every ℓ_0

$$\int_{\Omega_{\ell_0}} A(x)\nabla u_\ell \cdot \nabla v dx = \langle f, v \rangle \quad \forall v \in H_0^1(\Omega_{\ell_0})$$

and passing to the limit in ℓ we get

$$\int_{\Omega_{\ell_0}} A(x)\nabla u_\infty \cdot \nabla v dx = \langle f, v \rangle \quad \forall v \in H_0^1(\Omega_{\ell_0}).$$

e) Estimate of $\left\|\nabla u_\infty\right\|_{2,\Omega_\ell}$

From (2.11) one has

$$|u_\ell - u_{\ell+t}|_{V_{\frac{\ell}{2}}} \leq C'e^{-\beta\ell}$$

i.e.

$$|u_{2\ell} - u_{2\ell+t}|_{V_\ell} \leq C'e^{-2\beta\ell}.$$

Passing to the limit in t we obtain

$$|u_{2\ell} - u_\infty|_{V_\ell} \leq C'e^{-2\beta\ell}$$

which implies by (2.10)

$$|u_\infty|_{V_\ell} \leq C'e^{-2\beta\ell} + |u_{2\ell}|_{V_\ell} \leq C'e^{-2\beta\ell} + |u_{2\ell}|_{V_{2\ell}}$$
$$\leq C'e^{-\beta\ell} + C''(2\ell)^\gamma = O(\ell^\gamma).$$

f) Uniqueness

Suppose that u_∞, u_∞' are two solutions to (2.5). Then one has

$$\int_{\Omega_\ell} A(x)\nabla(u_\infty - u_\infty') \cdot \nabla v dx = 0 \quad \forall v \in H_0^1(\Omega_\ell).$$

Taking $v = (u_\infty - u_\infty')\rho_{\ell_1}$ and arguing as in the step a) above will lead to

$$\int_{\Omega_{\frac{\ell}{2}}} |\nabla(u_\infty - u_\infty')|^2 dx \leq C^2 e^{-2\beta\ell}.$$

Letting $\ell \to \infty$ leads to $u_\infty = u_\infty'$. This completes the proof of the theorem.

$\square$

Remark 2.2. One can replace the assumption $|f|_{V'_\ell} = O(\ell^\gamma)$ by $|f|_{V'_\ell} = O(e^{\gamma\ell})$ for some small enough γ.

Note that we are also recovering here the Theorem 1.2. Indeed in the case of Theorem 1.2, f is independent of x_1 and

$$f \in L^2(\omega_2).$$

Due to the Poincaré inequality, we have

$$\langle f, v \rangle = \int_{\Omega_\ell} fv\,dx \leq |f|_{2,\Omega_\ell}|v|_{2,\Omega_\ell} \leq \sqrt{2}|f|_{2,\Omega_\ell}||\nabla v||_{2,\Omega_\ell}$$

i.e.

$$|f|_{V'_\ell} \leq \sqrt{2}|f|_{2,\Omega_\ell} = \sqrt{2}\Big(\int_{-\ell}^{\ell}\int_{-1}^{1} f^2(x_2)\,dx_2\,dx_1\Big)^{\frac{1}{2}} = 2\sqrt{\ell}|f|_{2,\omega_2}$$

and

$$|f|_{V'_\ell} = O(\ell^{\frac{1}{2}}).$$

Thus applying Theorem 2.1 we see that u_ℓ converges toward the unique solution to (2.5). But it is easy to see that u_∞ solution to (1.6) satisfies (2.5) hence the recovery of Theorem 1.2. One can generalize this. Suppose indeed that $\Omega = \mathbb{R}^p \times \omega_2$ in such a way that

$$\Omega_\ell = \ell\omega_1 \times \omega_2$$

(ω_1 is as before a convex bounded open subset of $\mathbb{R}^p$ containing 0). Consider

$$f = f(X_2) \in H^{-1}(\omega_2).$$

Suppose that A is an elliptic matrix of the type

$$A(x) = \begin{pmatrix} A_{11}(x) & A_{12}(X_2) \\ A_{21}(x) & A_{22}(X_2) \end{pmatrix}$$

where A_{11}, A_{22} are respectively $p \times p$ and $(n-p) \times (n-p)$-matrices. One can define a continuous linear form on V_ℓ by setting

$$\langle f, v \rangle = \int_{\ell\omega_1} \langle f, v \rangle_{-1,1}\,dX_1 \tag{2.12}$$

(we denoted by $\langle, \rangle$ the V'_ℓ, V_ℓ duality and by $\langle, \rangle_{-1,1}$ the $H^{-1}(\omega_2), H^1_0(\omega_2)$ duality, $|\;|_{-1,\omega_2}$ will denote the strong dual norm in $H^{-1}(\omega_2)$). Indeed this follows from

$$|\langle f, v \rangle_{-1,1}| \leq |f|_{-1,\omega_2}||\nabla_{X_2}v(X_1,.)||_{2,\omega_2}.$$

In addition one has

$$
|\langle f, v \rangle| \leq |f|_{-1,\omega_2} \int_{\ell\omega_1} \left\| \nabla_{X_2} v(X_1, .) \right\|_{2,\omega_2} dX_1
$$

$$
\leq |f|_{-1,\omega_2} |\ell\omega_1|^{\frac{1}{2}} \left(\int_{\ell\omega_1} \left\| \nabla_{X_2} v(X_1, .) \right\|_{2,\omega_2}^2 dX_1 \right)^{\frac{1}{2}}
$$

$$
= \ell^{\frac{p}{2}} |\omega_1|^{\frac{1}{2}} |f|_{-1,\omega_2} \left\| \nabla_{X_2} v \right\|_{2,\Omega_\ell}
$$

and thus

$$
|f|_{V'_\ell} \leq \ell^{\frac{p}{2}} |\omega_1|^{\frac{1}{2}} |f|_{-1,\omega_2}
$$

where $|\omega_1|$ denotes the measure of ω_1. It follows that one can apply Theorem 2.1 and u_ℓ solution to (2.3) with f defined as (2.12) converges toward the unique solution to (2.5). It is a straightforward computation to check that this solution is also u_∞ solution to

$$
\begin{cases}
u_\infty \in H_0^1(\omega_2), \\[2mm]
\int_{\omega_2} A_{22}(X_2) \nabla_{X_2} u_\infty \cdot \nabla_{X_2} v \, dX_2 = \langle f, v \rangle_{-1,1} \quad \forall v \in H_0^1(\omega_2),
\end{cases}
\tag{2.13}
$$

i.e. f independent of X_1 forces in this case the limit solution to be independent of X_1 as well. (Note that u_∞ solution to (2.13) exists by the Lax-Milgram theorem and it satisfies also (2.5)).

Remark 2.3. For simplicity we took

$$
V_\ell = \{ v \in H^1(\Omega_\ell) \mid v = 0 \text{ on } \partial\Omega_\ell \}.
$$

One can replace V_ℓ by any subspace $\hat{V}_\ell$ of

$$
\{ v \in H^1(\Omega_\ell) \mid v = 0 \text{ on } \partial\Omega_\ell \cap \partial\Omega \}
$$

containing V_ℓ and get convergence of u_ℓ - solution to (2.3) where V_ℓ is replaced by $\hat{V}_\ell$ - toward the solution to (2.5). Indeed in the step a) of the proof we would have also

$$
(u_\ell - u_{\ell+r})\rho \in V_\ell \cap V_{\ell+r} \subset \hat{V}_\ell \cap \hat{V}_{\ell+r}
$$

and one can then proceed as there. This shows that only the lateral boundary conditions are here playing a role.

Remark 2.4. Instead of the operator

$$-\nabla \cdot (A(x)\nabla u)$$

one could add a lower order term i.e. consider the operator

$$-\nabla \cdot (A(x)\nabla u) + a(x)u$$

where a is some bounded function. When $a \geq 0$ the analysis above goes through with no modification. We refer the reader to Section 6.4 for the case where a is allowed to change sign.

Remark 2.5. We supposed above that Ω_ℓ is bounded in one direction. This can be relaxed to treat more complicated situation like for instance the one depicted in the figure below in dimension 2.

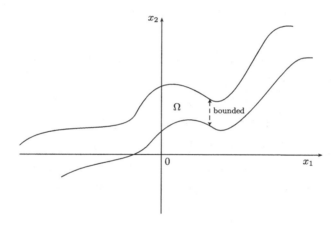

Fig. 2.2   Another possible unbounded situation.

2.2   A quasilinear case

As above we consider a domain Ω such that

$$\Omega \subset \mathbb{R}^p \times \omega_2 = S \subset \mathbb{R}^n$$

where ω_2 is a bounded domain of $\mathbb{R}^{n-p}$. We denote by $A(x,u)$ a $n \times n$-matrix where the entries $A_{i,j}$ are Carathéodory functions -i.e. measurable in x for every $u \in \mathbb{R}$ and continuous in u for almost every $x \in S$. Moreover we assume that A satisfies

$$\lambda|\xi|^2 \leq A(x,u)\xi \cdot \xi , \ |A(x,u)\xi| \leq \Lambda|\xi| \ \forall \xi \in \mathbb{R}^n, \ a.e. \ x \in S, \ \forall u \in \mathbb{R} \quad (2.14)$$

and (with the summation convention in j)

$$|\partial_{x_j} A_{i,j}(x,u)| \leq \Lambda \quad a.e. \ x \in S, \quad \forall u \in \mathbb{R}, \quad \forall i \tag{2.15}$$

for some positive constants λ and Λ -i.e. A is uniformly elliptic and bounded. We will use also $|\ |$ to denote the operator norm of matrices defined as

$$|A| = \sup_{\xi \neq 0} \frac{|A\xi|}{|\xi|}$$

and we will assume that there exists a non decreasing function ω such that

$$|A(x,u) - A(x,v)| \leq \omega(|u-v|) \quad a.e. \ x \in S, \quad \forall u,v \in \mathbb{R} \tag{2.16}$$

where ω satisfies

$$\int_{0+} \frac{ds}{\omega^2(s)} = +\infty. \tag{2.17}$$

Remark 2.6. The assumptions (2.16), (2.17) are satisfied for instance for $A(x,\cdot)$ Hölder continuous of order $1/2$.

We would like then to find u solution to

$$\begin{cases} -\nabla \cdot (A(x,u)\nabla u) = f & \text{in } \Omega, \\ u = 0 & \text{on } \partial\Omega, \end{cases}$$

for some class of functions or distributions f. If ω_1 is a bounded, convex open set containing 0 we denote by Ω_ℓ the open set

$$\Omega_\ell = (\ell\omega_1 \times \omega_2) \cap \Omega$$

and we set

$$V_\ell = \{v \in H^1(\Omega_\ell) \mid v = 0 \text{ on } \partial\Omega_\ell \cap \partial\Omega\}.$$

We suppose V_ℓ equipped with the usual Dirichlet norm and denote by V_ℓ^* the dual of V_ℓ. One can easily show, using the Schauder fixed point theorem (see [13]) that for $f \in V_\ell^*$ there exists a unique u_ℓ solution to

$$\begin{cases} u_\ell \in V_\ell, \\ \int_{\Omega_\ell} A(x,u_\ell)\nabla u_\ell \cdot \nabla v \ dx = \langle f,v \rangle & \forall v \in V_\ell, \end{cases} \tag{2.18}$$

where $\langle \ , \ \rangle$ denotes the duality bracket between V_ℓ^* and V_ℓ. Then we have:

Theorem 2.2. *Suppose that for some $\gamma > 0$*

$$|f|_{V_\ell^*} = O(\ell^\gamma) \tag{2.19}$$

where $|\ |_{V_{\ell}^*}$ is the strong dual norm in V_{ℓ}^*. Then u_{ℓ} converges when $\ell \to +\infty$ toward the unique weak solution to

$$\begin{cases} u_{\infty} \in H^1_{loc}(\overline{\Omega}), \quad u_{\infty} = 0 \quad on \ \partial\Omega, \\ -\nabla \cdot (A(x, u_{\infty})\nabla u_{\infty}) = f \ in \ \Omega, \ |u_{\infty}|_{1,\Omega_{\ell}} = O(\ell^{\gamma + \frac{p}{2}}) \ \forall \ell > 0. \end{cases} \tag{2.20}$$

Moreover one has the following estimate

$$|u_{\ell} - u_{\infty}|_{1,\Omega_{\frac{\ell}{2}}} \leq Ce^{-\beta\ell} \tag{2.21}$$

for some positive constants C and β.

We will need several preliminary lemmas using arguments previously introduced for instance in [25]–[37].

Lemma 2.1. *For $r > 0$ let u_{ℓ}, $u_{\ell+r}$ be the solutions to (2.18) corresponding to Ω_{ℓ}, $\Omega_{\ell+r}$ respectively. One has*

$$\int_{\{u_{\ell} > u_{\ell+r}\}} \{A(x, u_{\ell})\nabla u_{\ell} - A(x, u_{\ell+r})\nabla u_{\ell+r}\} \cdot \nabla\zeta \ dx \leq 0 \tag{2.22}$$

$$\forall \zeta \in H^1(\ell\omega_1 \times \omega_2), \quad 0 \leq \zeta \leq M, \quad \zeta = 0 \ on \ \partial(\ell\omega_1) \times \omega_2.$$

($\{u_{\ell} > u_{\ell+r}\} = \{x \in \Omega_{\ell} \mid u_{\ell}(x) > u_{\ell+r}(x)\}$, M is a constant).

Proof. By subtraction of the equations in (2.18) one has

$$\int_{\Omega_{\ell}} \{A(x, u_{\ell})\nabla u_{\ell} - A(x, u_{\ell+r})\nabla u_{\ell+r}\} \cdot \nabla v \ dx = 0 \quad \forall v \in H^1_0(\Omega_{\ell}). \tag{2.23}$$

Let us set

$$F_{\epsilon}(z) = \begin{cases} \dfrac{1}{I_{\epsilon}} \int_{\epsilon}^{z} \dfrac{ds}{\omega^2(s)}, \quad I_{\epsilon} = \int_{\epsilon}^{+\infty} \dfrac{ds}{\omega^2(s)}, \quad for \ z > \epsilon, \\[2mm] 0 \quad for \ z \leq \epsilon. \end{cases}$$

Note that there is no loss of generality in assuming that the integral I_{ϵ} converges at $+\infty$. Then

$$v = \zeta \ F_{\epsilon}(u_{\ell} - u_{\ell+r})$$

is a suitable test function for (2.23) and we derive

$$\int_{\Omega_{\ell}} \{A(x, u_{\ell})\nabla u_{\ell} - A(x, u_{\ell+r})\nabla u_{\ell+r}\} \cdot \nabla\zeta \ F_{\epsilon}(u_{\ell} - u_{\ell+r}) \ dx$$

$$= -\int_{\Omega_{\ell}} \{A(x, u_{\ell})\nabla u_{\ell} - A(x, u_{\ell+r})\nabla u_{\ell+r}\} \cdot \nabla F_{\epsilon}(u_{\ell} - u_{\ell+r})\zeta \ dx.$$

Denote by J_ϵ the right hand side integral. One has, writing F_ϵ for $F_\epsilon(u_\ell - u_{\ell+r})$:

$$J_\epsilon = -\int_{\Omega_\ell} A(x, u_\ell)\nabla(u_\ell - u_{\ell+r}) \cdot \nabla F_\epsilon \zeta \, dx$$

$$+ \int_{\Omega_\ell} \{A(x, u_{\ell+r}) - A(x, u_\ell)\}\nabla u_{\ell+r} \cdot \nabla F_\epsilon \zeta \, dx.$$

Noting that

$$\nabla F_\epsilon = \frac{1}{I_\epsilon} \frac{\nabla(u_\ell - u_{\ell+r})}{\omega^2(|u_\ell - u_{\ell+r}|)}\chi_{\{u_\ell - u_{\ell+r} > \epsilon\}}$$

(χ_A = the characteristic function of A), using (2.14), (2.16) we derive

$$J_\epsilon \leq -\frac{\lambda}{I_\epsilon}\int_{\{u_\ell - u_{\ell+r} > \epsilon\}} \frac{|\nabla(u_\ell - u_{\ell+r})|^2}{\omega^2(|u_\ell - u_{\ell+r}|)}\zeta \, dx$$

$$+ \frac{1}{I_\epsilon}\int_{\{u_\ell - u_{\ell+r} > \epsilon\}} |\nabla u_{\ell+r}|\frac{|\nabla(u_\ell - u_{\ell+r})|}{\omega(|u_\ell - u_{\ell+r}|)}\zeta \, dx$$

$$\leq \frac{1}{4\lambda I_\epsilon}\int_{\{u_\ell - u_{\ell+r} > \epsilon\}} |\nabla u_{\ell+r}|^2\zeta \, dx$$

$$\leq \frac{1}{4\lambda I_\epsilon}\int_{\Omega_\ell} |\nabla u_{\ell+r}|^2\zeta \, dx \leq \frac{M}{4\lambda I_\epsilon}\int_{\Omega_\ell} |\nabla u_{\ell+r}|^2 \, dx$$

by using in the second integral the inequality $ab \leq \lambda a^2 + \frac{1}{4\lambda}b^2$. Since $I_\epsilon \to +\infty$ when $\epsilon \to 0$, $J_\epsilon \to 0$ and since a.e.

$$F_\epsilon(u_\ell - u_{\ell+r}) \to \chi_{\{u_\ell - u_{\ell+r} > 0\}}$$

one gets (2.22). This completes the proof of the lemma. $\square$

We can now show:

Lemma 2.2. *For $0 \leq r \leq 1$ there exist constants C, β such that*

$$|u_\ell - u_{\ell+r}|_{1,\Omega_{\frac{\ell}{2}}} \leq Ce^{-\beta\ell}. \tag{2.24}$$

Proof. Let us set for $i, j = 1, \cdots, n$

$$\mathcal{A}_{i,j}(x, u) = \int_0^u A_{i,j}(x, s) \, ds.$$

If $u = u(x)$ is a function in V_ℓ one derives easily with the summation convention in j that

$$\partial_{x_j}\mathcal{A}_{i,j}(x, u) = A_{i,j}(x, u)\partial_{x_j}u + \int_0^u \partial_{x_j}A_{i,j}(x, s) \, ds$$

and (2.22) can be written as

$$
\int_{\{u_\ell - u_{\ell+r} > 0\}} \{\partial_{x_j} \big(\mathcal{A}_{i,j}(x, u_\ell) - \mathcal{A}_{i,j}(x, u_{\ell+r}) \big)
$$
$$
- \int_{u_{\ell+r}}^{u_\ell} \partial_{x_j} A_{i,j}(x, s)\, ds \} \partial_{x_i} \zeta\, dx \le 0. \tag{2.25}
$$

For $\ell_1 < \ell - 1$ denote by $\rho = \rho_{\ell_1}$ a smooth function of X_1 only such that

$$
0 \le \rho \le 1, \quad \rho = 1 \text{ on } \ell_1 \omega_1, \quad \rho = 0 \text{ outside } (\ell_1 + 1)\omega_1,
$$
$$
|\nabla_{X_1} \rho|, \ |\nabla_{X_1}^2 \rho| \le C \tag{2.26}
$$

for some constant C ($\nabla_{X_1} = (\partial_{x_1}, \cdots, \partial_{x_p})$, $\nabla_{X_1}^2$ denotes the matrix of the second derivatives). Then in (2.25) we choose

$$
\zeta = (M - e^{\alpha|X_2|^2}) \rho(X_1)
$$

(recall that ω_2 is bounded and this function is nonnegative and bounded for M large enough). One has integrating by parts

$$
\int_{\{u_\ell - u_{\ell+r} > 0\}} \partial_{x_j} \big(\mathcal{A}_{i,j}(x, u_\ell) - \mathcal{A}_{i,j}(x, u_{\ell+r}) \big) \partial_{x_i} \zeta\, dx
$$
$$
= \int_{\{u_\ell - u_{\ell+r} > 0\}} \partial_{x_j} \{ \big(\mathcal{A}_{i,j}(x, u_\ell) - \mathcal{A}_{i,j}(x, u_{\ell+r}) \big) \partial_{x_i} \zeta \}\, dx
$$
$$
- \int_{\{u_\ell - u_{\ell+r} > 0\}} \big(\mathcal{A}_{i,j}(x, u_\ell) - \mathcal{A}_{i,j}(x, u_{\ell+r}) \big) \partial_{x_i x_j}^2 \zeta\, dx
$$
$$
= - \int_{\{u_\ell - u_{\ell+r} > 0\}} \big(\mathcal{A}_{i,j}(x, u_\ell) - \mathcal{A}_{i,j}(x, u_{\ell+r}) \big) \partial_{x_i x_j}^2 \zeta\, dx
$$
$$
= - \int_{\{u_\ell - u_{\ell+r} > 0\}} \int_{u_{\ell+r}}^{u_\ell} A_{i,j}(x, s)\, ds\, \partial_{x_i x_j}^2 \zeta\, dx.
$$

Going back to (2.25) and setting $P_{\ell_1} = \Omega_{\ell_1+1} \cap \{u_\ell - u_{\ell+r} > 0\}$ we obtain since $\rho = 0$ outside $(\ell_1 + 1)\omega_1$

$$
- \int_{P_{\ell_1}} \int_{u_{\ell+r}}^{u_\ell} \{ \sum_{i,j>p} A_{i,j}(x, s)\, \partial_{x_i x_j}^2 \zeta + \sum_{i>p} \partial_{x_j} A_{i,j}(x, s) \partial_{x_i} \zeta \} ds\, dx
$$
$$
= \int_{P_{\ell_1}} \int_{u_{\ell+r}}^{u_\ell} \{ \sum_{i \text{ or } j \le p} A_{i,j}(x, s)\, \partial_{x_i x_j}^2 \zeta + \sum_{i \le p} \partial_{x_j} A_{i,j}(x, s) \partial_{x_i} \zeta \} ds\, dx.
$$

For $i, j > p$ one has

$$
\partial_{x_i} \zeta = -2\alpha x_i e^{\alpha|X_2|^2} \rho, \quad \partial_{x_i x_j} \zeta = -4\alpha^2 x_i x_j e^{\alpha|X_2|^2} \rho.
$$

Choosing if necessary the coordinates such that $|X_2| \geq 1$ on ω_2 (or replacing X_2 by $X_2 - X_2^0$ for some X_2^0) one gets

$$-\left(\sum_{i,j>p} A_{i,j}(x,s)\, \partial^2_{x_i x_j}\zeta + \sum_{i>p} \partial_{x_j}A_{i,j}(x,s)\partial_{x_i}\zeta \right)$$

$$\geq \{4\alpha^2\lambda|X_2|^2 - C\alpha|X_2|\}e^{\alpha|X_2|^2}\rho$$

$$= |X_2|\{4\alpha^2\lambda|X_2| - C\alpha\}e^{\alpha|X_2|^2}\rho \geq \rho$$

for α large enough. Since $\partial_{x_i}\zeta = \partial_{x_i}\rho = 0$ on Ω_{ℓ_1} for $i \leq p$ one derives from above that

$$\int_{\Omega_{\ell_1}} (u_\ell - u_{\ell+r})^+\, dx \leq \int_{\Omega_{\ell_1+1}} (u_\ell - u_{\ell+r})^+\rho\, dx$$

$$\leq C \int_{\Omega_{\ell_1+1}\setminus\Omega_{\ell_1}} (u_\ell - u_{\ell+r})^+\, dx,$$

for some constant C. Exchanging the roles of u_ℓ and $u_{\ell+r}$ (Cf (2.23)) one gets

$$\int_{\Omega_{\ell_1}} |u_\ell - u_{\ell+r}|\, dx \leq C \int_{\Omega_{\ell_1+1}\setminus\Omega_{\ell_1}} |u_\ell - u_{\ell+r}|\, dx$$

for some constant $C = C(A, \omega_2, \alpha)$ independent of ℓ_1 and ℓ. This last inequality can be written

$$\int_{\Omega_{\ell_1}} |u_\ell - u_{\ell+r}|\, dx \leq \delta \int_{\Omega_{\ell_1+1}} |u_\ell - u_{\ell+r}|\, dx$$

with $\delta = \frac{C}{C+1} < 1$. Iterating this formula $[\frac{\ell}{2}]$ times we arrive easily since $\frac{\ell}{2} - 1 < [\frac{\ell}{2}] \leq \frac{\ell}{2}$ to

$$\int_{\Omega_{\frac{\ell}{2}}} |u_\ell - u_{\ell+r}|\, dx \leq \delta^{\frac{\ell}{2}-1} \int_{\Omega_\ell} |u_\ell - u_{\ell+r}|\, dx. \qquad (2.27)$$

Taking now $v = u_\ell$ in (2.18) one derives also

$$\lambda \int_{\Omega_\ell} |\nabla u_\ell|^2\, dx \leq \int_{\Omega_\ell} A\nabla u_\ell \cdot \nabla u_\ell\, dx \leq |f|_{V_\ell^*}||\nabla u_\ell||_{2,\Omega_\ell}$$

and thus

$$||\nabla u_\ell||_{2,\Omega_\ell} \leq \frac{|f|_{V_\ell^*}}{\lambda} \leq C\ell^\gamma. \qquad (2.28)$$

It follows by the Poincaré inequality that

$$\int_{\Omega_\ell} |u_\ell|\, dx \leq \{\int_{\Omega_\ell} |u_\ell|^2\, dx\}^{\frac{1}{2}} |\Omega_\ell|^{\frac{1}{2}} \leq C||\nabla u_\ell||_{2,\Omega_\ell} \ell^{\frac{p}{2}} \leq C\ell^{\gamma+\frac{p}{2}} \qquad (2.29)$$

and thus from (2.27) for different constants C

$$\int_{\Omega_{\frac{\ell}{2}}} |u_\ell - u_{\ell+r}| \, dx \le \frac{1}{\delta} e^{-\frac{\ell}{2} \ln(\frac{1}{\delta})} \{ \int_{\Omega_\ell} |u_\ell| \, dx + \int_{\Omega_\ell} |u_{\ell+r}| \, dx \}$$

$$\le \frac{1}{\delta} e^{-\frac{\ell}{2} \ln(\frac{1}{\delta})} C \{ \ell^{\gamma+\frac{p}{2}} + (\ell+r)^{\gamma+\frac{p}{2}} \}$$

$$\le \frac{1}{\delta} e^{-\frac{\ell}{2} \ln(\frac{1}{\delta})} C \ell^{\gamma+\frac{p}{2}} \{ 1 + (1 + \frac{r}{\ell})^{\gamma+\frac{p}{2}} \} \le C e^{-\beta \ell},$$

provided we choose $\beta < \frac{1}{2} \ln(\frac{1}{\delta})$. This completes the proof of the lemma. $\square$

Lemma 2.3. *For any fixed $\ell_0 < \ell - 1$ one has for some constant $K(\ell_0)$ independent of ℓ*

$$\int_{\Omega_{\ell_0}} |\nabla u_\ell|^2 \, dx \le K(\ell_0). \tag{2.30}$$

Proof. One considers $\rho = \rho_{\ell_0}$ defined in (2.26). One derives from (2.18)

$$\int_{\Omega_{\ell_0+1}} A(x, u_\ell) \nabla u_\ell \cdot \nabla(\rho_{\ell_0} u_\ell) \, dx = \langle f, \rho_{\ell_0} u_\ell \rangle$$

which implies

$$\int_{\Omega_{\ell_0+1}} \{ A(x, u_\ell) \nabla u_\ell \cdot \nabla u_\ell \} \rho_{\ell_0} \, dx$$

$$= - \int_{\Omega_{\ell_0+1}} \{ A(x, u_\ell) \nabla u_\ell \cdot \nabla \rho_{\ell_0} \} u_\ell \, dx + \langle f, \rho_{\ell_0} u_\ell \rangle .$$

Since $\nabla \rho_{\ell_0} = 0$ on Ω_{ℓ_0} we derive easily

$$\lambda \int_{\Omega_{\ell_0+1}} |\nabla u_\ell|^2 \rho_{\ell_0} \, dx$$

$$\le \int_{\Omega_{\ell_0+1} \backslash \Omega_{\ell_0}} C\Lambda |\nabla u_\ell| |u_\ell| \, dx + |f|_{V^*_{\ell_0+1}} \| \nabla(\rho_{\ell_0} u_\ell) \|_{2, \Omega_{\ell_0+1}}$$

where C is the constant in (2.26). Setting $D_{\ell_0} = \Omega_{\ell_0+1} \backslash \Omega_{\ell_0}$ we obtain, applying Young's inequality:

$$\lambda \int_{\Omega_{\ell_0+1}} |\nabla u_\ell|^2 \rho_{\ell_0} \, dx \le C\Lambda \int_{D_{\ell_0}} |\nabla u_\ell| |u_\ell| \, dx$$

$$+ |f|_{V^*_{\ell_0+1}} \{ \| \rho_{\ell_0} \nabla u_\ell \|_{2, \Omega_{\ell_0+1}} + \| u_\ell \nabla \rho_{\ell_0} \|_{2, \Omega_{\ell_0+1}} \}$$

$$\le C\Lambda \int_{D_{\ell_0}} |\nabla u_\ell| |u_\ell| \, dx + \frac{\lambda}{2} \int_{\Omega_{\ell_0+1}} |\nabla u_\ell|^2 \rho_{\ell_0} \, dx$$

$$+ \frac{1}{2\lambda} |f|^2_{V^*_{\ell_0+1}} + \frac{C^2}{2} \int_{D_{\ell_0}} |u_\ell|^2 \, dx + \frac{1}{2} |f|^2_{V^*_{\ell_0+1}} .$$

Thus it comes after applying the Poincaré inequality:

$$\int_{\Omega_{\ell_0}} |\nabla u_\ell|^2 \, dx \leq \int_{\Omega_{\ell_0+1}} |\nabla u_\ell|^2 \rho_{\ell_0} \, dx \leq C \int_{D_{\ell_0}} |\nabla u_\ell|^2 \, dx + C|f|^2_{V^*_{\ell_0+1}}$$

$$\leq C \int_{D_{\ell_0}} |\nabla u_\ell|^2 \, dx + C(\ell_0 + 1)^{2\gamma}$$

for some constant C independent of ℓ and ℓ_0. This leads to

$$(1+C) \int_{\Omega_{\ell_0}} |\nabla u_\ell|^2 \, dx \leq C \int_{\Omega_{\ell_0+1}} |\nabla u_\ell|^2 \, dx + C(\ell_0 + 1)^{2\gamma}$$

which is

$$\int_{\Omega_{\ell_0}} |\nabla u_\ell|^2 \, dx \leq \delta \int_{\Omega_{\ell_0+1}} |\nabla u_\ell|^2 \, dx + \delta(\ell_0 + 1)^{2\gamma}$$

for $\delta = \frac{C}{C+1} < 1$. Iterating this formula $[\ell - \ell_0 - 1]$ times, we get easily

$$\int_{\Omega_{\ell_0}} |\nabla u_\ell|^2 \, dx \leq \delta\{\delta \int_{\Omega_{\ell_0+2}} |\nabla u_\ell|^2 \, dx + \delta(\ell_0 + 2)^{2\gamma}\} + \delta(\ell_0 + 1)^{2\gamma}$$

$$= \delta^2 \int_{\Omega_{\ell_0+2}} |\nabla u_\ell|^2 \, dx + \delta^2(\ell_0 + 2)^{2\gamma} + \delta(\ell_0 + 1)^{2\gamma}$$

$$\leq \delta^{[\ell-\ell_0-1]} \int_{\Omega_\ell} |\nabla u_\ell|^2 \, dx + \sum_{k=1}^{+\infty} \delta^k (\ell_0 + k)^{2\gamma}$$

$$\leq \frac{1}{\delta^{\ell_0+2}} \delta^\ell \int_{\Omega_\ell} |\nabla u_\ell|^2 \, dx + \sum_{k=1}^{+\infty} \delta^k (\ell_0 + k)^{2\gamma}$$

where we have used the fact that $\ell - \ell_0 - 2 < [\ell - \ell_0 - 1] \leq \ell - \ell_0 - 1$. Noting that the above series converges and recalling (2.28) we obtain

$$\int_{\Omega_{\ell_0}} |\nabla u_\ell|^2 \, dx \leq K(\ell_0)$$

which completes the proof of the lemma. $\square$

We can now give a proof of the theorem.

Proof. Let $t > 0$ arbitrary. From the estimate above the Lemma 2.3 one deduces

$$|u_\ell - u_{\ell+t}|_{1,\Omega_{\frac{\ell}{2}}} \leq |u_\ell - u_{\ell+1}|_{1,\Omega_{\frac{\ell}{2}}} +$$

$$|u_{\ell+1} - u_{\ell+2}|_{1,\Omega_{\frac{\ell}{2}}} + \cdots + |u_{\ell+[t]} - u_{\ell+t}|_{1,\Omega_{\frac{\ell}{2}}}$$

$$\leq |u_\ell - u_{\ell+1}|_{1,\Omega_{\frac{\ell}{2}}} +$$

$$|u_{\ell+1} - u_{\ell+2}|_{1,\Omega_{\frac{\ell+1}{2}}} + \cdots + |u_{\ell+[t]} - u_{\ell+t}|_{1,\Omega_{\frac{\ell+[t]}{2}}} \qquad (2.31)$$

$$\leq Ce^{-\beta\ell} + Ce^{-\beta(\ell+1)} + \cdots + Ce^{-\beta(\ell+[t])}$$

$$\leq \frac{C}{1 - e^{-\beta}} e^{-\beta\ell}.$$

Thus for any $\ell_0 \leq \frac{\ell}{2}$, u_ℓ is a Cauchy-sequence in $L^1(\Omega_{\ell_0})$ and converges toward some function u_∞ in $L^1(\Omega_{\ell_0})$. From Lemma 2.3 one has also

$$u_\ell \to u_\infty \text{ in } L^2(\Omega_{\ell_0}), \quad u_\ell \rightharpoonup u_\infty \text{ in } H^1(\Omega_{\ell_0})$$

which proves that $u_\infty \in H^1_{loc}(\overline{\Omega})$ and $u_\infty = 0$ on $\partial\Omega$. Now from

$$\int_{\Omega_{\ell_0}} A(x, u_\ell) \nabla u_\ell \cdot \nabla v \, dx = \langle f, v \rangle \quad \forall v \in H^1_0(\Omega_{\ell_0})$$

one can pass to the limit. Indeed up to a subsequence $u_\ell \to u_\infty$ a.e. and thus - denoting by T the transposition - $A(x, u_\ell)^T \nabla v \to A(x, u_\infty)^T \nabla v$ strongly in $L^2(\Omega_{\ell_0})$. So we obtain

$$\int_{\Omega_{\ell_0}} A(x, u_\infty) \nabla u_\infty \cdot \nabla v \, dx = \langle f, v \rangle \quad \forall v \in H^1_0(\Omega_{\ell_0}).$$

Now (2.31) written for 2ℓ gives

$$|u_{2\ell} - u_{2\ell+t}|_{1,\Omega_\ell} \leq \frac{C}{1 - e^{-\beta}} e^{-2\beta\ell}.$$

Letting $t \to +\infty$ we get

$$|u_{2\ell} - u_\infty|_{1,\Omega_\ell} \leq \frac{C}{1 - e^{-\beta}} e^{-2\beta\ell}$$

hence

$$|u_\infty|_{1,\Omega_\ell} \leq \frac{C}{1 - e^{-\beta}} e^{-2\beta\ell} + |u_{2\ell}|_{1,\Omega_\ell}$$

$$\leq \frac{C}{1 - e^{-\beta}} e^{-2\beta\ell} + |u_{2\ell}|_{1,\Omega_{2\ell}} = O(\ell^{\gamma + \frac{p}{2}})$$

by (2.29). Finally if u_∞ and u'_∞ are two solutions of (2.20) one would get (2.22) with u_ℓ, $u_{\ell+r}$ replaced by u_∞ and u'_∞ and (2.24) will become - following the same arguments:

$$|u_\infty - u'_\infty|_{1,\Omega_{\frac{\ell}{2}}} \leq Ce^{-\beta\ell}$$

which implies $u_\infty = u'_\infty$ by letting $\ell \to +\infty$. This completes the proof of the theorem. $\square$

Remark 2.7. It is possible to replace (2.19) by

$$|f|_{V^*_\ell} = O(e^{\gamma\ell})$$

for γ small enough. The reader will adapt the proof easily. When

$$A = \begin{pmatrix} A_{11}(x, u) & A_{12}(X_2, u) \\ A_{21}(x, u) & A_{22}(X_2, u) \end{pmatrix}$$

where $A_{11}(x, u)$ is a $p \times p$-matrix and $A_{22}(X_2, u)$ is a $(n-p) \times (n-p)$-matrix and

$$f = f(X_2) \in L^2(\omega_2)$$

the assumption (2.19) is clearly satisfied. In this case (see [15]) there exists a unique u_∞ weak solution to

$$\begin{cases} -\text{div}_{X_2}(A_{22}(X_2, u_\infty)\nabla_{X_2}u_\infty) = f(X_2) & \text{in } \omega_2, \\ u_\infty = 0 & \text{on } \partial\omega_2. \end{cases}$$

We have denoted by div_{X_2} the divergence operator in X_2 in such a way that the first equation above reads

$$-\sum_{i=p+1}^{n}\sum_{j=p+1}^{n} \partial_{x_i}(A_{i,j}(X_2, u_\infty)\partial_{x_j}u_\infty) = f.$$

It is clear that u_∞ satisfies (2.20). Thus in this case, when the data are depending only on the "section" ω_2 the solution u_ℓ converges toward u_∞ solution of the same problem set on the section.

2.3   Pointwise convergence

Consider a domain Ω_ℓ defined as

$$\Omega_\ell = \ell\omega_1 \times \omega_2$$

where ω_1 is a bounded convex open set of $\mathbb{R}^p$ containing 0 and ω_2 a bounded open subset of $\mathbb{R}^{n-p}$. Denote by $x = (X_1, X_2)$ the points in Ω_ℓ and by $A = A(x)$ the matrix

$$A = \begin{pmatrix} A_{11}(X_1) & 0 \\ 0 & A_{22}(X_2) \end{pmatrix}$$

satisfying (2.1). Let $f = f(X_2) \in H^{-1}(\omega_2)$. By the remark 2.2 if u_ℓ denotes the solution to

$$\begin{cases} u_\ell \in H_0^1(\Omega_\ell), \\ \int_{\Omega_\ell} A\nabla u_\ell \cdot \nabla v \, dx = \langle f, v \rangle & \forall v \in H_0^1(\Omega_\ell), \end{cases} \tag{2.32}$$

then u_ℓ converges toward u_∞ the solution to (2.13) when $\ell \to \infty$. More precisely one has for some constants C, β

$$\left\lVert\nabla(u_\ell - u_\infty)\right\rVert_{2,\Omega_{\frac{\ell}{2}}} \leq Ce^{-\beta\ell}.$$

This estimate does not give any indication on the pointwise convergence of u_ℓ toward u_∞. This is what we would like to analyze here. We will consider

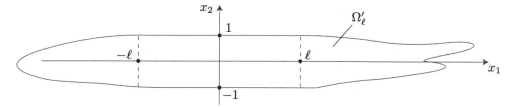

Fig. 2.3   The two dimensional situation.

a slightly more general situation. Indeed let Ω'_ℓ be a domain in $\mathbb{R}^n$ such that

$$\Omega_\ell \subset \Omega'_\ell \subset \mathbb{R}^p \times \omega_2.$$

Figure 2.3 represents such a domain Ω'_ℓ in dimension two when $\omega_1 = \omega_2 = (-1, 1)$.

For $f \in L^2(\omega_2)$ consider u'_ℓ the solution to

$$\begin{cases} u'_\ell \in H^1_0(\Omega'_\ell), \\ \int_{\Omega'_\ell} A\nabla u'_\ell \cdot \nabla v \, dx = \int_{\Omega'_\ell} fv \, dx \quad \forall v \in H^1_0(\Omega'_\ell). \end{cases} \tag{2.33}$$

We are going to show that u'_ℓ converges to u_∞ pointwise. First we have

Lemma 2.4. *Suppose that the entries of the matrix A_{11} belong to $W^{1,\infty}(\mathbb{R}^p)$ where $W^{1,\infty}(A)$ denotes the space of uniformly Lipschitz continuous functions on A. For $\lambda_1 > 0$ let w_ℓ be the weak solution to*

$$\begin{cases} -\nabla_{X_1} \cdot (A_{11}\nabla_{X_1} w_\ell) + \lambda_1 w_\ell = 0 \quad in \ \ell\omega_1, \\ w_\ell = 1 \quad on \ \partial(\ell\omega_1). \end{cases}$$

Then $w_\ell \to 0$ pointwise exponentially quickly.

Proof. By definition one has $w_\ell - 1 \in H^1_0(\ell\omega_1)$ and

$$\int_{\ell\omega_1} A_{11}\nabla_{X_1} w_\ell \cdot \nabla_{X_1} v + \lambda_1 w_\ell v \, dX_1 = 0 \quad \forall v \in H^1_0(\ell\omega_1). \tag{2.34}$$

Taking $v = w_\ell^-$ one derives

$$\int_{\ell\omega_1} A_{11}\nabla_{X_1} w_\ell \cdot \nabla_{X_1} w_\ell^- + \lambda_1 w_\ell w_\ell^- \, dX_1 = 0$$

which is equivalent to

$$\int_{\ell\omega_1} A_{11}\nabla_{X_1} w_\ell^- \cdot \nabla_{X_1} w_\ell^- + \lambda_1 w_\ell^- w_\ell^- \, dX_1 = 0$$

i.e. one has $w_\ell^- = 0$ which implies $w_\ell \geq 0$. Taking now $v = (w_\ell - 1)^+$ leads to

$$\int_{\ell\omega_1} A_{11} \nabla_{X_1}(w_\ell - 1) \cdot \nabla_{X_1}(w_\ell - 1)^+ + \lambda_1 w_\ell(w_\ell - 1)^+ \, dX_1 = 0$$

and thus

$$\int_{\ell\omega_1} A_{11} \nabla_{X_1}(w_\ell - 1)^+ \cdot \nabla_{X_1}(w_\ell - 1)^+ \, dX_1 \leq 0.$$

This implies by (2.1) that $(w_\ell - 1)^+ = 0$ and we have proved that

$$0 \leq w_\ell \leq 1 \quad \text{in } \ell\omega_1.$$

By our choice of ω_1 there exists $a > 0$ such that

$$(-a, a)^p \subset \omega_1.$$

On $Q_\ell = (-\ell a, \ell a)^p$ consider the function ρ defined by

$$\rho = \sum_{i=1}^{p} \frac{\cosh(\mu x_i)}{\cosh(\mu a \ell)}$$

where $\mu > 0$ will be fixed later. Clearly, since the functions in the sum above are positive, one has

$$\rho \geq 1 \geq w_\ell \geq 0 \quad \text{on } \partial Q_\ell.$$

Taking $v = (w_\ell - \rho)^+$ in (2.34) we get since $v = 0$ outside Q_ℓ

$$\int_{Q_\ell} A_{11} \nabla_{X_1} w_\ell \cdot \nabla_{X_1}(w_\ell - \rho)^+ + \lambda_1 w_\ell(w_\ell - \rho)^+ \, dX_1 = 0.$$

This implies

$$\begin{aligned}
\int_{Q_\ell} &A_{11} \nabla_{X_1}(w_\ell - \rho) \cdot \nabla_{X_1}(w_\ell - \rho)^+ + \lambda_1(w_\ell - \rho)(w_\ell - \rho)^+ \, dX_1 \\
&= -\int_{Q_\ell} A_{11} \nabla_{X_1}\rho \cdot \nabla_{X_1}(w_\ell - \rho)^+ + \lambda_1 \rho(w_\ell - \rho)^+ \, dX_1 \qquad (2.35) \\
&= -\int_{Q_\ell} \{-\nabla_{X_1} \cdot (A_{11}\nabla_{X_1}\rho) + \lambda_1\rho\}(w_\ell - \rho)^+ \, dX_1.
\end{aligned}$$

Denoting by $A_{i,j}$ the entries of the matrix A one has with the summation convention from 1 to p

$$\begin{aligned}
-\nabla_{X_1} \cdot (A_{11}\nabla_{X_1}\rho) + \lambda_1\rho &= -\partial_{x_i}(A_{i,j}\partial_{x_j}\rho) + \lambda_1\rho \\
&= -A_{i,j}\partial^2_{x_i x_j}\rho - \partial_{x_i}A_{i,j}\partial_{x_j}\rho + \lambda_1\rho.
\end{aligned}$$

Since

$$\partial_{x_j}\rho = \mu\frac{\sinh(\mu x_j)}{\cosh(\mu a\ell)}, \quad \partial^2_{x_i x_j}\rho = \mu^2\delta_{i,j}\frac{\cosh(\mu x_j)}{\cosh(\mu a\ell)}$$

we get (note that $|\frac{\sinh}{\cosh}| \leq 1$)

$$-\nabla_{X_1}\cdot(A_{11}\nabla_{X_1}\rho)+\lambda_1\rho$$

$$= -\mu^2 A_{i,i}\frac{\cosh(\mu x_i)}{\cosh(\mu a\ell)} - \mu\partial_{x_i}A_{i,j}\frac{\sinh(\mu x_j)}{\cosh(\mu a\ell)} + \lambda_1\rho$$

$$\geq -\mu^2 A_{i,i}\rho - \mu|\sum_{j=1}^{p}\partial_{x_i}A_{i,j}|\rho + \lambda_1\rho$$

$$= (-\mu^2 A_{i,i} - \mu|\sum_{j=1}^{p}\partial_{x_i}A_{i,j}| + \lambda_1)\rho \geq 0$$

for μ small enough. Going back to (2.35) we deduce

$$0 \leq w_\ell \leq \rho = \sum_{i=1}^{p}\frac{\cosh(\mu x_i)}{\cosh(\mu a\ell)} \quad \text{on } Q_\ell.$$

This completes the proof of the lemma. □

Then we can show:

Theorem 2.3. *Let λ_1, v_1 be respectively the first eigenvalue and a positive eigenfunction of the problem*

$$\begin{cases} -\nabla_{X_2}\cdot(A_{22}\nabla_{X_2}v_1) = \lambda_1 v_1 & in \ \omega_2, \\ v_1 = 0 & on \ \partial\omega_2. \end{cases}$$

Suppose that for some constant C one has

$$|u_\infty| \leq Cv_1 \ in \ \omega_2. \tag{2.36}$$

Then if u_ℓ is the solution to (2.32) one has

$$|u_\ell - u_\infty| \leq Cv_1\sum_{i=1}^{p}\frac{\cosh(\mu x_i)}{\cosh(\mu a\ell)} \quad on \ Q_\ell\times\omega_2. \tag{2.37}$$

(In particular $u_\ell \to u_\infty$ pointwise and with an exponential speed on $\mathbb{R}^p\times\omega_2$.)

Proof. For $v \in H^1_0(\Omega_\ell)$, $v(X_1,\cdot) \in H^1_0(\omega_2)$ for a.e. $X_1 \in \ell\omega_1$. Thus by (2.13)

$$\int_{\omega_2} A_{22}\nabla_{X_2}u_\infty\cdot\nabla_{X_2}v\,dX_2 = \int_{\omega_2} fv\,dX_2.$$

Integrating in X_1 and using the fact that u_∞ is independent of X_1, we obtain

$$\int_{\Omega_\ell} A\nabla u_\infty \cdot \nabla v \, dx = \int_{\Omega_\ell} fv \, dx.$$

Combining this with (2.32), we deduce

$$\int_{\Omega_\ell} A\nabla(u_\ell - u_\infty) \cdot \nabla v \, dx = 0 \quad \forall v \in H_0^1(\Omega_\ell).$$

One has also if w_ℓ is the function introduced in Lemma 2.4

$$\int_{\Omega_\ell} A\nabla(w_\ell v_1) \cdot \nabla v \, dx = \int_{\Omega_\ell} \begin{pmatrix} A_{11} & 0 \\ 0 & A_{22} \end{pmatrix} \begin{pmatrix} v_1 \nabla_{X_1} w_\ell \\ w_\ell \nabla_{X_2} v_1 \end{pmatrix} \cdot \begin{pmatrix} \nabla_{X_1} v \\ \nabla_{X_2} v \end{pmatrix} dx$$

$$= \int_{\Omega_\ell} v_1 A_{11} \nabla_{X_1} w_\ell \cdot \nabla_{X_1} v + w_\ell A_{22} \nabla_{X_2} v_1 \cdot \nabla_{X_2} v \, dx$$

$$= \int_{\omega_2} v_1 \int_{\ell\omega_1} A_{11} \nabla_{X_1} w_\ell \cdot \nabla_{X_1} v dx + \int_{\ell\omega_1} w_\ell \int_{\omega_2} A_{22} \nabla_{X_2} v_1 \cdot \nabla_{X_2} v \, dx$$

$$= \int_{\omega_2} v_1 \int_{\ell\omega_1} -\lambda_1 w_\ell v dx + \int_{\ell\omega_1} w_\ell \int_{\omega_2} \lambda_1 v_1 v \, dx = 0.$$

Thus we derive

$$\int_{\Omega_\ell} A\nabla(Cw_\ell v_1 \pm (u_\ell - u_\infty)) \cdot \nabla v \, dx = 0 \quad \forall v \in H_0^1(\Omega_\ell).$$

If C is chosen such that (2.36) holds then

$$Cw_\ell v_1 \pm (u_\ell - u_\infty) \geq 0 \quad \text{on } \partial\Omega_\ell.$$

By the weak maximum principle it follows that

$$\pm(u_\ell - u_\infty) \leq Cw_\ell v_1 \quad \text{on } \Omega_\ell$$

and the result follows. $\qquad\square$

Remark 2.8. For smooth data thanks to the strong maximum principle it is easy to see that (2.36) holds if $u_\infty \in W^{1,\infty}(\omega_2)$.

As a consequence we have

Theorem 2.4. *Suppose that A, ω_2 and f are smooth and let u'_ℓ, u_∞ be the solutions to (2.33), (2.13). If $u_\infty \in W^{1,\infty}(\omega_2)$ there exists positive constants $C = C(\ell_0, f)$ and α such that*

$$|u'_\ell - u_\infty|_{\infty,\Omega_{\ell_0}} \leq Ce^{-\alpha\ell}$$

($|\ |_{\infty,O}$ denotes the essential supremum norm on O).

Proof. Assuming first $f \geq 0$ by the weak maximum principle, it is easy to show that

$$0 \leq u_\ell \leq u'_\ell \leq u_\infty \quad \text{on } \Omega_\ell$$

and thus by (2.37)

$$|u'_\ell - u_\infty| \leq |u_\ell - u_\infty| \leq C v_1 \sum_{i=1}^{p} \frac{\cosh(\mu a \ell_1)}{\cosh(\mu a \ell)} \quad \text{on } Q_{\ell_1} \times \omega_2.$$

It is clear that for $\ell_1 = \ell_1(\ell_0)$ large enough one has $\ell_0 \omega_1 \subset Q_{\ell_1}$ and the result follows in this case. If $u'_{\ell,\pm}$, $u_{\infty,\pm}$ denote the solutions to (2.33), (2.13) corresponding to f^+ and f^- one has

$$|u'_{\ell,\pm} - u_{\infty,\pm}|_{\infty,\Omega_{\ell_0}} \leq C e^{-\alpha \ell}.$$

The result is then a consequence from

$$u'_\ell = u'_{\ell,+} - u'_{\ell,-}, \quad u_\infty = u_{\infty,+} - u_{\infty,-}$$

since we are dealing with linear problems (note however that $u'_{\ell,\pm}, u'_{\infty,\pm}$ are not the positive or negative part of u'_ℓ, u'_∞). $\qquad\square$

Chapter 3

The pure Neumann problem

3.1 Introduction

Let $\omega' \subset \mathbb{R}^{n-1}$ be a bounded Lipschitz domain. For $\ell \in (0, +\infty)$, we define

$$\Omega_\ell := (-\ell, \ell) \times \omega', \quad \Omega_\infty := \mathbb{R} \times \omega'.$$

Denote by $A = A(x) = (A_{i,j}(x))_{1 \leq i,j \leq n}$ a $n \times n$ uniformly bounded elliptic matrix defined on Ω_∞, i.e. such that there exist $\lambda, \Lambda > 0$ such that

$$A(x)\xi \cdot \xi \geq \lambda|\xi|^2 \quad \text{for all} \quad \xi \in \mathbb{R}^n, \text{ for a. e. } \quad x \in \Omega_\infty, \qquad (3.1)$$

$$|A(x)\xi| \leq \Lambda|\xi| \quad \text{for all} \quad \xi \in \mathbb{R}^n, \text{ for a. e. } \quad x \in \Omega_\infty. \qquad (3.2)$$

The right hand side of the problem is given by a function $f \in L^2_{\text{loc}}(\overline{\Omega}_\infty)$ (see below) satisfying

$$\int_{\Omega_\ell} f(x)\,dx = 0 \quad \text{for all} \quad \ell > 0. \qquad (3.3)$$

Then for each $\ell > 0$, we consider the following Neumann problem on Ω_ℓ (Cf. [27]):

$$\begin{cases} -\operatorname{div}(A\nabla u_\ell) = f & \text{in} \quad \Omega_\ell, \\ A\nabla u_\ell \cdot \nu = 0 & \text{on} \quad \partial\Omega_\ell, \end{cases} \qquad (3.4)$$

where $\nu : \partial\Omega_\ell \to \mathbb{R}^n$ is the outward unit normal to Ω_ℓ.

Thanks to the Lax-Milgram theorem, we know that up to an additive constant the problem (3.4) has a unique weak solution in $H^1(\Omega_\ell)$. We would like to show, under a weak growth condition on f, that u_ℓ converges to the solution u_∞ of a Neumann problem in the infinite cylinder Ω_∞. Moreover, the rate of this convergence is exponential, hence we retrieve the same type of result as for the Dirichlet problem.

In this special setting a generic point in $\mathbb{R}^n$, $n \geq 2$ is denoted $x = (x_1, x_2, \ldots, x_n) = (x_1, x')$, where

$$x' = (x_2, \ldots, x_n) \in \mathbb{R}^{n-1}.$$

We also introduce the operators ∇', Δ', div$'$ defined by

$$\nabla' = (\partial_{x_2}, \ldots, \partial_{x_n}), \quad \Delta' = \partial^2_{x_2} + \cdots + \partial^2_{x_n}, \quad \text{div}\,'v = \partial_{x_2} v^2 + \cdots + \partial_{x_n} v^n$$

for $v = (v^2, \cdots, v^n)$. For a set $E \subset \mathbb{R}^n$, we define the subsets E^+ and E^- by

$$E^+ := \{x = (x_1, x') \in E \mid x_1 > 0\}, \quad E^- := \{x = (x_1, x') \in E \mid x_1 < 0\}.$$

The outward unit normal to a Lipschitz domain in $\mathbb{R}^n$ is denoted by ν and the normal derivative is denoted by $\frac{\partial}{\partial \nu}$.

Recall that for an open set $O \subset \mathbb{R}^k$ ($k \in \mathbb{N}^*$), the space of infinitely differentiable functions with compact support in O is denoted by $\mathcal{D}(O)$. The notation $|\cdot|_{2,O}$ designates the usual L^2-norm on the space $L^2(O)$, while $|\cdot|$ is the euclidean norm in $\mathbb{R}^k$. Hence, if $v \in H^1(O)$, the notation $||\nabla v||_{2,O}$ is the L^2-norm of $|\nabla v|$. We define as usual $|v|_{H^1(O)} = ||(v, \nabla v)||_{2,O}$ for any $v \in H^1(O)$. If O is bounded, then

$$|v|_{H^1(O)/\mathbb{R}} = \inf_{c \in \mathbb{R}} |v + c|_{H^1(O)}.$$

If $O \subset \mathbb{R}^n$ is an unbounded open set, then we define

$$L^2_{\text{loc}}(\overline{O}) := \{v \mid v \in L^2(U) \text{ for all bounded open set } U \subset O\}$$

and

$$H^1_{\text{loc}}(\overline{O}) := \{v \mid v \in H^1(U) \text{ for all bounded open set } U \subset O\}.$$

On Ω_ℓ ($0 < \ell < +\infty$), we introduce the space

$$V_0(\Omega_\ell) := \{v \in H^1(\Omega_\ell) \mid v = 0 \text{ on } \{-\ell, \ell\} \times \omega'\}.$$

For any $v \in V_0(\Omega_\ell)$ we denote also by v its extension by 0 outside Ω_ℓ - note that $v \in H^1(\Omega_\infty)$.

We denote by A_1 the first row of the matrix A, hence $A_1 = (A_{1,j})_{1 \leq j \leq n}$. For $\beta \in \mathbb{R}$ we define the following space of functions on Ω_∞:

$$W_\beta(\Omega_\infty) := \{f \in L^2_{\text{loc}}(\overline{\Omega}_\infty) \mid \int_{\omega'} f(x_1, x')\, dx' = 0 \text{ for a.e. } x_1 \in \mathbb{R}$$

$$\text{and } \exists C \geq 0 \text{ such that } |f|_{2,\Omega_\ell} \leq Ce^{\beta \ell} \,\forall \ell > 0\}.$$

Note that for all $\beta > 0$, $W_\beta(\Omega_\infty)$ contains all the functions of sectional average 0 with polynomial growth at infinity for the L^2-norm. More precisely, if $f \in L^2_{loc}(\overline{\Omega}_\infty)$ and there exist constants $C \geq 0$, $\gamma \in \mathbb{R}$ such that

$$|f|_{2,\Omega_\ell} \leq C(1 + \ell^\gamma) \quad \text{for all} \quad \ell > 0,$$

then, if its average is 0 on ω', $f \in W_\beta(\Omega_\infty)$ for all $\beta > 0$.

In our convergence result, the assumption on the data f is that it belongs to $W_\beta(\Omega_\infty)$ for a small positive β. The necessity of the growth condition in the definition of $W_\beta(\Omega_\infty)$ appears clearly in the proof of our result, but as noticed above, this is a very mild restriction for a function in $L^2_{loc}(\overline{\Omega}_\infty)$. However, the assumption

$$\int_{\omega'} f(x_1, x') \, dx' = 0 \text{ for a.e. } x_1 \in \mathbb{R} \tag{3.5}$$

is stronger than the necessary condition (3.3) (in order to have existence for the problem (3.4)). There are at least two reasons to consider the stronger condition (3.5) instead of the necessary one (3.3).

One is that we would like to be able to extend our convergence result to a more general case where the cylinders approaching Ω_∞ do not necessarily have the same hyperplane of symmetry. We only chose the cylinders Ω_ℓ to be defined by $(-\ell, \ell) \times \omega'$ for the simplicity of the presentation, but we could consider the more general case of cylinders of the type $(a_\ell, b_\ell) \times \omega'$ where $(-c_1\ell^{s_1}, c_1\ell^{s_1}) \subset (a_\ell, b_\ell) \subset (-c_2\ell^{s_2}, c_2\ell^{s_2})$, where $0 < s_1 \leq s_2$ and c_1, c_2 are positive constants.

The other reason is more serious. Indeed the convergence result that we will establish in the next section under the stronger assumption (3.5) may not be true under the hypothesis (3.3). Indeed, consider for instance the case $n = 2$, $A = I_2$ the identity 2×2 matrix, $\omega' = (-1, 1)$ and $f : \Omega_\infty \to \mathbb{R}$ defined by

$$f(x_1, x_2) = \begin{cases} -2 & \text{if } x_1 < 0, \\ 2 & \text{if } x_1 > 0. \end{cases}$$

It is clear that the condition (3.3) is satisfied by the function f above. The solution to (3.4) (note that in this case the equation on Ω_ℓ is $-\Delta u_\ell = f$) is then given up to a constant by

$$u_\ell(x_1, x_2) = \begin{cases} (x_1 + \ell)^2 - \ell^2 & \text{if } x_1 < 0, \\ -(x_1 - \ell)^2 + \ell^2 & \text{if } x_1 > 0. \end{cases}$$

Hence the gradient of u_ℓ is given by

$$\nabla u_\ell(x_1, x_2) = \begin{cases} (2(\ell + x_1), 0) & \text{if } x_1 < 0, \\ (2(\ell - x_1), 0) & \text{if } x_1 > 0. \end{cases}$$

Making the computations, we can see that $\left\|\nabla u_\ell\right\|_{2,\Omega_1} \to +\infty$, so the restrictions of u_ℓ to Ω_1 cannot converge with respect to the norm of the quotient space $H^1(\Omega_1)/\mathbb{R}$.

Therefore, it is more natural to consider the hypothesis $\int_{(a,b)\times\omega'} f \, dx = 0$ for all $a, b \in \mathbb{R}$ $(a < b)$ which is in fact equivalent to (3.5).

In the remaining part of this section, we give two preliminary results that will be useful in the proof of our convergence result.

First, note that the variational formulation of the Neumann problem (3.4) is the following:

$$
\begin{cases}
\text{Find } u_\ell \in H^1(\Omega_\ell) \text{ such that} \\
\displaystyle\int_{\Omega_\ell} A\nabla u_\ell \cdot \nabla v \, dx = \int_{\Omega_\ell} fv \, dx \quad \text{for all} \quad v \in H^1(\Omega_\ell).
\end{cases}
\tag{3.6}
$$

The next lemma gives a very useful property of the solution to such a problem.

Lemma 3.1. *If u_ℓ is a solution to problem* (3.6), *then u_ℓ satisfies*

$$
\int_{\omega'} (A_1 \cdot \nabla u_\ell)(x_1, x') \, dx' = 0 \ \text{ for a.e. } \ x_1 \in (-\ell, \ell) \,.
$$

Proof. In the equation (3.6) satisfied by u_ℓ, taking as test function $v = \varphi(x_1)$, where $\varphi \in H^1((-\ell, \ell))$, we get

$$
\int_{\Omega_\ell} (A_1 \cdot \nabla u_\ell)(x_1, x')\varphi'(x_1) \, dx = \int_{\Omega_\ell} f\varphi \, dx.
$$

Using Fubini's theorem, this implies

$$
\int_{-\ell}^{\ell} \left(\int_{\omega'} (A_1 \cdot \nabla u_\ell)(x_1, x') \, dx' \right)\varphi'(x_1) \, dx_1
$$
$$
= \int_{-\ell}^{\ell} \left(\int_{\omega'} f(x_1, x') \, dx' \right)\varphi(x_1) \, dx_1 = 0
\tag{3.7}
$$

for all $\varphi \in H^1((-\ell, \ell))$, thanks to the property (3.5) satisfied by f. Since (3.7) holds for every φ in $\mathcal{D}((-\ell, \ell))$, we derive

$$
\int_{\omega'} (A_1 \cdot \nabla u_\ell)(x_1, x') \, dx' = K \quad \text{for a.e.} \quad x_1 \in (-\ell, \ell),
$$

where K is a constant. Finally, taking $\varphi = x_1$ in (3.7), we obtain that $K = 0$. $\qquad\square$

Remark 3.1. 1. Note that the property of u_ℓ is a consequence of the assumption (3.5) that we made on f.

2. A consequence of Lemma 3.1 is that a solution u_ℓ to problem (3.6) also satisfies

$$\int_{(a,b)\times\omega'} A_1 \cdot \nabla u_\ell \, dx = 0 \quad \text{for all} \quad a, b \in (-\ell, \ell) \quad (a < b). \tag{3.8}$$

The second preliminary result is in fact the Poincaré-Wirtinger inequality on Ω_ℓ, but we are interested here in the dependence with respect to ℓ of the constant appearing in the inequality.

Lemma 3.2. *There exists a constant C independent of ℓ such that for all $\ell > 0$,*

$$|u|_{2,\Omega_\ell} \le C(\ell+1)||\nabla u||_{2,\Omega_\ell} \quad \text{for all} \quad u \in H^1(\Omega_\ell) \text{ satisfying } \int_{\Omega_\ell} u \, dx = 0.$$

Proof. We use a scaling argument. Let $u \in H^1(\Omega_\ell)$ such that $\int_{\Omega_\ell} u \, dx = 0$. Then $\tilde{u} : \Omega_1 \to \mathbb{R}$ defined by

$$\tilde{u}(x_1, x') = u(\ell x_1, x')$$

belongs to $H^1(\Omega_1)$ and

$$\int_{\Omega_1} \tilde{u} \, dx = \int_{\Omega_1} u(\ell x_1, x') dx = \frac{1}{\ell} \int_{\Omega_\ell} u(y_1, y') \, dy = 0$$

by the change of variable $y = (y_1, y') = (\ell x_1, x')$.

We also have

$$|\tilde{u}|_{2,\Omega_1}^2 = \int_{\Omega_1} u^2(\ell x_1, x') \, dx = \frac{1}{\ell} \int_{\Omega_\ell} u^2(y_1, y') \, dy = \frac{1}{\ell} |u|_{2,\Omega_\ell}^2$$

and

$$||\nabla \tilde{u}||_{2,\Omega_1}^2 = \int_{\Omega_1} \left(\ell^2 (\partial_{x_1} u(\ell x_1, x'))^2 + |\nabla' u(\ell x_1, x')|^2 \right) dx$$

$$= \frac{1}{\ell} \int_{\Omega_\ell} \left(\ell^2 (\partial_{x_1} u(y_1, y'))^2 + |\nabla' u(y_1, y')|^2 \right) dy$$

$$\le \frac{1}{\ell} \int_{\Omega_\ell} (\ell^2 + 1) |\nabla u(y_1, y')|^2 \, dy \le \frac{(\ell+1)^2}{\ell} ||\nabla u||_{2,\Omega_\ell}^2.$$

Using the Poincaré-Wirtinger inequality for $\tilde{u}$ on the fixed domain Ω_1 (see, e.g. [16], [52]), we get

$$\frac{1}{\ell} |u|_{2,\Omega_\ell}^2 = |\tilde{u}|_{2,\Omega_1}^2 \le C^2 ||\nabla \tilde{u}||_{2,\Omega_1}^2 \le C^2 \frac{(\ell+1)^2}{\ell} ||\nabla u||_{2,\Omega_\ell}^2$$

and the desired inequality follows. $\qquad\qquad\qquad\qquad\qquad\qquad\square$

Remark 3.2. An obvious consequence of Lemma 3.2 is that $\left\| \cdot \right\|_{H^1(\Omega_\ell)/\mathbb{R}}$ and $\left\| \nabla \cdot \right\|_{2,\Omega_\ell}$ are two equivalent norms on the space $H^1(\Omega_\ell)/\mathbb{R}$. More precisely, for any $\ell > 0$, we have

$$|u|_{H^1(\Omega_\ell)/\mathbb{R}} \le C(\ell+1)\left\| \nabla u \right\|_{2,\Omega_\ell} \quad \text{for all} \quad u \in H^1(\Omega_\ell),$$

with a constant C independent of ℓ.

3.2 Construction of a solution to the pure Neumann problem

In this section, we prove the convergence of u_ℓ. We will prove in particular that the restriction of u_ℓ to a fixed cylinder Ω_{ℓ_0} converges with respect to the norm of the space $H^1(\Omega_{\ell_0})/\mathbb{R}$, the rate of the convergence being exponential.

Theorem 3.1. *Let $f \in W_\beta(\Omega_\infty)$ for $\beta > 0$ small enough and $u_\ell \in H^1(\Omega_\ell)$ be the solution (unique up to an additive constant) of the problem (3.6). Then for any $\ell_0 > 0$,*

$$u_\ell \to u_\infty \quad \text{strongly in } H^1(\Omega_{\ell_0})/\mathbb{R}$$

as $\ell \to \infty$, where $u_\infty \in H^1_{\text{loc}}(\overline{\Omega}_\infty)$ is the weak solution (unique up to an additive constant) to the following Neumann problem in the infinite strip Ω_∞:

$$
\begin{cases}
-\text{div}\,(A\nabla u_\infty) = f & in \quad \Omega_\infty, \\[2mm]
A\nabla u_\infty \cdot \nu = 0 & on \quad \partial\Omega_\infty, \\[2mm]
\displaystyle\int_{\omega'} (A_1 \cdot \nabla u_\infty)(x^1, x')\,dx' = 0 \text{ for a.e. } x^1 \in \mathbb{R}, \\[2mm]
\left\| \nabla u_\infty \right\|_{2,\Omega_\ell} \le C_\infty(\ell+1)e^{2\beta\ell} \text{ for all } \ell > 0,
\end{cases}
\tag{3.9}
$$

the last inequality being satisfied for some positive constant C_∞ independent of ℓ.

Moreover, the following estimate holds:

$$\left\| \nabla(u_\ell - u_\infty) \right\|_{2,\Omega_{\ell/2}} \le Ce^{-\alpha\ell} \quad \text{for all} \quad \ell > 0, \tag{3.10}$$

where $C \ge 0$, $\alpha > 0$ are constants depending only on ω', λ and Λ.

Remark 3.3. 1. The variational formulation corresponding to the first two equations of (3.9) is the following:

$$\int_{\Omega_\infty} A\nabla u_\infty \cdot \nabla v \, dx = \int_{\Omega_\infty} fv \, dx \quad \forall \, v \in V_0(\Omega_\ell), \quad \forall \ell > 0. \tag{3.11}$$

2. The existence of a solution to (3.9) will be obtained by passing to the limit in the variational equations satisfied by u_ℓ and the uniqueness will be established in the end of the proof of Theorem 3.1 by repeating the arguments employed to prove the convergence of u_ℓ.

Note that without any one of the last two conditions in (3.9), the uniqueness - even up to an additive constant - of the solution to the remaining part of the problem can be lost.

Let us assume for instance that the last condition of the problem (3.9) was not required and consider the case $f = 0$ and $A = I_n$, the identity $n \times n$ matrix. Let $\mu > 0$ be any eigenvalue and $u_\mu \in H^1(\omega')$ be an associated eigenfunction for the problem

$$\begin{cases} -\Delta' u_\mu = \mu u_\mu & \text{in } \omega', \\ \dfrac{\partial u_\mu}{\partial \nu} = 0 & \text{on } \partial \omega', \\ \displaystyle\int_{\omega'} u_\mu \, dx' = 0. \end{cases}$$

Then $u(x_1, x') = e^{\sqrt{\mu} x_1} u_\mu(x')$ is a solution to the problem corresponding to the first three equations of (3.9).

If the third condition in problem (3.9) is missing, then for any constants $a, b \in \mathbb{R}$, the function $u = u(x_1) = ax_1 + b$ is a solution to the problem

$$\begin{cases} -\Delta u = 0 & \text{in } \Omega_\infty, \\ \dfrac{\partial u}{\partial \nu} = 0 & \text{on } \partial\Omega_\infty, \\ \left\| \nabla u \right\|_{2,\Omega_\ell} \le C\sqrt{\ell} & \text{for all } \ell > 0. \end{cases}$$

Proof. We proceed in six steps. In the first three steps we will prove that (u_ℓ) is a Cauchy "sequence" with respect to the norm of $H^1(\Omega_{\ell_0})/\mathbb{R}$ for all $\ell_0 > 0$, then we will prove that the limit of u_ℓ is a solution to the problem (3.9) and in the final step of the proof we establish the uniqueness of the solution to problem (3.9).

(i) *There exists a constant $a \in (0,1)$ depending only on ω', λ and Λ
such that*

$$\int_{\Omega_{\ell_1}} |\nabla(u_\ell - u_{\ell+r})|^2 \, dx \leq a \int_{\Omega_{\ell_1+1}} |\nabla(u_\ell - u_{\ell+r})|^2 \, dx, \qquad (3.12)$$

for all $\ell_1 \leq \ell - 1$ and all $r \geq 0$.

In this part of the proof, we consider $\ell > 1$ and $r \geq 0$ is fixed.

Let $v \in V_0(\Omega_\ell)$. Then $v \in H^1(\Omega_\infty)$ and in particular belongs to
$H^1(\Omega_{\ell+r})$. Therefore, we can use v as test function for the equations satisfied by u_ℓ and $u_{\ell+r}$. Subtracting the two equations, we obtain

$$\int_{\Omega_\ell} A\nabla(u_\ell - u_{\ell+r}) \cdot \nabla v \, dx = 0 \quad \text{for all} \quad v \in V_0(\Omega_\ell), \qquad (3.13)$$

since v and ∇v vanish on $\Omega_{\ell+r} \setminus \Omega_\ell$.

Let now ℓ_1 be an arbitrary real number such that $0 < \ell_1 \leq \ell - 1$.
Denote then by ρ the function whose graph is given by Figure 3.1.

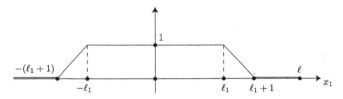

Fig. 3.1   The function ρ.

Since $\ell_1 + 1 \leq \ell$, one has clearly $v = \rho(x_1)(u_\ell - u_{\ell+r}) \in V_0(\Omega_\ell)$ and it
is a test function for (3.13). Since

$$\nabla v = \rho\nabla(u_\ell - u_{\ell+r}) + \big(\rho'(x_1)(u_\ell - u_{\ell+r}), 0, \dots, 0\big)$$

we have

$$A\nabla(u_\ell - u_{\ell+r}) \cdot \nabla v = \rho A\nabla(u_\ell - u_{\ell+r}) \cdot \nabla(u_\ell - u_{\ell+r})$$
$$+ A_1 \cdot \nabla(u_\ell - u_{\ell+r})\rho'(x_1)(u_\ell - u_{\ell+r}).$$

From (3.13), we obtain since v and ∇v vanish on $\Omega_\ell \setminus \Omega_{\ell_1+1}$

$$\int_{\Omega_{\ell_1+1}} \rho A\nabla(u_\ell - u_{\ell+r}) \cdot \nabla(u_\ell - u_{\ell+r}) \, dx$$

$$= -\int_{\Omega_{\ell_1+1}} A_1 \cdot \nabla(u_\ell - u_{\ell+r})\rho'(u_\ell - u_{\ell+r}) \, dx$$

$$= -\int_{\Omega_{\ell_1+1} \setminus \Omega_{\ell_1}} A_1 \cdot \nabla(u_\ell - u_{\ell+r})\rho'(u_\ell - u_{\ell+r}) \, dx,$$

since $\rho' = 0$ outside $D_{\ell_1} = \Omega_{\ell_1+1} \setminus \Omega_{\ell_1}$. $\rho' = 1$ on $D_{\ell_1}^-$ and $\rho' = -1$ on $D_{\ell_1}^+$, therefore, the equality above becomes

$$\int_{\Omega_{\ell_1+1}} \rho A \nabla(u_\ell - u_{\ell+r}) \cdot \nabla(u_\ell - u_{\ell+r}) \, dx$$

$$= -\int_{D_{\ell_1}^-} A_1 \cdot \nabla(u_\ell - u_{\ell+r})(u_\ell - u_{\ell+r}) \, dx$$

$$+ \int_{D_{\ell_1}^+} A_1 \cdot \nabla(u_\ell - u_{\ell+r})(u_\ell - u_{\ell+r}) \, dx \, .$$

From the ellipticity of A (see (3.1)), the positivity of ρ and the fact that $\rho = 1$ on Ω_{ℓ_1}, we derive

$$\int_{\Omega_{\ell_1+1}} \rho A \nabla(u_\ell - u_{\ell+r}) \cdot \nabla(u_\ell - u_{\ell+r}) \, dx \geq \lambda \int_{\Omega_{\ell_1+1}} \rho |\nabla(u_\ell - u_{\ell+r})|^2 \, dx$$

$$\geq \lambda \int_{\Omega_{\ell_1}} |\nabla(u_\ell - u_{\ell+r})|^2 \, dx.$$

Combining this inequality with the last equality above, we get

$$\lambda \int_{\Omega_{\ell_1}} |\nabla(u_\ell - u_{\ell+r})|^2 \, dx \leq -\int_{D_{\ell_1}^-} A_1 \cdot \nabla(u_\ell - u_{\ell+r})(u_\ell - u_{\ell+r}) \, dx$$

$$+ \int_{D_{\ell_1}^+} A_1 \cdot \nabla(u_\ell - u_{\ell+r})(u_\ell - u_{\ell+r}) \, dx.$$

Using the property (3.8) for u_ℓ and $u_{\ell+r}$, we have that

$$\int_{D_{\ell_1}^-} A_1 \cdot \nabla(u_\ell - u_{\ell+r}) \, dx = \int_{D_{\ell_1}^+} A_1 \cdot \nabla(u_\ell - u_{\ell+r}) \, dx = 0,$$

so for any constants $c_+, c_- \in \mathbb{R}$, the last inequality is equivalent to

$$\lambda \int_{\Omega_{\ell_1}} |\nabla(u_\ell - u_{\ell+r})|^2 \, dx \leq \int_{D_{\ell_1}^+} A_1 \cdot \nabla(u_\ell - u_{\ell+r})(u_\ell - u_{\ell+r} - c_+) \, dx$$

$$- \int_{D_{\ell_1}^-} A_1 \cdot \nabla(u_\ell - u_{\ell+r})(u_\ell - u_{\ell+r} - c_-) \, dx \, .$$

From (3.2), we derive

$$|A_1 \cdot \nabla(u_\ell - u_{\ell+r})| \leq |A \nabla(u_\ell - u_{\ell+r})| \leq \Lambda |\nabla(u_\ell - u_{\ell+r})| \, .$$

Thus, using the Cauchy-Schwarz inequality, we get

$$\lambda \int_{\Omega_{\ell_1}} |\nabla(u_\ell - u_{\ell+r})|^2 \, dx$$

$$\leq \Lambda \int_{D_{\ell_1}^+} |\nabla(u_\ell - u_{\ell+r})| |u_\ell - u_{\ell+r} - c_+| \, dx$$

$$+ \Lambda \int_{D_{\ell_1}^-} |\nabla(u_\ell - u_{\ell+r})| |u_\ell - u_{\ell+r} - c_-| \, dx \qquad (3.14)$$

$$\leq \Lambda \big\| |\nabla(u_\ell - u_{\ell+r})| \big\|_{2,D_{\ell_1}^+} \big| u_\ell - u_{\ell+r} - c_+ \big|_{2,D_{\ell_1}^+}$$

$$+ \Lambda \big\| |\nabla(u_\ell - u_{\ell+r})| \big\|_{2,D_{\ell_1}^-} \big| u_\ell - u_{\ell+r} - c_- \big|_{2,D_{\ell_1}^-}.$$

Let us now choose

$$c_+ = \fint_{D_{\ell_1}^+} (u_\ell - u_{\ell+r}) \, dx \quad \text{and} \quad c_- = \fint_{D_{\ell_1}^-} (u_\ell - u_{\ell+r}) \, dx \,,$$

i.e., c_+ and c_- are the mean values of $u_\ell - u_{\ell+r}$ on $D_{\ell_1}^+$ and respectively $D_{\ell_1}^-$. Then $\int_{D_{\ell_1}^+} (u_\ell - u_{\ell+r} - c_+) \, dx = 0$, and by the Poincaré-Wirtinger inequality, we have

$$\big| u_\ell - u_{\ell+r} - c_+ \big|_{2,D_{\ell_1}^+} \leq C_P \big\| |\nabla(u_\ell - u_{\ell+r})| \big\|_{2,D_{\ell_1}^+} \,,$$

where C_P is a constant depending only on ω', since it is the Poincaré-Wirtinger constant for the space $H^1((0,1) \times \omega')$ (note that $D_{\ell_1}^+$ is just a translated set of $(0,1) \times \omega'$). We obtain a similar inequality for $u_\ell - u_{\ell+r} - c_-$ on $D_{\ell_1}^-$ with exactly the same constant.

Combining the last inequality with (3.14), we get

$$\lambda \big\| |\nabla(u_\ell - u_{\ell+r})| \big\|_{2,\Omega_{\ell_1}}^2$$

$$\leq \Lambda C_P \Big(\big\| |\nabla(u_\ell - u_{\ell+r})| \big\|_{2,D_{\ell_1}^+}^2 + \big\| |\nabla(u_\ell - u_{\ell+r})| \big\|_{2,D_{\ell_1}^-}^2 \Big)$$

$$= \Lambda C_P \big\| |\nabla(u_\ell - u_{\ell+r})| \big\|_{2,D_{\ell_1}}^2$$

$$= \Lambda C_P \Big(\big\| |\nabla(u_\ell - u_{\ell+r})| \big\|_{2,\Omega_{\ell_1+1}}^2 - \big\| |\nabla(u_\ell - u_{\ell+r})| \big\|_{2,\Omega_{\ell_1}}^2 \Big)$$

and consequently,

$$\big\| |\nabla(u_\ell - u_{\ell+r})| \big\|_{2,\Omega_{\ell_1}}^2 \leq \frac{\Lambda C_P}{\lambda + \Lambda C_P} \big\| |\nabla(u_\ell - u_{\ell+r})| \big\|_{2,\Omega_{\ell_1+1}}^2 \,,$$

which proves the estimate (3.12) with $a = \dfrac{\Lambda C_P}{\lambda + \Lambda C_P} \in (0,1)$.

(ii) *There exist positive constants C, α depending only on ω', λ, Λ and β such that*

$$\left\|\nabla(u_\ell - u_{\ell+r})\right\|_{2,\Omega_{\ell/2}} \le Ce^{-\alpha\ell} \quad \text{for all} \quad \ell > 0 \quad \text{and} \quad r \in [0,1]. \quad (3.15)$$

Let $\ell > 0$ and $r \in [0,1]$. Starting with $\ell_1 = \dfrac{\ell}{2}$, we iterate the inequality (3.12) $\left[\dfrac{\ell}{2}\right]$ times (where $\left[\dfrac{\ell}{2}\right]$ denotes the integer part of $\dfrac{\ell}{2}$). We obtain

$$\int_{\Omega_{\ell/2}} |\nabla(u_\ell - u_{\ell+r})|^2 \, dx \le a^{\left[\frac{\ell}{2}\right]} \int_{\Omega_{\ell/2+[\ell/2]}} |\nabla(u_\ell - u_{\ell+r})|^2 \, dx.$$

Noting that

$$\frac{\ell}{2} - 1 \le \left[\frac{\ell}{2}\right] \le \frac{\ell}{2} \quad \text{and} \quad 0 < a < 1,$$

we have $\Omega_{\ell/2+[\ell/2]} \subset \Omega_\ell$ and consequently,

$$\int_{\Omega_{\ell/2}} |\nabla(u_\ell - u_{\ell+r})|^2 \, dx \le a^{\frac{\ell}{2}-1} \int_{\Omega_\ell} |\nabla(u_\ell - u_{\ell+r})|^2 \, dx.$$

Thus,

$$\left\|\nabla(u_\ell - u_{\ell+r})\right\|_{2,\Omega_{\ell/2}} \le ce^{-\alpha'\ell}\left\|\nabla(u_\ell - u_{\ell+r})\right\|_{2,\Omega_\ell}, \quad (3.16)$$

with $c = a^{-1/2}$ and $\alpha' = \frac{1}{4}\ln(\frac{1}{a}) > 0$.

We need then to estimate $\left\|\nabla(u_\ell - u_{\ell+r})\right\|_{2,\Omega_\ell}$. Taking $\tilde{u}_\ell = u_\ell - \fint_{\Omega_\ell} u_\ell \, dx$ as test function in the equation satisfied by u_ℓ and using (3.1) and Lemma 3.2, we obtain

$$\lambda \int_{\Omega_\ell} |\nabla u_\ell|^2 \, dx \le \int_{\Omega_\ell} A\nabla u_\ell \cdot \nabla u_\ell \, dx = \int_{\Omega_\ell} f\tilde{u}_\ell \, dx \le |f|_{2,\Omega_\ell} |\tilde{u}_\ell|_{2,\Omega_\ell}$$

$$\le C(\ell+1)e^{\beta\ell}\left\|\nabla u_\ell\right\|_{2,\Omega_\ell},$$

since $f \in W_\beta(\Omega_\infty)$. Therefore,

$$\left\|\nabla u_\ell\right\|_{2,\Omega_\ell} \le C(\ell+1)e^{\beta\ell} \quad (3.17)$$

for a constant C depending only on λ, the constant from Lemma 3.2 and the constant appearing in the definition of the space $W_\beta(\Omega_\infty)$.

In the same manner,

$$\left\|\nabla u_{\ell+r}\right\|_{2,\Omega_\ell} \le \left\|\nabla u_{\ell+r}\right\|_{2,\Omega_{\ell+r}} \le C(\ell+r+1)e^{\beta(\ell+r)} \le Ce^{\beta}(\ell+2)e^{\beta\ell}$$

since $r \in [0,1]$. Here the constant C is exactly the same as the one in (3.17).

Consequently,

$$\big\|\nabla(u_\ell - u_{\ell+r})\big\|_{2,\Omega_\ell} \leq \big\|\nabla u_\ell\big\|_{2,\Omega_\ell} + \big\|\nabla u_{\ell+r}\big\|_{2,\Omega_\ell} \leq C(\ell+1)e^{\beta\ell}, \quad (3.18)$$

where the last constant depends on the same parameters as the one appearing in the inequalities (3.17) and β, hence is independent of $\ell > 0$ and $r \in [0,1]$.

Combining (3.18) with (3.16), we get

$$\big\|\nabla(u_\ell - u_{\ell+r})\big\|_{2,\Omega_{\ell/2}} \leq C(\ell+1)e^{-(\alpha'-\beta)\ell}.$$

Finally, if β from the definition of the space $W_\beta(\Omega_\infty)$ is such that $\beta < \alpha'$ then we have

$$\big\|\nabla(u_\ell - u_{\ell+r})\big\|_{2,\Omega_{\ell/2}} \leq Ce^{-\alpha\ell},$$

for all $\ell > 0$ and all $r \in [0,1]$, where α can be any real number satisfying $0 < \alpha < \alpha' - \beta$.

(iii) *There exist constants* $C, \alpha > 0$ *depending only on* ω', λ, Λ *and* β *such that*

$$\big\|\nabla(u_\ell - u_{\ell+t})\big\|_{2,\Omega_{\ell/2}} \leq Ce^{-\alpha\ell} \qquad (3.19)$$

for all $\ell > 0$ *and* $t \geq 0$.

Indeed,

$$\big\|\nabla(u_\ell - u_{\ell+t})\big\|_{2,\Omega_{\ell/2}}$$

$$\leq \sum_{i=0}^{[t]-1} \big\|\nabla(u_{\ell+i} - u_{\ell+i+1})\big\|_{2,\Omega_{\ell/2}} + \big\|\nabla(u_{\ell+[t]} - u_{\ell+t})\big\|_{2,\Omega_{\ell/2}}$$

$$\leq \sum_{i=0}^{[t]-1} \big\|\nabla(u_{\ell+i} - u_{\ell+i+1})\big\|_{2,\Omega_{(\ell+i)/2}} + \big\|\nabla(u_{\ell+[t]} - u_{\ell+t})\big\|_{2,\Omega_{(\ell+[t])/2}}$$

$$\leq \sum_{i=0}^{[t]} Ce^{-\alpha(\ell+i)} = Ce^{-\alpha\ell}\sum_{i=0}^{[t]} e^{-\alpha i} \leq C\frac{1}{1-e^{-\alpha}}e^{-\alpha\ell},$$

so we obtain (3.19) with the same constant α as in the inequality (3.15) and a modified C, but depending on the same parameters as the constant C of (3.15).

(iv) *There exists* $u_\infty \in H^1_{\mathrm{loc}}(\overline{\Omega}_\infty)$ *such that for all* $\ell_0 > 0$, $u_\ell \to u_\infty$ *in* $H^1(\Omega_{\ell_0})/\mathbb{R}$ *and* $u_\ell - u_\infty$ *satisfies the estimate* (3.10).

A trivial consequence of (3.19) and the Poincaré-Wirtinger inequality - see Remark 3.2 - is that for a fixed $\ell_0 > 0$,

$$|u_\ell - u_{\ell+t}|_{H^1(\Omega_{\ell_0})/\mathbb{R}} \leq C(\ell_0 + 1)||\nabla(u_\ell - u_{\ell+t})||_{2,\Omega_{\ell_0}} \leq Ce^{-\alpha\ell}$$

for all $\ell \geq 2\ell_0$ and all $t \geq 0$. This implies that $(u_\ell)_{\ell>0}$ is a Cauchy "sequence" with respect to the norm of the quotient space $H^1(\Omega_{\ell_0})/\mathbb{R}$. But $H^1(\Omega_{\ell_0})/\mathbb{R}$ is a Banach space. Thus there exists $u_\infty^{\ell_0} \in H^1(\Omega_{\ell_0})$ such that $u_\ell \to u_\infty^{\ell_0}$ in $H^1(\Omega_{\ell_0})/\mathbb{R}$. For $\ell_0 \geq 1$, let us choose the representative in $H^1(\Omega_{\ell_0})/\mathbb{R}$ satisfying $\int_{\Omega_1} u_\infty^{\ell_0} \, dx = 0$.

Let $\ell_0 = k$ vary in $\mathbb{N}^*$. For all $k \in \mathbb{N}^*$, we have that $\nabla u_\infty^k = \nabla u_\infty^{k+1}$ a.e. in Ω_k, since ∇u_∞^k and ∇u_∞^{k+1} are both strong limits of ∇u_ℓ (as $\ell \to +\infty$) in $(L^2(\Omega_k))^n$. By the choice of our representatives, we also have $\int_{\Omega_1} (u_\infty^k - u_\infty^{k+1}) \, dx = 0$ and therefore (recall that Ω_k is connected) $u_\infty^k = u_\infty^{k+1}$ a.e. in Ω_k.

Consequently, we can construct a function $u_\infty \in H^1_{\text{loc}}(\overline{\Omega}_\infty)$ such that $u_\infty = u_\infty^k$ a.e. in Ω_k for all $k \in \mathbb{N}^*$. It is easy to see that this function satisfies

$$u_\ell \to u_\infty \quad \text{in} \quad H^1(\Omega_{\ell_0})/\mathbb{R} \quad \text{for all} \quad \ell_0 > 0 \,.$$

In order to obtain the estimate (3.10) for $u_\ell - u_\infty$, we consider $\ell > 0$ fixed (but otherwise arbitrary) and let $t \to +\infty$ in the inequality (3.19).

(v) *The limit u_∞ is a solution to problem* (3.9).

For $\ell_0 > 0$ fixed, let $v \in V_0(\Omega_{\ell_0})$. Then $v \in H^1(\Omega_\infty)$, hence $v \in H^1(\Omega_\ell)$ for all $\ell > 0$. Consequently, v is a good test function for the variational problems satisfied by all u_ℓ for $\ell > 0$. Since $v = 0$ outside Ω_{ℓ_0}, we have for all $\ell \geq \ell_0$,

$$\int_{\Omega_{\ell_0}} A\nabla u_\ell \cdot \nabla v \, dx = \int_{\Omega_\ell} A\nabla u_\ell \cdot \nabla v \, dx = \int_{\Omega_\ell} fv \, dx = \int_{\Omega_{\ell_0}} fv \, dx \,.$$

Since $\nabla u_\ell \to \nabla u_\infty$ strongly in $(L^2(\Omega_{\ell_0}))^n$, by letting ℓ go to $+\infty$, we get

$$\int_{\Omega_\infty} A\nabla u_\infty \cdot \nabla v \, dx = \int_{\Omega_{\ell_0}} A\nabla u_\infty \cdot \nabla v \, dx = \int_{\Omega_{\ell_0}} fv \, dx = \int_{\Omega_\infty} fv \, dx \,,$$

that is (3.11), taking into account the fact that ℓ_0 was arbitrarily chosen. Therefore u_∞ satisfies the first two equations of problem (3.9).

Once again, let $\ell_0 > 0$ be fixed. Then, for any $\ell \geq \ell_0$, we have that

$$\int_{\omega'} (A_1 \cdot \nabla u_\ell)(x_1, x') \, dx' = 0 \quad \text{for a.e.} \quad x_1 \in (-\ell_0, \ell_0) \qquad (3.20)$$

thanks to Lemma 3.1. Since $\nabla u_\ell \to \nabla u_\infty$ strongly in $(L^2(\Omega_{\ell_0}))^n$, we have that, up to a "subsequence", $\nabla u_\ell(x_1, \cdot) \to \nabla u_\infty(x_1, \cdot)$ strongly in $(L^2(\omega'))^n$, for a.e. $x_1 \in (-\ell_0, \ell_0)$. Letting $\ell \to +\infty$ in (3.20) and taking into account that ℓ_0 is arbitrary, we prove that u_∞ satisfies the third equation of problem (3.9).

Using the estimate (3.10) proved in the previous step and the inequality (3.17), we get

$$
\begin{aligned}
\left\|\nabla u_\infty\right\|_{2,\Omega_\ell} &\leq \left\|\nabla(u_\infty - u_{2\ell})\right\|_{2,\Omega_\ell} + \left\|\nabla u_{2\ell}\right\|_{2,\Omega_\ell} \\
&\leq \left\|\nabla(u_\infty - u_{2\ell})\right\|_{2,\Omega_\ell} + \left\|\nabla u_{2\ell}\right\|_{2,\Omega_{2\ell}} \\
&\leq C\left(e^{-2\alpha\ell} + (2\ell+1)e^{2\beta\ell}\right) \\
&\leq C_\infty(\ell+1)e^{2\beta\ell},
\end{aligned}
$$

i.e., exactly the inequality appearing in problem (3.9).

(vi) *There exists a unique solution (up to an additive constant) to problem (3.9).*

The existence was established in the previous step. For the uniqueness, consider u_∞, u'_∞ two solutions to problem (3.9). Then, for any $\ell_1 > 0$, the arguments used in step (i) are valid with u_ℓ, $u_{\ell+r}$ replaced by u_∞, respectively u'_∞. Therefore, we obtain the inequality

$$
\int_{\Omega_{\ell_1}} |\nabla(u_\infty - u'_\infty)|^2 \, dx \leq a \int_{\Omega_{\ell_1+1}} |\nabla(u_\infty - u'_\infty)|^2 \, dx,
$$

for all $\ell_1 > 0$ with exactly the same constant a as the one obtained at step (i). Iterating k-times the inequality above, we get

$$
\begin{aligned}
\left\|\nabla(u_\infty - u'_\infty)\right\|_{2,\Omega_{\ell_1}} &\leq a^{\frac{k}{2}} \left\|\nabla(u_\infty - u'_\infty)\right\|_{2,\Omega_{\ell_1+k}} \\
&= e^{-2\alpha'k} \left\|\nabla(u_\infty - u'_\infty)\right\|_{2,\Omega_{\ell_1+k}},
\end{aligned}
$$

since $\alpha' = \frac{1}{4}\ln(\frac{1}{a})$ (see step (ii)).

Using the inequality in (3.9) satisfied by u_∞ and u'_∞, we derive

$$
\begin{aligned}
\left\|\nabla(u_\infty - u'_\infty)\right\|_{2,\Omega_{\ell_1}} &\leq 2C_\infty(\ell_1+k+1)e^{2\beta(\ell_1+k)}e^{-2\alpha'k} \\
&= 2C_\infty e^{2\beta\ell_1}(\ell_1+k+1)e^{-2(\alpha'-\beta)k}.
\end{aligned}
$$

Keeping ℓ_1 fixed and letting k go to $+\infty$, we obtain $\left\|\nabla(u_\infty - u'_\infty)\right\|_{2,\Omega_{\ell_1}} = 0$, since $\beta < \alpha'$. Therefore, $\nabla u_\infty = \nabla u'_\infty$ a.e. in Ω_{ℓ_1} and consequently, $\nabla u_\infty = \nabla u'_\infty$ a.e. in Ω_∞, the result above being valid for any $\ell_1 > 0$. This proves that $u_\infty = u'_\infty$ up to an additive constant, since Ω_∞ is connected. □

The key ingredient in the proof of Theorem 3.1 is the introduction of the constants c_+ and c_-, which allows to apply the Poincaré-Wirtinger inequality on $D_{\ell_1}^+$ and $D_{\ell_1}^-$ with a constant independent of ℓ and ℓ_1. This is specific to the case of the usual cylinders studied here and is no longer true for instance for generalized cylinders of the form $(-\ell, \ell)^k \times \omega'$, if $k \geq 2$ (here ω' is a bounded Lipschitz domain in $\mathbb{R}^{n-k}$).

One can consider a slightly different equation (3.4), the data f having a different form than the one considered above. More precisely one can consider the following equation:

$$\begin{cases} -\operatorname{div}(A\nabla u_\ell) = -\operatorname{div} f & \text{in} \quad \Omega_\ell, \\ A\nabla u_\ell \cdot \nu = f \cdot \nu & \text{on} \quad \partial\Omega_\ell, \end{cases} \quad (3.21)$$

where this time $f = (f_1, \ldots, f_n)$ is a vector field in $(L^2_{\text{loc}}(\overline{\Omega}_\infty))^n$.

The variational formulation of problem (3.21) is the following:

$$\begin{cases} \text{Find } u_\ell \in H^1(\Omega_\ell) \text{ such that} \\ \displaystyle\int_{\Omega_\ell} A\nabla u_\ell \cdot \nabla v \, dx = \int_{\Omega_\ell} f \cdot \nabla v \, dx \quad \text{for all} \quad v \in H^1(\Omega_\ell). \end{cases} \quad (3.22)$$

Note that no compatibility condition for f is needed in this case, i.e., the problem (3.22) admits a unique solution - up to an additive constant - for any $f \in (L^2(\Omega_\ell))^n$.

Let us notice that the results of Theorem 3.1 can be extended to the problem (3.21) under the assumption that $f \in (W_\beta(\Omega_\infty))^n$ for some small enough $\beta > 0$. The problem satisfied by the limit u_∞ is the following:

$$\begin{cases} -\operatorname{div}(A\nabla u_\infty) = -\operatorname{div} f & \text{in} \quad \Omega_\infty, \\ A\nabla u_\infty \cdot \nu = f \cdot \nu & \text{on} \quad \partial\Omega_\infty, \\ \displaystyle\int_{\omega'} (A_1 \cdot \nabla u_\infty)(x^1, x') \, dx' = \int_{\omega'} f_1(x_1, x') \, dx' & \text{for a.e. } x^1 \in \mathbb{R}, \\ \|\nabla u_\infty\|_{2,\Omega_\ell} \leq C_\infty e^{2\beta\ell} & \text{for all } \ell > 0. \end{cases} \quad (3.23)$$

Indeed, using the same arguments as in the proof of Lemma 3.1, we prove that a solution u_ℓ of (3.22) satisfies

$$\int_{\omega'} (A_1 \cdot \nabla u_\ell)(x_1, x') \, dx' = \int_{\omega'} f_1(x_1, x') \, dx' \quad \text{for a.e. } x_1 \in (-\ell, \ell).$$

In particular, for all $a, b \in (-\ell, \ell)$ with $a < b$, we have that

$$\int_{(a,b)\times\omega'} A_1 \cdot \nabla u_\ell \, dx = \int_{(a,b)\times\omega'} f_1 \, dx.$$

Therefore, for any $\ell \geq 1$ and $r \geq 0$ and for any $\ell_1 \leq \ell - 1$, we get

$$\int_{D_{\ell_1}^-} A_1 \cdot \nabla(u_\ell - u_{\ell+r}) \, dx = \int_{D_{\ell_1}^-} A_1 \cdot \nabla u_\ell \, dx - \int_{D_{\ell_1}^-} A_1 \cdot \nabla u_{\ell+r} \, dx$$

$$= \int_{D_{\ell_1}^-} f_1 \, dx - \int_{D_{\ell_1}^-} f_1 \, dx = 0$$

and the corresponding equality on $D_{\ell_1}^+$. Keeping these equalities in mind, we can follow the proof of Theorem 3.1 and see that all the arguments used there remain valid if we replace the problem (3.6) by the problem (3.22) in the cylinder Ω_ℓ. Note that in this case, we do not need to use the Poincaré-Wirtinger inequality in order to estimate $\left\|\nabla u_\ell\right\|_{2,\Omega_\ell}$. Hence, instead of (3.17), we obtain $\left\|\nabla u_\ell\right\|_{2,\Omega_\ell} \leq Ce^{\beta\ell}$ for all $\ell > 0$, since

$$\lambda \int_{\Omega_\ell} |\nabla u_\ell|^2 \, dx \leq \int_{\Omega_\ell} A\nabla u_\ell \cdot \nabla u_\ell = \int_{\Omega_\ell} f \cdot \nabla u_\ell \, dx$$

$$\leq |f|_{2,\Omega_\ell} \left\|\nabla u_\ell\right\|_{2,\Omega_\ell} \leq Ce^{\beta\ell} \left\|\nabla u_\ell\right\|_{2,\Omega_\ell}.$$

This estimate explains the disappearance of the multiplying factor $(\ell+1)$ in the inequality giving the estimate of $\left\|\nabla u_\infty\right\|_{2,\Omega_\ell}$ in problem (3.23). Thanks to the same estimate, we can even take $\alpha = \alpha' - \beta$ instead of $0 < \alpha < \alpha' - \beta$, where α' is the constant appearing in the proof of Theorem 3.1.

We end this section with an obvious corollary of Theorem 3.1.

Corollary 3.1. *Let $f \in W_\beta(\Omega_\infty)$ for $\beta > 0$ small enough. Let $u_\ell \in H^1(\Omega_\ell)$ be the solution (unique up to an additive constant) to problem (3.6) and $u_\infty \in H^1_{loc}(\overline{\Omega}_\infty)$ be the weak solution (unique up to an additive constant) to problem (3.9) . Then for any $\ell_0 > 0$,*

$$u_\ell \to u_\infty \quad \text{strongly in } H^1(\Omega_{\ell_0})/\mathbb{R}$$

as $\ell \to \infty$ and the following estimate holds:

$$|u_\ell - u_\infty|_{H^1(\Omega_{\ell_0})/\mathbb{R}} \leq Ce^{-\alpha\ell} \quad \text{for all} \quad \ell \geq \ell_0,$$

where $C \geq 0$ is a constant depending only on ω', λ, Λ, ℓ_0 and $\alpha > 0$ is a constant depending only on ω', λ, Λ.

3.3 The case of data independent of x_1

We would like to know if data independent of x_1 force the solution to (3.6) to be independent of x_1 at the limit. Hence we suppose $A = A(x')$ and $f = f(x')$ and we want to know if the limit u_∞ is also independent of x_1. We will see that this is not true but that u_∞ can be written as the sum (up to an additive constant) of a function independent of x_1 and a linear function of x_1.

In order to fix the ideas, we give here the precise assumptions satisfied by the data A and f in this section. For the measurable matrix field $A = A(x')$, we assume that

$$A(x')\xi \cdot \xi \geq \lambda|\xi|^2, \ |A(x')\xi| \leq \Lambda|\xi| \quad \text{for a.e.} \quad x' \in \omega' \tag{3.24}$$

for some positive constants $\lambda, \Lambda > 0$. For the function f, we suppose

$$f = f(x') \quad \text{with} \quad f \in L^2(\omega') \quad \text{and} \quad \int_{\omega'} f \, dx' = 0. \tag{3.25}$$

Let us introduce the following notation: A'_1 and $A'_{,1}$ denote the vector fields with values in $\mathbb{R}^{n-1}$ defined by the last $n-1$ components of the first row and respectively first column of A, i.e.,

$$A'_1 = (A_{1,j})_{2\leq j\leq n} \ , \quad A'_{,1} = (A_{i,1})_{2\leq i\leq n}.$$

Define also A' to be the $(n-1) \times (n-1)$ matrix given by

$$A' = (A_{i,j})_{2\leq i,j\leq n} \ .$$

Finaly, we use the notation ν' to designate the outward unit normal to ω'.

We have:

Theorem 3.2. *Let A and f satisfy (3.24) and (3.25). Let $u_\ell \in H^1(\Omega_\ell)$ be the solution (unique up to an additive constant) of the problem (3.6). Then for any $\ell_0 > 0$,*

$$u_\ell \to u_\infty \quad \text{strongly in } H^1(\Omega_{\ell_0})/\mathbb{R}$$

as $\ell \to \infty$, the limit $u_\infty \in H^1_{\text{loc}}(\overline{\Omega}_\infty)$ being given by the formula

$$u_\infty(x_1, x') = \tilde{u}_\infty(x') + kx_1,$$

where $(\tilde{u}_\infty, k) \in H^1(\omega') \times \mathbb{R}$ is the weak solution (unique up to an additive constant for the first component) to the following problem in ω':

$$\begin{cases} -\mathrm{div}\,'(A'\nabla'\tilde{u}_\infty) = f + k\,\mathrm{div}\,'A'_{.,1} & in\ \ \omega', \\[2mm] A'\nabla'\tilde{u}_\infty \cdot \nu' = -kA'_{.,1}\cdot\nu' & on\ \ \partial\omega', \\[2mm] \displaystyle\int_{\omega'}(A'_1\cdot\nabla'\tilde{u}_\infty)\,dx' = -k\int_{\omega'}A_{1,1}\,dx'. \end{cases} \qquad (3.26)$$

Moreover, the following estimate holds:

$$\big\|\nabla(u_\ell - u_\infty)\big\|_{2,\Omega_{\ell/2}} \leq Ce^{-\alpha\ell} \quad \textit{for all}\ \ \ell > 0, \qquad (3.27)$$

where $C \geq 0$, $\alpha > 0$ are constants depending only on ω', λ and Λ.

Remark 3.4. The weak formulation corresponding to the first two equations of (3.26) is the following:

$$\begin{aligned} \int_{\omega'} A'\nabla'\tilde{u}_\infty \cdot \nabla'v\,dx' \\ = \int_{\omega'} fv\,dx' - k\int_{\omega'}A'_{.,1}\cdot\nabla'v\,dx' \quad \text{for all}\ \ v \in H^1(\omega'). \end{aligned} \qquad (3.28)$$

Proof. First we remark that $\big|f\big|_{2,\Omega_\ell} = \big|f\big|_{2,\omega'}\sqrt{2\ell}$ for all $\ell > 0$, hence $f \in W_\beta(\Omega_\infty)$ for all $\beta > 0$. Therefore, all the hypotheses of Theorem 3.1 are satisfied and consequently u_ℓ converges (in the sense of Theorem 3.1) to the limit u_∞, which is the weak solution to the problem (3.9) in the infinite cylinder Ω_∞.

Let us assume for a moment that we have proved the existence of a solution $(\tilde{u}_\infty, k) \in H^1(\omega') \times \mathbb{R}$ to problem (3.26). Then, thanks to the uniqueness of the solution to problem (3.9), it is enough to prove that the function u_∞ defined by

$$u_\infty(x_1, x') = \tilde{u}_\infty(x') + kx_1,$$

is a solution to problem (3.9).

We begin by verifying the first two equations, i.e., the variational equation (3.11). Let $v \in V_0(\Omega_\ell)$ for some $\ell > 0$. Then we have (note that $\nabla u_\infty = (k, \nabla'\tilde{u}_\infty(x'))$ is independent of x_1):

$$\int_{\Omega_\infty} A\nabla u_\infty \cdot \nabla v \, dx = \int_{\Omega_\ell} A\nabla u_\infty \cdot \nabla v \, dx$$

$$= \int_{\Omega_\ell} \left((A_1 \cdot \nabla u_\infty)\partial_{x_1} v + (kA'_{,1} + A'\nabla'\tilde{u}_\infty)\cdot \nabla'v \right) dx$$

$$= \int_{\omega'} \left((A_1 \cdot \nabla u_\infty)(x') \int_{-\ell}^{\ell} (\partial_{x_1} v)(x_1, x') \, dx_1 \right) dx' \qquad (3.29)$$

$$+ \int_{-\ell}^{\ell} \left(\int_{\omega'} (kA'_{,1} + A'\nabla'\tilde{u}_\infty)(x')\cdot (\nabla'v)(x_1, x') \, dx' \right) dx_1 \,.$$

Since $v \in H^1(\Omega_\ell)$, we have that $v(x_1, \cdot) \in H^1(\omega')$ and $\nabla'\big(v(x_1, \cdot)\big) = (\nabla'v)(x_1, \cdot)$ for a.e. $x_1 \in (-\ell, \ell)$. Therefore, using (3.28), we derive

$$\int_{\omega'} (kA'_{,1} + A'\nabla'\tilde{u}_\infty)(x') \cdot (\nabla'v)(x_1, x') \, dx' = \int_{\omega'} f(x')v(x_1, x') \, dx'$$

for a.e. $x_1 \in (-\ell, \ell)$.

Moreover, since $v = 0$ on $\{-\ell, \ell\} \times \omega'$, we deduce that

$$\int_{-\ell}^{\ell} (\partial_{x_1} v)(x_1, x') \, dx_1 = v(\ell, x') - v(-\ell, x') = 0$$

for a.e. $x' \in \omega'$.

Using the last two equalities in (3.29), we get

$$\int_{\Omega_\infty} A\nabla u_\infty \cdot \nabla v \, dx = \int_{-\ell}^{\ell} \int_{\omega'} f(x')v(x_1, x') \, dx' \, dx_1 = \int_{\Omega_\ell} fv \, dx = \int_{\Omega_\infty} fv \, dx$$

which proves that u_∞ satisfies the variational equation (3.11).

The third equation in (3.9) is a trivial consequence of the last equation in (3.26), taking into account the fact that $A_1 = (A_{1,1}, A'_1)$ and $\nabla u_\infty = (k, \nabla'\tilde{u}_\infty)$.

Finally, since $\nabla u_\infty = \nabla u_\infty(x')$, we have that

$$||\nabla u_\infty||_{2,\Omega_\ell} = ||\nabla u_\infty||_{2,\omega'}\sqrt{2\ell} \quad \text{for all} \quad \ell > 0 \,.$$

Thus, u_∞ also satisfies the inequality in (3.9) for any $\beta > 0$. We have therefore proved that $u_\infty = \tilde{u}_\infty + kx_1$ is the solution to the problem (3.9), provided that $(\tilde{u}_\infty, k)$ is a solution to problem (3.26).

To complete the proof, we only have to justify the existence of a solution to the problem (3.26). In fact, we will also prove the uniqueness up to an additive constant for the first component.

Let $u_f, u_A \in H^1(\omega')$ be the solutions (unique up to additive constants) to the following problems:

$$\int_{\omega'} A' \nabla' u_f \cdot \nabla' v \, dx' = \int_{\omega'} fv \, dx' \quad \text{for all} \quad v \in H^1(\omega') \tag{3.30}$$

and respectively

$$\int_{\omega'} A' \nabla' u_A \cdot \nabla' v \, dx' = \int_{\omega'} A'_{\cdot,1} \cdot \nabla' v \, dx' \quad \text{for all} \quad v \in H^1(\omega'). \tag{3.31}$$

These solutions exist thanks to the Lax-Milgram theorem (the matrix A being uniformly elliptic on ω', so is A').

For a fixed k, the solution $\tilde{u}_\infty$ to the problem (3.28) is unique, up to an additive constant. Therefore, thanks to the linearity of the problem, we must have

$$\tilde{u}_\infty = u_f - k u_A. \tag{3.32}$$

Replacing $\tilde{u}_\infty$ in the last equation of (3.26), we obtain

$$k \int_{\omega'} (A_{1,1} - A'_1 \cdot \nabla' u_A) \, dx' = - \int_{\omega'} A'_1 \cdot \nabla' u_f \, dx'.$$

Let us prove that $\int_{\omega'} (A_{1,1} - A'_1 \cdot \nabla' u_A) \, dx' \neq 0$. Taking u_A as test function in (3.31), we derive

$$\int_{\omega'} \left(-A'_{\cdot,1} \cdot \nabla' u_A + A' \nabla' u_A \cdot \nabla' u_A \right) dx' = 0.$$

Hence we can write

$$\int_{\omega'} (A_{1,1} - A'_1 \cdot \nabla' u_A) \, dx'$$

$$= \int_{\omega'} \left(A_{1,1} - A'_1 \cdot \nabla' u_A - A'_{\cdot,1} \cdot \nabla' u_A + A' \nabla' u_A \cdot \nabla' u_A \right) dx'.$$

One has

$$\nabla(-x_1 + u_A) = \begin{pmatrix} -1 \\ \nabla' u_A \end{pmatrix} \quad , \quad A\nabla(-x_1 + u_A) = \begin{pmatrix} -A_{1,1} + A'_1 \nabla' u_A \\ -A'_{\cdot,1} + A' \nabla' u_A \end{pmatrix}$$

and thus

$$A\nabla(-x_1 + u_A) \cdot \nabla(-x_1 + u_A) = A_{1,1} - A'_1 \cdot \nabla' u_A - A'_{\cdot,1} \cdot \nabla' u_A + A' \nabla' u_A \cdot \nabla' u_A.$$

One derives from above

$$\int_{\omega'} (A_{1,1} - A'_1 \cdot \nabla' u_A) \, dx'$$

$$= \int_{\omega'} A\nabla(-x_1 + u_A) \cdot \nabla(-x_1 + u_A) \, dx'$$

$$\geq \lambda \int_{\omega'} |\nabla(-x_1 + u_A)|^2 \, dx' = \lambda \int_{\omega'} \left(1 + |\nabla' u_A|^2 \right) dx'$$

$$\geq \lambda \int_{\omega'} dx' > 0.$$

Consequently, the quantity $\int_{\omega'}(A_{1,1} - A_1' \cdot \nabla' u_A)\,dx'$ does not vanish, so k is uniquely determined by the formula

$$k = -\frac{\int_{\omega'} A_1' \cdot \nabla' u_f\,dx'}{\int_{\omega'}(A_{1,1} - A_1' \cdot \nabla' u_A)\,dx'}.$$

Replacing this k in the formula (3.32) will give us the couple $(\tilde{u}_\infty, k) \in H^1(\omega') \times \mathbb{R}$ as the solution to the problem (3.26). $\quad\square$

Remark 3.5. If the solution u_f of the problem (3.30) satisfies the equality

$$\int_{\omega'} A_1' \cdot \nabla' u_f\,dx' = 0, \tag{3.33}$$

then $k = 0$ and $u_\infty = \tilde{u}_\infty = u_f$. Hence in this case the limit of u_ℓ is independent of x_1, i.e., $u_\infty = u_\infty(x')$ and is the solution - unique up to an additive constant - to the problem

$$\int_{\omega'} A' \nabla' u_\infty \cdot \nabla' v\,dx' = \int_{\omega'} f v\,dx' \quad \text{for all} \quad v \in H^1(\omega'). \tag{3.34}$$

In fact, in this case, the result remains valid even if we allow the first column of A to be dependent of x_1. More precisely, if we only have

$$A_{i,j} = A_{i,j}(x') \quad \text{for all} \quad i \in \{1,\ldots,n\} \quad \text{and} \quad j \in \{2,\ldots,n\} \tag{3.35}$$

- the other assumptions of Theorem 3.2 remaining unchanged - then the limit u_∞ is independent of x_1 and is the solution to the problem (3.34).

Note that the equality (3.33) is automatically satisfied if $A_1' = 0$. This is a very special case, since we have in fact

$$u_\infty = u_\ell \quad \text{a.e. in} \quad \Omega_\ell, \quad \text{for all} \quad \ell > 0, \tag{3.36}$$

where u_∞ is the solution to (3.34). Indeed, it is easy to verify that in this case u_∞ also satisfies the variational problem (3.6) for all $\ell > 0$.

One can ask if this is really a special case, or if that can happen in a more general situation. Intuitively, there is no reason to have $u_\infty = u_\ell$ in the general case, because of the boundary condition that u_ℓ must satisfy on $\{-\ell, \ell\} \times \omega'$. We can provide a whole class of examples where we can be sure that the statement (3.36) is false.

Let us consider the situation where $A_1' = 0$, A satisfies (3.35) and $f \in W_\beta(\Omega_\infty)$ for some $\beta > 0$ small enough. Then the limit u_∞ satisfies (3.36) if and only if $f = f(x')$.

We just saw that $f = f(x')$ implies that $u_\infty = u_\ell$. Conversely if $u_\infty = u_\ell$ for all $\ell > 0$, then, since $A_1' = 0$, $A_{1,1}\partial_{x_1}u_\infty = A_1 \cdot \nabla u_\infty = A_1 \cdot \nabla u_\ell = 0$ a.e. on $\{-\ell, \ell\} \times \omega'$, for a.e. $\ell \in \mathbb{R}$. Hence $\partial_{x_1}u_\infty = 0$ a.e. in Ω_∞, since $A_{1,1} \geq \lambda > 0$ a.e. in Ω_∞. Consequently, $u_\infty = u_\infty(x')$ and thus $\nabla u_\infty = (0, \nabla' u_\infty(x'))$, which implies

$$A\nabla u_\infty = (0, A'\nabla' u_\infty(x')).$$

Therefore $A\nabla u_\infty$ depends only on x', which means that $-\text{div}\,(A\nabla u_\infty)$ depends only on x', i.e. $f = f(x')$, taking into account the equation satisfied by u_∞ in the infinite cylinder Ω_∞.

Thus in order to construct an example of Neumann problem in Ω_ℓ such that (3.36) is not satisfied, we can choose data that fulfill the following conditions: the measurable matrix field A satisfies (3.1), (3.2), (3.35) and $A_1' = 0$ (for instance $A = I_n$, the $n \times n$ identity matrix) and the function f belongs to some space $W_\beta(\Omega_\infty)$ for a small enough $\beta > 0$ and depends on the variable x_1.

3.4   The case with a lower order term

The case where we have a lower order term in the equation is simpler than the case that we have seen above and we are going to consider it briefly. The notation is slightly different than the one of the preceding section and we have

$$\Omega_\ell = \ell\omega_1 \times \omega_2, \quad \Omega_\infty = \mathbb{R}^p \times \omega_2$$

with ω_1 a convex bounded domain of $\mathbb{R}^p$ containing 0 and ω_2 a bounded domain in $\mathbb{R}^{n-p}$. We would like to find a solution to

$$\begin{cases} -\text{div}\,(A(x)\nabla u) + a(x)u = f \text{ in } \Omega_\infty, \\ A(x)\nabla u \cdot \nu = 0 \text{ on } \partial\Omega_\infty, \end{cases} \tag{3.37}$$

where ν denotes the outward unit normal to $\partial\Omega_\infty$ the boundary of Ω_∞ that we will assume to be piecewise Lipschitz continuous which guarantees the existence of ν. For the reader convenience we recall our assumptions on A i.e. for some positive constants λ, Λ we have

$$\lambda|\xi|^2 \leq A(x)\xi \cdot \xi \ \forall \xi \in \mathbb{R}^n, \quad \text{a. e.} \quad x \in \Omega_\infty, \tag{3.38}$$

$$|A(x)\xi| \leq \Lambda|\xi| \ \forall \xi \in \mathbb{R}^n, \quad \text{a. e.} \quad x \in \Omega_\infty. \tag{3.39}$$

Regarding the function a we will assume that there exists a constant Λ such that

$$0 \leq a(x) \leq \Lambda \quad \text{a. e.} \quad x \in \Omega_\infty. \tag{3.40}$$

(Note that there is no loss of generality in taking the same constant as in (3.39)). We will use the following lemma introduced in [8]:

Lemma 3.3. *Let a be a function satisfying for some positive constants μ, Λ*

$$0 \leq a(X_2) \leq \Lambda \quad a.e. \ X_2 \in \omega_2, \quad \int_{\omega_2} a(X_2)dX_2 \geq \mu. \tag{3.41}$$

Then there exists a constant $\delta = \delta(\Lambda, \mu) > 0$ such that

$$\delta \int_{\omega_2} v^2(X_2)dX_2$$
$$\leq \int_{\omega_2} |\nabla v(X_2)|^2 + a(X_2)v^2(X_2)dX_2 \quad \forall v \in H^1(\omega_2). \tag{3.42}$$

Proof. If (3.42) fails then there exist a sequence a_n satisfying (3.41), a sequence $v_n \in H^1(\omega_2)$ such that

$$\int_{\omega_2} |\nabla v_n(X_2)|^2 + a_n(X_2)v_n^2(X_2)dX_2 \leq \frac{1}{n} \int_{\omega_2} v_n^2(X_2)dX_2. \tag{3.43}$$

Considering possibly $\pm v_n/|v_n|_{2,\omega_2}$ one can assume

$$\int_{\omega_2} v_n^2(X_2)dX_2 = 1, v_n > 0. \tag{3.44}$$

Then it follows from (3.43), (3.44) that v_n is bounded in $H^1(\omega_2)$ and –up to a subsequence– there exists $v \in H^1(\omega_2)$ such that

$$v_n \rightharpoonup v \text{ in } H^1(\omega_2), \quad v_n \to v \text{ in } L^2(\omega_2).$$

From (3.43), (3.44) one deduces that $\nabla v = 0$, i.e. v is constant and from (3.44)

$$v = \frac{1}{\sqrt{|\omega_2|}}$$

where $|\omega_2|$ denotes the measure of ω_2. Going back to (3.41), (3.43) one has

$$\frac{\mu}{|\omega_2|} \leq \int_{\omega_2} \frac{a_n}{|\omega_2|}dX_2 = \int_{\omega_2} a_n\left(\frac{1}{|\omega_2|} - v_n^2\right) + a_n v_n^2 dX_2$$
$$\leq \Lambda\left|\frac{1}{\sqrt{|\omega_2|}} - v_n\right|_{2,\omega_2}\left|\frac{1}{\sqrt{|\omega_2|}} + v_n\right|_{2,\omega_2} + \frac{1}{n}$$
$$\leq \Lambda\left|\frac{1}{\sqrt{|\omega_2|}} - v_n\right|_{2,\omega_2}\left\{\left|\frac{1}{\sqrt{|\omega_2|}}\right|_{2,\omega_2} + |v_n|_{2,\omega_2}\right\} + \frac{1}{n}$$

by Cauchy-Schwarz inequality, i.e. –see (3.44)

$$\frac{\mu}{|\omega_2|} \leq 2\Lambda |\frac{1}{\sqrt{|\omega_2|}} - v_n|_{2,\omega_2} + \frac{1}{n} \to 0$$

and a contradiction. This completes the proof of the lemma.　　　□

Remark 3.6. It follows, if (3.41) holds, that for some positive constants μ_1, μ_2 one has

$$\mu_1 |v|^2_{H^1(\omega_2)} \leq \int_{\omega_2} |\nabla v|^2 + av^2 dX_2 \leq \mu_2 |v|^2_{H^1(\omega_2)} \quad \forall v \in H^1(\omega_2) \quad (3.45)$$

with

$$|v|^2_{H^1(\omega_2)} = \int_{\omega_2} |\nabla v|^2 + v^2 dX_2.$$

Indeed

$$\int_{\omega_2} |\nabla v|^2 + av^2 dX_2 \leq \int_{\omega_2} |\nabla v|^2 dX_2 + \Lambda \int_{\omega_2} v^2 dX_2 \leq (1 \vee \Lambda)|v|^2_{H^1(\omega_2)}$$

where $\vee$ denotes the maximum of two numbers and

$$\int_{\omega_2} |\nabla v|^2 + av^2 dX_2 \geq \frac{1}{2}\int_{\omega_2} |\nabla v|^2 dX_2 + \frac{1}{2}\int_{\omega_2} |\nabla v|^2 + av^2 dX_2$$

$$\geq \frac{1}{2}\int_{\omega_2} |\nabla v|^2 dX_2 + \frac{\delta}{2}\int_{\omega_2} v^2 dX_2 \geq \frac{1 \wedge \delta}{2}|v|^2_{H^1(\omega_2)}$$

where $\wedge$ denotes the minimum of two numbers.

As a corollary we have

Corollary 3.2. *Suppose that a satisfies*

$$0 < \lambda \leq a(x) \leq \Lambda \ a.e. \ x \in \Omega_\infty \quad (3.46)$$

or

$$0 \leq a(x) \leq \Lambda \ a.e. \ x \in \Omega_\infty \ and \ there \ exists \ \mu > 0 \ such \ that$$
$$\int_{\omega_2} a(X_1, X_2)dX_2 \geq \mu \ a.e. \ X_1 \in \mathbb{R}^p. \quad (3.47)$$

Then there exists two positive constants c_1, c_2 independent of ℓ such that

$$c_1 |v|^2_{H^1(\Omega_\ell)} \leq \int_{\Omega_\ell} |\nabla v|^2 + av^2 dx \leq c_2 |v|^2_{H^1(\Omega_\ell)} \quad \forall v \in H^1(\Omega_\ell). \quad (3.48)$$

Proof. Since in both cases $0 \leq a \leq \Lambda$ one has

$$\int_{\Omega_\ell} |\nabla v|^2 + av^2 dx \leq \int_{\Omega_\ell} |\nabla v|^2 + \Lambda v^2 dx \leq (1 \vee \Lambda)|v|^2_{H^1(\Omega_\ell)}.$$

This shows the right hand side inequality. In the case where (3.46) holds, one has for almost every $X_1 \in \ell\omega_1$ and $v \in H^1(\Omega_\ell)$

$$\int_{\omega_2} |\nabla_{X_2} v|^2 + av^2 dX_2 \geq (\lambda \wedge 1) \int_{\omega_2} |\nabla_{X_2} v|^2 + v^2 dX_2$$

and in the case of (3.47) by (3.45)

$$\int_{\omega_2} |\nabla_{X_2} v|^2 + av^2 dX_2 \geq \mu_1 \int_{\omega_2} |\nabla_{X_2} v|^2 + v^2 dX_2.$$

Integrating in X_1 we get for some constant c independent of ℓ

$$\int_{\Omega_\ell} |\nabla_{X_2} v|^2 + av^2 dx \geq c \int_{\Omega_\ell} |\nabla_{X_2} v|^2 + v^2 dx.$$

Thus

$$\int_{\Omega_\ell} |\nabla v|^2 + av^2 dx \geq \frac{1}{2} \int_{\Omega_\ell} |\nabla v|^2 dx + \frac{1}{2} \int_{\Omega_\ell} |\nabla_{X_2} v|^2 + av^2 dx$$

$$\geq \frac{1}{2} \int_{\Omega_\ell} |\nabla v|^2 dx + \frac{c}{2} \int_{\Omega_\ell} |\nabla_{X_2} v|^2 + v^2 dx$$

$$\geq \frac{1 \wedge c}{2} \int_{\Omega_\ell} |\nabla v|^2 + v^2 dx.$$

This completes the proof of the corollary. □

Remark 3.7. Note that in fact (3.46) implies (3.47) with $\mu = \lambda|\omega_2|$. In the case where (3.46) holds then (3.48) holds for $\Omega_\ell = (\ell\omega_1 \times \omega_2) \cap \Omega$ where Ω is an unbounded open set such that $\Omega \subset \Omega_\infty$.

Suppose now that

$$f \in L^2(\Omega_\ell) \quad \forall \ell.$$

Assuming (3.46) or (3.47) it follows from (3.48) and the Lax-Milgram theorem that for every $\ell > 0$ there exists a unique u_ℓ solution to

$$\begin{cases} u_\ell \in H^1(\Omega_\ell), \\ \int_{\Omega_\ell} A(x)\nabla u_\ell \cdot \nabla v + a(x)u_\ell v dx = \int_{\Omega_\ell} fv dx \quad \forall v \in H^1(\Omega_\ell). \end{cases} \quad (3.49)$$

Let us set

$$V_0^\ell = \{v \in H^1(\Omega_\ell) \mid v = 0 \text{ on } \partial\ell\omega_1 \times \omega_2\}. \quad (3.50)$$

One has

Theorem 3.3. *Suppose that* (3.46) *or* (3.47) *holds and that for some non-negative constant* γ

$$|f|_{2,\Omega_\ell} = O(\ell^\gamma). \tag{3.51}$$

Then there exists a unique u_∞ *solution to*

$$\begin{cases} u_\infty \in H^1_{loc}(\overline{\Omega}_\infty), \\ \int_{\Omega_\ell} A(x)\nabla u_\infty \cdot \nabla v + a(x)u_\infty v dx = \int_{\Omega_\ell} f v dx \quad \forall v \in V_0^\ell, \quad \forall \ell, \\ |u_\infty|_{H^1(\Omega_\ell)} = O(\ell^\gamma). \end{cases} \tag{3.52}$$

Moreover

$$|u_\ell - u_\infty|_{H^1(\Omega_{\frac{\ell}{2}})} \leq C e^{-\alpha \ell} \tag{3.53}$$

for two positive constants C *and* α.

Proof. Since the proof is very similar to the one developed in the previous chapter we will only underline the important steps. We want to show that u_ℓ is a Cauchy sequence and thus we start first to estimate $u_\ell - u_{\ell+r}$ for $r \in [0,1]$. From (3.49) we derive

$$\int_{\Omega_\ell} A(x)\nabla(u_\ell - u_{\ell+r}) \cdot \nabla v + a(x)(u_\ell - u_{\ell+r})v dx = 0 \quad \forall v \in V_0^\ell$$

(one assumes v extended by 0 outside of Ω_ℓ). Then if $\rho = \rho_{\ell_1}(X_1)$ is the function defined in (2.7) one has

$$\rho(u_\ell - u_{\ell+r}) \in V_0^\ell$$

and it follows that

$$\int_{\Omega_\ell} A(x)\nabla(u_\ell - u_{\ell+r}) \cdot \nabla[\rho(u_\ell - u_{\ell+r})] + a(x)\rho(u_\ell - u_{\ell+r})^2 dx = 0.$$

Since it is enough to integrate on Ω_{ℓ_1+1} we deduce

$$\int_{\Omega_{\ell_1+1}} A(x)\nabla(u_\ell - u_{\ell+r}) \cdot \nabla(u_\ell - u_{\ell+r})\rho + a(x)\rho(u_\ell - u_{\ell+r})^2 dx$$

$$= -\int_{\Omega_{\ell_1+1}} A(x)\nabla(u_\ell - u_{\ell+r}) \cdot \nabla \rho (u_\ell - u_{\ell+r}) dx.$$

Since $\rho = 1$ and $\nabla \rho = 0$ on Ω_{ℓ_1} one obtains easily (see (3.38), (3.39), (2.7))

$$\int_{\Omega_{\ell_1}} \lambda |\nabla(u_\ell - u_{\ell+r})|^2 + a(x)(u_\ell - u_{\ell+r})^2 dx$$

$$\leq C \int_{D_{\ell_1}} \Lambda |\nabla(u_\ell - u_{\ell+r})||u_\ell - u_{\ell+r}| dx$$

$(D_{\ell_1} = \Omega_{\ell_1+1}\backslash\Omega_{\ell_1})$. Then by (3.48) and the Cauchy-Young inequality one deduces

$$c_1(\lambda \wedge 1)|u_\ell - u_{\ell+r}|^2_{H^1(\Omega_{\ell_1})} \leq \frac{C\Lambda}{2}\int_{D_{\ell_1}}|\nabla(u_\ell - u_{\ell+r})|^2 + (u_\ell - u_{\ell+r})^2 dx$$

i.e. for $\kappa = \frac{C\Lambda}{2c_1(\lambda\wedge1)}$

$$|u_\ell - u_{\ell+r}|^2_{H^1(\Omega_{\ell_1})} \leq \kappa|u_\ell - u_{\ell+r}|^2_{H^1(\Omega_{\ell_1+1})} - \kappa|u_\ell - u_{\ell+r}|^2_{H^1(\Omega_{\ell_1})}$$

which implies

$$|u_\ell - u_{\ell+r}|^2_{H^1(\Omega_{\ell_1})} \leq a|u_\ell - u_{\ell+r}|^2_{H^1(\Omega_{\ell_1+1})}, \quad a = \frac{\kappa}{\kappa+1} < 1.$$

From this last inequality, performing $[\frac{\ell}{2}]$ iterations starting from $\ell_1 = \frac{\ell}{2}$, one ends up with

$$|u_\ell - u_{\ell+r}|^2_{H^1(\Omega_{\frac{\ell}{2}})} \leq a^{[\frac{\ell}{2}]}|u_\ell - u_{\ell+r}|^2_{H^1(\Omega_{\frac{\ell}{2}+[\frac{\ell}{2}]})}.$$

Recalling that

$$\frac{\ell}{2} - 1 \leq [\frac{\ell}{2}] \leq \frac{\ell}{2}$$

it comes

$$|u_\ell - u_{\ell+r}|^2_{H^1(\Omega_{\frac{\ell}{2}})} \leq \frac{1}{a}e^{-\frac{\ell}{2}\ln(\frac{1}{a})}|u_\ell - u_{\ell+r}|^2_{H^1(\Omega_\ell)}$$

i.e.

$$|u_\ell - u_{\ell+r}|_{H^1(\Omega_{\frac{\ell}{2}})} \leq \frac{1}{\sqrt{a}}e^{-\frac{\ell}{4}\ln(\frac{1}{a})}\{|u_\ell|_{H^1(\Omega_\ell)} + |u_{\ell+r}|_{H^1(\Omega_{\ell+r})}\}. \quad (3.54)$$

To estimate the right hand side of this inequality one takes $v = u_\ell$ in (3.49) to get

$$\int_{\Omega_\ell} \lambda|\nabla u_\ell|^2 + au_\ell^2 dx \leq \int_{\Omega_\ell} fu_\ell dx \leq |f|_{2,\Omega_\ell}|u_\ell|_{2,\Omega_\ell}$$

and by (3.48)

$$c_1(\lambda \wedge 1)\int_{\Omega_\ell} |\nabla u_\ell|^2 + u_\ell^2 dx \leq |f|_{2,\Omega_\ell}|u_\ell|_{H^1(\Omega_\ell)}.$$

It follows then by (3.51) that one has

$$|u_\ell|_{H^1(\Omega_\ell)} \leq \frac{|f|_{2,\Omega_\ell}}{c_1(\lambda \wedge 1)} \leq C\ell^\gamma. \quad (3.55)$$

Going back to (3.54) we obtain for different constants C

$$|u_\ell - u_{\ell+r}|_{H^1(\Omega_{\frac{\ell}{2}})} \leq C(\ell^\gamma + (\ell+r)^\gamma)e^{-\frac{\ell}{4}\ln(\frac{1}{a})}$$

$$\leq C\ell^\gamma(1 + (1 + \frac{r}{\ell})^\gamma)e^{-\frac{\ell}{4}\ln(\frac{1}{a})} \leq Ce^{-\alpha\ell}$$

for any $\alpha < \frac{1}{4}\ln(\frac{1}{a})$. By the triangular inequality one has then for any t

$$
\begin{aligned}
|u_\ell - u_{\ell+t}|_{H^1(\Omega_{\frac{\ell}{2}})} &\leq |u_\ell - u_{\ell+1}|_{H^1(\Omega_{\frac{\ell}{2}})} \\
&+ |u_{\ell+1} - u_{\ell+2}|_{H^1(\Omega_{\frac{\ell}{2}})} + \cdots + |u_{\ell+[t]} - u_{\ell+t}|_{H^1(\Omega_{\frac{\ell}{2}})} \\
&\leq Ce^{-\alpha\ell} + Ce^{-\alpha(\ell+1)} + \cdots + Ce^{-\alpha(\ell+[t])} \\
&\leq C\frac{1}{1-e^{-\alpha}}e^{-\alpha\ell} \leq Ce^{-\alpha\ell}.
\end{aligned}
\tag{3.56}
$$

Thus for any $\ell_0 \leq \frac{\ell}{2}$, u_ℓ is a Cauchy sequence in $H^1(\Omega_{\ell_0})$. Let us denote by u_∞ the element of $H^1(\Omega_{\ell_0})$ such that

$$u_\ell \to u_\infty \text{ in } H^1(\Omega_{\ell_0})$$

when $\ell \to \infty$. For any $v \in V_0^{\ell_0}$ one has

$$\int_{\Omega_{\ell_0}} A(x)\nabla u_\ell \cdot \nabla v + a(x)u_\ell v dx = \int_{\Omega_{\ell_0}} fv dx.$$

Passing to the limit in ℓ we obtain

$$\int_{\Omega_{\ell_0}} A(x)\nabla u_\infty \cdot \nabla v + a(x)u_\infty v dx = \int_{\Omega_{\ell_0}} fv dx \quad \forall v \in V_0^{\ell_0}.$$

This gives us the first two properties of (3.52). Passing to the limit in t in (3.56) we get

$$|u_\ell - u_\infty|_{H^1(\Omega_{\frac{\ell}{2}})} \leq Ce^{-\alpha\ell}$$

i.e.

$$|u_{2\ell} - u_\infty|_{H^1(\Omega_\ell)} \leq Ce^{-2\alpha\ell}.$$

This implies

$$
\begin{aligned}
|u_\infty|_{H^1(\Omega_\ell)} &\leq Ce^{-2\alpha\ell} + |u_{2\ell}|_{H^1(\Omega_\ell)} \\
&\leq Ce^{-2\alpha\ell} + |u_{2\ell}|_{H^1(\Omega_{2\ell})} = O(\ell^\gamma)
\end{aligned}
$$

by (3.55). The uniqueness of the solution to (3.52) follows easily since for two solutions u_∞, u'_∞ of (3.52) one has

$$\int_{\Omega_\ell} A(x)\nabla(u_\infty - u'_\infty) \cdot \nabla v + a(x)(u_\infty - u'_\infty)v dx = 0 \quad \forall v \in V_0^\ell$$

and arguing as above with $u_\ell - u_{\ell+r}$ one gets

$$|u_\infty - u'_\infty|_{H^1(\Omega_{\frac{\ell}{2}})} \leq Ce^{-\alpha\ell}.$$

Letting $\ell \to \infty$ the uniqueness of the solution to (3.52) follows. This completes the proof of the theorem. $\qquad\square$

Remark 3.8. In the case where

$$f = f(X_2) \in L^2(\omega_2), \qquad A = \begin{pmatrix} A_{11}(x) & A_{12}(X_2) \\ A_{21}(x) & A_{22}(X_2) \end{pmatrix}$$

and

$$0 \leq a = a(X_2) \leq \Lambda, \quad a \not\equiv 0,$$

it is easy to see (Cf. Lemma 3.3 and Remark 3.6) that there exists a unique solution to

$$\begin{cases} u_\infty \in H^1(\omega_2), \\ \int_{\omega_2} A_{22} \nabla_{X_2} u_\infty \cdot \nabla_{X_2} v + a u_\infty v dX_2 = \int_{\omega_2} f v dX_2 \quad \forall v \in H^1(\omega_2). \end{cases}$$

Clearly this function satisfies all the properties of (3.52). This follows from the fact that

$$\begin{aligned} A\nabla u_\infty \cdot \nabla v &= \begin{pmatrix} A_{11}(x) & A_{12}(X_2) \\ A_{21}(x) & A_{22}(X_2) \end{pmatrix} \begin{pmatrix} 0 \\ \nabla_{X_2} u_\infty \end{pmatrix} \cdot \begin{pmatrix} \nabla_{X_1} v \\ \nabla_{X_2} v \end{pmatrix} \\ &= \begin{pmatrix} A_{12}(X_2) \nabla_{X_2} u_\infty \\ A_{22}(X_2) \nabla_{X_2} u_\infty \end{pmatrix} \cdot \begin{pmatrix} \nabla_{X_1} v \\ \nabla_{X_2} v \end{pmatrix} \\ &= A_{12}(X_2) \nabla_{X_2} u_\infty \cdot \nabla_{X_1} v + A_{22}(X_2) \nabla_{X_2} u_\infty \cdot \nabla_{X_2} v. \end{aligned}$$

Indeed for $v \in V_0^\ell$ one has then

$$\begin{aligned} &\int_{\Omega_\ell} A(x) \nabla u_\infty \cdot \nabla v + a u_\infty v dx \\ &= \int_{\Omega_\ell} A_{12}(X_2) \nabla_{X_2} u_\infty \cdot \nabla_{X_1} v + A_{22}(X_2) \nabla_{X_2} u_\infty \cdot \nabla_{X_2} v + a u_\infty v dx \\ &= \int_{\Omega_\ell} \nabla_{X_1} \cdot (A_{12}(X_2) \nabla_{X_2} u_\infty v) + A_{22}(X_2) \nabla_{X_2} u_\infty \cdot \nabla_{X_2} v + a u_\infty v dx \\ &= \int_{\ell\omega_1} \int_{\omega_2} A_{22}(X_2) \nabla_{X_2} u_\infty \cdot \nabla_{X_2} v + a u_\infty v dX_2 dX_1 \\ &= \int_{\Omega_\ell} f v dx. \end{aligned}$$

Thus u_∞ is the solution to (3.52) with $\gamma = p$.

Remark 3.9. In the case where (3.46) holds –recalling the remark 3.7– if Ω is an open subset of $\mathbb{R}^n$ such that

$$\Omega \subset \mathbb{R}^p \times \omega_2$$

and if u_ℓ is the solution to (3.49) with $\Omega_\ell = (\ell\omega_1 \times \omega_2) \cap \Omega$ then for $f \in L^2(\Omega_\ell)$ such that for some nonnegative constant γ

$$|f|_{2,\Omega_\ell} = O(\ell^\gamma)$$

there exists a unique u_∞ solution to

$$\begin{cases} u_\infty \in H^1(\overline{\Omega}), \\ \int_{\Omega_\ell} A(x)\nabla u_\infty \cdot \nabla v + a(x)u_\infty v dx = \int_{\Omega_\ell} f v dx \quad \forall v \in V_0^\ell, \quad \forall \ell > 0, \\ |u_\infty|_{H^1(\Omega_\ell)} = O(\ell^\gamma). \end{cases}$$

Moreover

$$|u_\ell - u_\infty|_{H^1(\Omega_{\frac{\ell}{2}})} \le C e^{-\alpha \ell}$$

for two positive constants C and α. The proof follows the one of the theorem above having in mind the Remark 3.7. Note that when $f \in L^2(\Omega)$ the existence of a unique solution to (3.52) in $H^1(\Omega)$ follows from the Lax-Milgram theorem. This corresponds to the case where $\gamma = 0$.

Chapter 4

Periodic problems

In the previous chapters we have addressed in particular the issue of the convergence of problems set in cylinders becoming large in some directions and for data independent of these directions. As we already pointed out to be independent of one direction could also be interpreted as to be periodic in this direction for any period. Then the question arises for various problems to see if periodic data can force at the limit the solution of a problem to be periodic. This is the kind of issue that we would like to address now.

4.1 A general theory

For the sake of simplicity we consider here only the case of periodic functions with period

$$T = \prod_{i=1}^{n}(0, T_i) \tag{4.1}$$

where T_i are for $i = 1, \ldots, n$ positive numbers. If $f \in L^1_{\text{loc}}(\mathbb{R}^n)$ we will say that f is periodic of period T iff

$$f(x + T_i e_i) = f(x) \quad \text{a.e. } x \in \mathbb{R}^n, \ \forall\, i = 1, \ldots, n. \tag{4.2}$$

e_i denotes here the i-th vector of the canonical basis in $\mathbb{R}^n$.

Let us then consider a matrix A and a function a such that

$$A(x), \ a(x) \text{ are } T - \text{periodic} \tag{4.3}$$

(this means for the matrix $A = (A_{i,j})$ that all the entries are T periodic). We also assume that for some positive constants λ, Λ we have

$$\lambda|\xi|^2 \leq A(x)\xi \cdot \xi \quad \text{a.e. } x \in T, \quad \forall\, \xi \in \mathbb{R}^n, \tag{4.4}$$

$$|A(x)\xi| \leq \Lambda|\xi| \quad \text{a.e. } x \in T, \quad \forall\, \xi \in \mathbb{R}^n, \tag{4.5}$$

$$\lambda \leq a(x) \leq \Lambda \quad \text{a.e. } x \in T. \tag{4.6}$$

Note that in (4.4) and (4.6) we can assume the constants λ to be the same at the expense of taking the minimum of the two. The same remark applies to Λ in (4.5), (4.6).

Let us denote by $\beta(x, u)$ a $\mathbb{R}$-valued function defined on $\mathbb{R}^n \times \mathbb{R}$ such that

$$x \to \beta(x, u) \text{ is measurable and } T\text{-periodic } \forall u, \tag{4.7}$$

$$\beta(x, 0) = 0 \quad \text{a.e. } x \in \mathbb{R}^n, \tag{4.8}$$

$$|\beta(x, u) - \beta(x, v)| \leq C|u - v| \text{ a.e. } x \in \mathbb{R}^n, \quad \forall u, v \tag{4.9}$$

for some constant C. We suppose also that $\beta(x, u)$ is monotone in the sense that

$$(\beta(x, u) - \beta(x, v))(u - v) \geq 0 \text{ a.e. } x \in \mathbb{R}^n, \quad \forall u, v. \tag{4.10}$$

We define for $\ell > 0$

$$\Omega_\ell = \prod_{i=1}^{n} (-\ell T_i, \ell T_i), \tag{4.11}$$

and consider a bounded open set Ω'_ℓ such that

$$\Omega_\ell \subset \Omega'_\ell. \tag{4.12}$$

Let us denote by V_ℓ a subspace of $H^1(\Omega'_\ell)$ such that

$$H_0^1(\Omega_\ell) \subset V_\ell \subset H^1(\Omega'_\ell), \qquad V_\ell \text{ is closed in } H^1(\Omega'_\ell). \tag{4.13}$$

(As before for a domain O, $H_0^1(O)$ stands for the set of $H^1(O)$-functions vanishing on the boundary of O. We suppose that the elements of $H_0^1(\Omega_\ell)$ are extended by 0 outside of Ω_ℓ).

Let us also denote by f a T-periodic function such that

$$f \in L^2(T). \tag{4.14}$$

Then one can show (Cf. Theorem A.1 of the appendix) that there exists a unique u_ℓ such that

$$\begin{cases} u_\ell \in V_\ell, \\ \displaystyle\int_{\Omega'_\ell} A(x)\nabla u_\ell \cdot \nabla v + a(x)u_\ell v + \beta(x, u_\ell)v \, dx = \int_{\Omega'_\ell} fv \, dx \quad \forall v \in V_\ell. \end{cases} \tag{4.15}$$

We denote by $H_{\mathrm{per}}^1(T)$ the space defined as

$$H_{\mathrm{per}}^1(T) = \text{ the closure of } \{ v \in C^\infty(\overline{T}) \mid v \ \ T\text{-periodic} \} \tag{4.16}$$

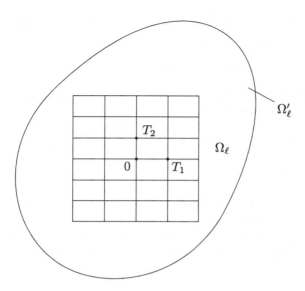

Fig. 4.1   The domain Ω'_ℓ.

where $C^\infty(\overline{T})$ denotes the space of restrictions of $C^\infty(\mathbb{R}^n)$-functions to $\overline{T}$. ($v|_T$ denotes the restriction of v to the cell T, the closure is understood in the $H^1(T)$ sense). $H^1_{\mathrm{per}}(T)$ stands here for the set of functions which are periodic on T – i.e. such that

$$v(x_1,\ldots,x_{i-1},0,x_{i+1},\ldots,x_n) = v(x_1,\ldots,x_{i-1},T_i,x_{i+1},\ldots,x_n)$$

for almost every $(x_1,\ldots,x_{i-1},0,x_{i+1},\ldots,x_n) \in (0,T_1) \times \cdots \times (0,T_{i-1}) \times \{0\} \times (0,T_{i+1}) \times \cdots \times (0,T_n)$. Note that by definition $H^1_{\mathrm{per}}(T)$ is a closed subspace of $H^1(T)$. Thus, see the appendix, there exists a unique u_∞ solution to

$$
\begin{cases}
u_\infty \in H^1_{\mathrm{per}}(T), \\
\displaystyle\int_T A(x)\nabla u_\infty \cdot \nabla v + a(x)u_\infty v + \beta(x,u_\infty)v \, dx \\
\hspace{4cm} = \int_T f v \, dx \; \forall \, v \in H^1_{\mathrm{per}}(T).
\end{cases}
\tag{4.17}
$$

Moreover we have:

Theorem 4.1. *For any $\ell > 0$ there exist constants $C, \sigma > 0$ independent of ℓ and β such that*

$$|u_\ell - u_\infty|^2_{H^1(\Omega_{\ell/2})} \le Ce^{-\sigma\ell}\{|f|^2_{2,\Omega'_\ell} + |u_\infty|^2_{H^1(\Omega_\ell)}\}. \tag{4.18}$$

Remark 4.1. Note that for instance $\beta \equiv 0$ verifies our assumptions and thus our theory holds for linear elliptic problems.

Let us first prove the following lemma.

Lemma 4.1. *Let us assume that u_∞ is extended by periodicity to the whole $\mathbb{R}^n$. Then we have for every $\Omega \subset \mathbb{R}^n$, Ω bounded*

$$\int_\Omega A(x)\nabla u_\infty \cdot \nabla v + a(x)u_\infty v + \beta(x, u_\infty)v \, dx$$

$$= \int_\Omega fv \, dx \quad \forall \, v \in H_0^1(\Omega). \tag{4.19}$$

Proof. In other words this theorem says that if a function is T-periodic and satisfies an elliptic equation in the period cell, then it satisfies this equation everywhere – of course for periodic data.

Let $v \in \mathcal{D}(\Omega)$ and suppose that v is extended by 0 outside Ω. One has if we denote by τ the vector of the period – i.e. $\tau = (T_1, T_2, \ldots, T_n)$

$$\int_\Omega A(x)\nabla u_\infty \cdot \nabla v + a(x)u_\infty v + \beta(x, u_\infty)v \, dx$$

$$= \int_{\mathbb{R}^n} A(x)\nabla u_\infty \cdot \nabla v + a(x)u_\infty v + \beta(x, u_\infty)v \, dx \tag{4.20}$$

$$= \sum_{z \in \mathbb{Z}^n} \int_{T_z} A(x)\nabla u_\infty \cdot \nabla v + a(x)u_\infty v + \beta(x, u_\infty)v \, dx$$

where $T_z = T + z\tau$, $z\tau$ being a shortcut for $(z_1 T_1, \ldots, z_n T_n)$. Note that in this sum only a finite number of terms are not equal to 0 since v is assumed to have a compact support.

Making the change of variable $x + z\tau \to x$ in T_z and using the T-periodicity of A, a, u_∞ and β – that is to say each of them satisfies like the function a

$$a(x + z\tau) = a(x) \quad \text{a.e. } x \in T \tag{4.21}$$

(see (4.2)) – we obtain

$$\int_\Omega A(x)\nabla u_\infty \cdot \nabla v + a(x)u_\infty v + \beta(x, u_\infty)v \, dx =$$

$$\sum_{z \in \mathbb{Z}^n} \int_T A(x)\nabla u_\infty \cdot \nabla v(x + z\tau) + \{a(x)u_\infty + \beta(x, u_\infty)\}v(x + z\tau) \, dx$$

$$= \int_T A(x)\nabla u_\infty \cdot \nabla \tilde{v} + a(x)u_\infty \tilde{v} + \beta(x, u_\infty)\tilde{v} \, dx \tag{4.22}$$

where we have set

$$\tilde{v}(x) = \sum_{z \in \mathbb{Z}^n} v(x + z\tau).$$

Now clearly this function is indefinitely differentiable and T-periodic. Thus from (4.17), (4.22) we derive

$$\int_{\Omega} A(x)\nabla u_{\infty} \cdot \nabla v + a(x)u_{\infty}v + \beta(x, u_{\infty})v \, dx = \int_{T} f\tilde{v} \, dx$$

$$= \sum_{z \in \mathbb{Z}^n} \int_{T} f(x)v(x + z\tau) \, dx$$

$$= \sum_{z \in \mathbb{Z}^n} \int_{T_z} f(x - z\tau)v(x) \, dx \qquad (4.23)$$

$$= \int_{\Omega} fv \, dx$$

by the periodicity of f. This completes the proof since $\mathcal{D}(\Omega)$ is dense in $H_0^1(\Omega)$. $\qquad \square$

Remark 4.2. One should note that the periodic extension of u_{∞} to Ω is such that

$$u_{\infty} \in H^1(\Omega). \qquad (4.24)$$

This is true for smooth functions of $H_{\text{per}}^1(T)$ and by density for any function of $H_{\text{per}}^1(T)$ – see (4.16).

End of the proof of Theorem 4.1. Let ℓ_0 be such that $0 < \ell_0 \leq \ell - 1$. There exists a Lipschitz continuous function ρ such that

$$0 \leq \rho \leq 1, \quad \rho = 1 \text{ on } \Omega_{\ell_0}, \quad \rho = 0 \text{ on } \mathbb{R}^n \backslash \Omega_{\ell_0+1}, \quad |\nabla \rho| \leq C_T. \quad (4.25)$$

(Note that this function belongs to $H_0^1(\Omega_{\ell_0+1}) \cap W^{1,\infty}(\Omega_{\ell_0+1})$). Then we have

$$(u_{\ell} - u_{\infty})\rho \in H_0^1(\Omega_{\ell_0+1}). \qquad (4.26)$$

Since V_{ℓ} is supposed to contain $H_0^1(\Omega_{\ell})$ – see (4.13) – combining (4.15), (4.19) we have

$$\int_{\Omega_{\ell}} A\nabla(u_{\ell} - u_{\infty}) \cdot \nabla v + a(u_{\ell} - u_{\infty})v + (\beta(x, u_{\ell}) - \beta(x, u_{\infty}))v \, dx$$

$$= 0 \quad \forall v \in H_0^1(\Omega_{\ell}). \qquad (4.27)$$

From (4.26) we derive

$$\int_{\Omega_\ell} A\nabla(u_\ell - u_\infty) \cdot \nabla\{(u_\ell - u_\infty)\rho\} + a(u_\ell - u_\infty)^2 \rho \, dx$$

$$+ \int_{\Omega_\ell} (\beta(x, u_\ell) - \beta(x, u_\infty))(u_\ell - u_\infty)\rho \, dx = 0. \tag{4.28}$$

Since β is monotone the last integral above is nonnegative and this implies

$$\int_{\Omega_\ell} A\nabla(u_\ell - u_\infty) \cdot \nabla(u_\ell - u_\infty)\rho + a(u_\ell - u_\infty)^2 \rho \, dx$$

$$\leq - \int_{\Omega_\ell} (A\nabla(u_\ell - u_\infty) \cdot \nabla\rho)(u_\ell - u_\infty) \, dx.$$

In the first integral above it is enough to integrate on Ω_{ℓ_0+1} and in the second one on $\Omega_{\ell_0+1} \backslash \Omega_{\ell_0}$. Moreover using the ellipticity conditions (4.4), (4.5) and (4.6) we derive easily

$$\lambda \int_{\Omega_{\ell_0+1}} \{|\nabla(u_\ell - u_\infty)|^2 + (u_\ell - u_\infty)^2\}\rho \, dx$$

$$\leq \int_{\Omega_{\ell_0+1} \backslash \Omega_{\ell_0}} \Lambda C_T |\nabla(u_\ell - u_\infty)| \, |u_\ell - u_\infty| \, dx$$

(we used (4.25)).

Since $\rho = 1$ on Ω_{ℓ_0} by the usual Young inequality

$$ab \leq \frac{1}{2}(a^2 + b^2)$$

we obtain

$$\int_{\Omega_{\ell_0}} |\nabla(u_\ell - u_\infty)|^2 + (u_\ell - u_\infty)^2 \, dx$$

$$\leq \frac{\Lambda C_T}{2\lambda} \int_{\Omega_{\ell_0+1} \backslash \Omega_{\ell_0}} |\nabla(u_\ell - u_\infty)|^2 + (u_\ell - u_\infty)^2 \, dx.$$

Setting $c = \frac{\Lambda C_T}{2\lambda}$ and noticing that

$$\int_{\Omega_{\ell_0+1} \backslash \Omega_{\ell_0}} |\nabla(u_\ell - u_\infty)|^2 + (u_\ell - u_\infty)^2 \, dx$$

$$= \int_{\Omega_{\ell_0+1}} |\nabla(u_\ell - u_\infty)|^2 + (u_\ell - u_\infty)^2 \, dx$$

$$- \int_{\Omega_{\ell_0}} |\nabla(u_\ell - u_\infty)|^2 + (u_\ell - u_\infty)^2 \, dx,$$

we derive then like in the proof of Theorem 2.1 that

$$\int_{\Omega_{\ell_0}} |\nabla(u_\ell - u_\infty)|^2 + (u_\ell - u_\infty)^2 \, dx$$

$$\leq \left(\frac{c}{1+c}\right) \int_{\Omega_{\ell_0+1}} |\nabla(u_\ell - u_\infty)|^2 + (u_\ell - u_\infty)^2 \, dx.$$

Iterating this formula from $\frac{\ell}{2}$, $\left[\frac{\ell}{2}\right]$ times ($\left[\frac{\ell}{2}\right]$ denotes the integer part of $\frac{\ell}{2}$) we obtain

$$\int_{\Omega_{\ell/2}} |\nabla(u_\ell - u_\infty)|^2 + (u_\ell - u_\infty)^2 \, dx$$

$$\leq \left(\frac{c}{1+c}\right)^{\left[\frac{\ell}{2}\right]} \int_{\Omega_{\frac{\ell}{2}+\left[\frac{\ell}{2}\right]}} |\nabla(u_\ell - u_\infty)|^2 + (u_\ell - u_\infty)^2 \, dx.$$

Noticing that

$$\frac{\ell}{2} - 1 \leq \left[\frac{\ell}{2}\right] \leq \frac{\ell}{2}$$

we get

$$\int_{\Omega_{\ell/2}} |\nabla(u_\ell - u_\infty)|^2 + (u_\ell - u_\infty)^2 \, dx$$

$$\leq c_0 e^{-\sigma \ell} \int_{\Omega_\ell} |\nabla(u_\ell - u_\infty)|^2 + (u_\ell - u_\infty)^2 \, dx \tag{4.29}$$

where $c_0 = \frac{1+c}{c}$, $\sigma = \frac{1}{2} \ln \frac{1+c}{c}$.

Next we need to estimate the integral on the right hand side of (4.29). For this, choosing $v = u_\ell$ in (4.15) we get

$$\int_{\Omega_\ell'} A\nabla u_\ell \cdot \nabla u_\ell + a u_\ell^2 + \beta(x, u_\ell) u_\ell \, dx = \int_{\Omega_\ell'} f u_\ell \, dx.$$

Due to (4.8), (4.10) we have

$$\int_{\Omega_\ell'} \beta(x, u_\ell) u_\ell \, dx = \int_{\Omega_\ell'} (\beta(x, u_\ell) - \beta(x, 0))(u_\ell - 0) \, dx \geq 0,$$

and thus

$$\int_{\Omega_\ell'} A\nabla u_\ell \cdot \nabla u_\ell + a u_\ell^2 \, dx \leq \int_{\Omega_\ell'} f u_\ell \, dx.$$

Using the ellipticity condition and the Cauchy–Schwarz inequality we have

$$\lambda \int_{\Omega_\ell'} |\nabla u_\ell|^2 + u_\ell^2 \, dx \leq |f|_{2,\Omega_\ell'} \left(\int_{\Omega_\ell'} u_\ell^2 \, dx\right)^{\frac{1}{2}}$$

$$\leq |f|_{2,\Omega_\ell'} \left\{\int_{\Omega_\ell'} |\nabla u_\ell|^2 + u_\ell^2 \, dx\right\}^{\frac{1}{2}}.$$

From this it follows that

$$\int_{\Omega'_\ell} |\nabla u_\ell|^2 + u_\ell^2 \, dx \leq \frac{|f|^2_{2,\Omega'_\ell}}{\lambda^2}. \qquad (4.30)$$

Going back to (4.29) we have

$$|u_\ell - u_\infty|^2_{H^1(\Omega_{\ell/2})} \leq 2c_0 e^{-\sigma\ell} \int_{\Omega_\ell} |\nabla u_\ell|^2 + |\nabla u_\infty|^2 + u_\ell^2 + u_\infty^2 \, dx$$

$$= 2c_0 e^{-\sigma\ell} \left\{ \int_{\Omega_\ell} |\nabla u_\ell|^2 + u_\ell^2 \, dx + \int_{\Omega_\ell} |\nabla u_\infty|^2 + u_\infty^2 \, dx \right\}$$

$$\leq 2c_0 e^{-\sigma\ell} \left\{ \frac{|f|^2_{2,\Omega'_\ell}}{\lambda^2} + |u_\infty|^2_{H^1(\Omega_\ell)} \right\}.$$

The result follows by choosing $C = 2c_0(\frac{1}{\lambda^2} \vee 1)$ where $\vee$ denotes the maximum of two numbers. This completes the proof of the theorem. $\qquad \square$

Remark 4.3. One should notice that the constants C, σ are depending only on λ, Λ and T. They are independent of the choice of V_ℓ – that is to say on the boundary conditions that we choose on Ω'_ℓ and on Ω'_ℓ itself.

For the estimate (4.29) we only used (4.27) which holds in fact if we only suppose $f = f_\ell$ periodic on Ω_ℓ or $\Omega_{[\ell]+1}$ (see the proof of (4.19)).

As a corollary of formula (4.18) we have

Theorem 4.2. *Suppose that $\Omega'_\ell = \Omega_\ell$. Let u_ℓ, u_∞ be the solutions to (4.15), (4.17). Then there exist two constants c_1, σ_1, depending on λ, Λ, n, T only such that*

$$|u_\ell - u_\infty|^2_{H^1(\Omega_{\ell/2})} \leq c_1 e^{-\sigma_1\ell} |f|^2_{2,T}. \qquad (4.31)$$

Proof. Of course we suppose that we are under the assumptions of Theorem 4.1. Then by (4.29) we have

$$|u_\ell - u_\infty|^2_{H^1(\Omega_{\ell/2})} \leq c_0 e^{-\sigma\ell} \int_{\Omega_\ell} |\nabla(u_\ell - u_\infty)|^2 + (u_\ell - u_\infty)^2 \, dx \qquad (4.32)$$

with $c_0 = \frac{c+1}{c}$, $\sigma = \frac{1}{2} \ln \frac{c+1}{c}$, $c = \frac{\Lambda C_T}{2\lambda}$. By (4.30) we have

$$\int_{\Omega_\ell} |\nabla u_\ell|^2 + u_\ell^2 \, dx \leq \frac{|f|^2_{2,\Omega_\ell}}{\lambda^2} \leq \frac{|f|^2_{2,\Omega_{[\ell]+1}}}{\lambda^2}$$

$$\leq \frac{2^n([\ell]+1)^n |f|^2_{2,T}}{\lambda^2} \leq \frac{2^n(\ell+1)^n |f|^2_{2,T}}{\lambda^2}. \qquad (4.33)$$

We used the periodicity of f. That is to say if $N \in \mathbb{N}$

$$\int_{\Omega_N} f^2 \, dx = \sum_z \int_{T_z} f^2 \, dx = \sum_z \int_T f^2 \, dx = (2N)^n |f|^2_{2,T}. \qquad (4.34)$$

(T_z are the T_z's covering Ω_N).

Taking now $v = u_\infty$ in (4.17) we derive easily

$$\int_T A(x) \nabla u_\infty \cdot \nabla u_\infty + a u_\infty^2 \, dx \le \int_T f u_\infty \, dx.$$

Using the ellipticity condition and Cauchy–Schwarz inequality we obtain

$$\lambda \int_T |\nabla u_\infty|^2 + u_\infty^2 \, dx \le |f|_{2,T} |u_\infty|_{2,T}$$

which implies

$$\int_T |\nabla u_\infty|^2 + u_\infty^2 \, dx \le \frac{|f|^2_{2,T}}{\lambda^2}.$$

Thus since u_∞ is periodic it comes

$$\int_{\Omega_\ell} |\nabla u_\infty|^2 + u_\infty^2 \, dx \le \int_{\Omega_{[\ell]+1}} |\nabla u_\infty|^2 + u_\infty^2 \, dx$$

$$= \{2([\ell]+1)\}^n \int_T |\nabla u_\infty|^2 + u_\infty^2 \, dx \qquad (4.35)$$

$$\le \frac{2^n (\ell+1)^n}{\lambda^2} |f|^2_{2,T}.$$

Combining then (4.33), (4.35) we get

$$|u_\ell - u_\infty|^2_{H^1(\Omega_{\ell/2})} \le 2c_0 e^{-\sigma\ell} \left\{ \int_{\Omega_\ell} |\nabla u_\ell|^2 + u_\ell^2 \, dx + \int_{\Omega_\ell} |\nabla u_\infty|^2 + u_\infty^2 \, dx \right\}$$

$$\le \frac{4c_0 2^n (\ell+1)^n}{\lambda^2} e^{-\sigma\ell} |f|^2_{2,T}.$$

The result follows with σ_1 taken as any positive constant less than σ. $\qquad \square$

For Ω'_ℓ arbitrary larger than Ω_ℓ we have

Theorem 4.3. *Under the assumptions of Theorem 4.1 we suppose that*

$$|f|^2_{2,\Omega'_\ell} \le C e^{(\sigma-\delta)\ell} \qquad (4.36)$$

for some $0 < \delta < \sigma$ and some positive constant C. Then we have for some constant $c' > 0$

$$|u_\ell - u_\infty|^2_{H^1(\Omega_{\ell/2})} \le c' e^{-\delta\ell} \qquad (4.37)$$

i.e. u_ℓ converges toward u_∞ at an exponential rate on $\Omega_{\ell/2}$.

Proof. Combining (4.18), (4.35) and (4.36) we get

$$|u_\ell - u_\infty|^2_{H^1(\Omega_{\ell/2})} \leq ce^{-\sigma\ell}\left\{Ce^{(\sigma-\delta)\ell} + \frac{2^n(\ell+1)^n}{\lambda^2}|f|^2_{2,T}\right\} \leq c'e^{-\delta\ell}$$

for some constant c'. □

Remark 4.4. Since f is periodic (4.36) holds for instance when

$$\Omega'_\ell \subset \Omega_{\ell'} \quad \text{with} \quad \ell' = O(e^{\frac{(\sigma-\delta)}{n}\ell}).$$

Note that (4.37) holds when (4.36) holds and thus f can be chosen arbitrary outside of Ω_ℓ (see also Remark 4.3) and V_ℓ can also be arbitrary provided it satisfies the assumptions of Theorem 4.1. Note that the Theorems 4.2 and 4.3 show the convergence of u_ℓ toward u_∞ in $H^1(\Omega)$ for any bounded open subset of $\mathbb{R}^n$ and this at an exponential rate of convergence. This is clear since for ℓ large enough we have

$$\Omega \subset \Omega_{\ell/2}.$$

4.2   Some degenerate case

We have addressed a non degenerate case that is to say a case where $a(x) \geq \lambda > 0$. The results developed in the preceding section can be extended easily in the case where

$$0 \leq a(x), \qquad a(x) \not\equiv 0, \tag{4.38}$$

(note that for some constant c one has then

$$\int_T A\nabla v \cdot \nabla v + av^2\,dx \geq c|v|^2_{H^1(T)} \quad \forall v \in H^1(T)).$$

In the case where

$$a \equiv 0 \tag{4.39}$$

the situation is quite different. Indeed for instance a solution to (4.17) with $\beta \equiv 0$ cannot be ensured by the Lax–Milgram theorem. Let us examine what can happen in dimension 1. For that let us consider

$$\Omega_\ell = (-\ell T_1, \ell T_1), \quad \ell \in \mathbb{N} \tag{4.40}$$

and u_ℓ the weak solution to

$$\begin{cases} -(A(x)u'_\ell)' = f & \text{in } \Omega_\ell, \\ u_\ell(-\ell T_1) = u_\ell(\ell T_1) = 0. \end{cases} \tag{4.41}$$

In (4.41) we suppose for instance that

$$f \in L^2(T) \tag{4.42}$$

where $T = (0, T_1)$ and A, f are T-periodic measurable functions such that for A we have

$$0 < \lambda \leq A(x) \leq \Lambda \quad \text{a.e. } x \in \mathbb{R}. \tag{4.43}$$

Then we have

Theorem 4.4. *If*

$$\int_T f(x)\, dx \neq 0, \tag{4.44}$$

the solution to (4.41) is unbounded when $\ell \to +\infty$.

(Before going into the details of the proof one has to note that the theorem above claims that in order to obtain convergence of u_ℓ some condition on f is necessary).

Proof. Let us set

$$F(x) = \int_0^x f(s)\, ds, \qquad c_0 = \int_T f(s)\, ds. \tag{4.45}$$

We have

$$
\begin{aligned}
F(x + T_1) &= \int_0^{x+T_1} f(s)\, ds \\
&= \int_0^x f(s)\, ds + \int_x^{x+T_1} f(s)\, ds = F(x) + c_0
\end{aligned} \tag{4.46}
$$

(note that by a change of variables one has $\int_x^{x+T_1} f(s)\, ds = \int_T f(s)\, ds$).

Integrating the first equation of (4.41) we have

$$-A(x)u'_\ell = F(x) + c_1$$

for some constant c_1 and thus

$$-u'_\ell = \frac{F(x)}{A(x)} + \frac{c_1}{A(x)}. \tag{4.47}$$

The value of c_1 is easily determined since

$$0 = \int_{-\ell T_1}^{\ell T_1} -u'_\ell(x)\, dx = \int_{-\ell T_1}^{\ell T_1} \frac{F(x)}{A(x)}\, dx + c_1 \int_{-\ell T_1}^{\ell T_1} \frac{dx}{A(x)}. \tag{4.48}$$

By periodicity of A we get

$$\int_{-\ell T_1}^{\ell T_1} \frac{dx}{A(x)} = 2\ell \int_T \frac{dx}{A(x)}. \tag{4.49}$$

Moreover by (4.46) for any integer k, we have

$$\int_{kT_1}^{(k+1)T_1} \frac{F(x)}{A(x)}\, dx = \int_{kT_1}^{(k+1)T_1} \frac{F(x+T_1) - c_0}{A(x)}\, dx$$

$$= \int_{(k+1)T_1}^{(k+2)T_1} \frac{F(x)}{A(x)}\, dx - c_0 \int_T \frac{dx}{A(x)}. \tag{4.50}$$

Setting

$$m_1 = \int_T \frac{dx}{A(x)}, \qquad m_2 = \int_T \frac{F(x)}{A(x)}\, dx$$

we derive – starting from $k = 0$ and iterating – that

$$\int_{kT_1}^{(k+1)T_1} \frac{F(x)}{A(x)}\, dx = \int_T \frac{F(x)}{A(x)}\, dx + c_0 k m_1 = m_2 + c_0 k m_1.$$

It follows that

$$\int_{-\ell T_1}^{\ell T_1} \frac{F(x)}{A(x)} = \sum_{k=-\ell}^{\ell-1} \int_{kT_1}^{(k+1)T_1} \frac{F(x)}{A(x)}\, dx = 2\ell m_2 - \ell c_0 m_1.$$

Going back to (4.47), (4.48) this implies that

$$2\ell m_1 c_1 + 2\ell m_2 - \ell c_0 m_1 = 0$$

$$\iff \qquad c_1 = \frac{c_0}{2} - \frac{m_2}{m_1}.$$

Then from (4.47) integrating between 0 and ℓT_1, we get

$$u_\ell(0) = \int_0^{\ell T_1} \frac{F(x)}{A(x)}\, dx + c_1 \int_0^{\ell T_1} \frac{dx}{A(x)}$$

$$= \sum_{k=0}^{\ell-1} (m_2 + c_0 k m_1) + c_1 \ell m_1$$

$$= \ell m_2 + c_0 m_1 \frac{(\ell-1)\ell}{2} + \left(\frac{c_0}{2} - \frac{m_2}{m_1}\right)\ell m_1$$

$$= c_0 m_1 \frac{\ell^2}{2} \to \infty \text{ when } c_0 \neq 0.$$

This completes the proof of the theorem.   □

The interested reader will find in [37] complements on these issues, i.e., when (4.39) holds.

4.3   Application to the periodic obstacle problem

Let $\varphi \in H^1(T)$ such that

$$\varphi \leq 0 \quad \text{on } \partial T, \quad \varphi \ \ T\text{-periodic} \tag{4.51}$$

(∂T denotes the boundary of T). Define K_φ as the set

$$K_\varphi = \{v \in H^1_{per}(T) \mid v(x) \geq \varphi(x) \text{ a.e. on } T\}. \tag{4.52}$$

Note that K_φ is not empty since $\varphi, \varphi^+ \in K_\varphi$. Then under the assumptions (4.4)–(4.6) (Cf. Theorem A.3 in the appendix) there exists a unique u_∞ solution to

$$\begin{cases} u_\infty \in K_\varphi, \\ \displaystyle\int_T A\nabla u_\infty \cdot \nabla(v - u_\infty) + au_\infty(v - u_\infty)\, dx \\ \qquad\qquad \geq \displaystyle\int_T f(v - u_\infty)\, dx \ \ \forall v \in K_\varphi. \end{cases} \tag{4.53}$$

u_∞ is the solution of the periodic obstacle problem. We will denote yet by u_∞ and φ the extension of these functions by periodicity to the whole $\mathbb{R}^n$.

We consider then a space V_ℓ satisfying (4.13) and such that $\varphi^+ \in V_\ell$. Then we set

$$K_\ell = \{v \in V_\ell \mid v(x) \geq \varphi(x) \text{ a.e. on } \Omega'_\ell\}. \tag{4.54}$$

It is clear that K_ℓ is not empty since φ^+ belongs to it. Then there exists a unique solution to

$$\begin{cases} u_\ell \in K_\ell, \\ \displaystyle\int_{\Omega'_\ell} A\nabla u_\ell \cdot \nabla(v - u_\ell) + au_\ell(v - u_\ell)\, dx \\ \qquad\qquad \geq \displaystyle\int_{\Omega'_\ell} f(v - u_\ell)\, dx \ \ \forall v \in K_\ell. \end{cases} \tag{4.55}$$

We would like to show that the periodicity of A, a, f and φ forces u_ℓ to converge toward u_∞. Let us first show that both problems above can be approximated by problems of the type introduced in the first section of this chapter. For that let us introduce a smooth monotone function such that

$$\beta(x) = 0 \ \ \forall x \geq 0 \ , \ \ \beta(x) < 0 \ \ \forall x < 0, \tag{4.56}$$

$$|\beta(x) - \beta(y)| \le C_\beta |x - y|, \quad \forall x, y. \tag{4.57}$$

i.e. β is uniformly Lipschitz continuous. Consider then u_∞^ε and u_ℓ^ε respectively solutions to

$$\begin{cases} u_\infty^\varepsilon \in H_{\mathrm{per}}^1(T), \\ \displaystyle\int_T A\nabla u_\infty^\varepsilon \cdot \nabla v + au_\infty^\varepsilon v + \frac{\beta}{\varepsilon}(u_\infty^\varepsilon - \varphi)v \, dx \\ \qquad\qquad\qquad = \displaystyle\int_T fv \, dx \quad \forall\, v \in H_{\mathrm{per}}^1(T). \end{cases} \tag{4.58}$$

$$\begin{cases} u_\ell^\varepsilon \in V_\ell, \\ \displaystyle\int_{\Omega_\ell'} A\nabla u_\ell^\varepsilon \cdot \nabla v + au_\ell^\varepsilon v + \frac{\beta}{\varepsilon}(u_\ell^\varepsilon - \varphi)v \, dx \\ \qquad\qquad\qquad = \displaystyle\int_{\Omega_\ell'} fv \, dx \quad \forall\, v \in V_\ell. \end{cases} \tag{4.59}$$

We have

Theorem 4.5. *Under the assumptions above*

$$u_\infty^\varepsilon \to u_\infty \quad \text{in } H^1(T), \quad u_\ell^\varepsilon \to u_\ell \quad \text{in } H^1(\Omega_\ell') \tag{4.60}$$

when $\varepsilon \to 0$.

Proof. We do the proof for the first convergence, the second being identical. It is clear that

$$u_\infty^\varepsilon - \varphi^+ \in H_{per}^1(T).$$

(We could here take φ instead of φ^+ but for the second convergence we would have to take φ^+ and thus to have a unified presentation we choose φ^+). Thus we have

$$\int_T A\nabla u_\infty^\varepsilon \cdot \nabla(u_\infty^\varepsilon - \varphi^+) + au_\infty^\varepsilon(u_\infty^\varepsilon - \varphi^+)$$
$$+ \frac{\beta}{\varepsilon}(u_\infty^\varepsilon - \varphi)(u_\infty^\varepsilon - \varphi^+) \, dx = \int_T f(u_\infty^\varepsilon - \varphi^+) \, dx. \tag{4.61}$$

One has $\frac{\beta}{\varepsilon}(u_\infty^\varepsilon - \varphi)(u_\infty^\varepsilon - \varphi^+) \ge 0$. Indeed this expression vanishes when $u_\infty^\varepsilon \ge \varphi$ and when $u_\infty^\varepsilon < \varphi \le \varphi^+$ both factors are negative. It follows that

$$\int_T A\nabla u_\infty^\varepsilon \cdot \nabla u_\infty^\varepsilon + au_\infty^\varepsilon u_\infty^\varepsilon \, dx \le \int_T A\nabla u_\infty^\varepsilon \cdot \nabla \varphi^+ + au_\infty^\varepsilon \varphi^+ \, dx$$
$$+ \int_T f(u_\infty^\varepsilon - \varphi^+) \, dx.$$

One can then deduce easily (recall (4.4)–(4.6)) that u_∞^ε is bounded in $H_{per}^1(T)$ independently of ε and thus, possibly for a "subsequence" one has

$$u_\infty^\varepsilon \rightharpoonup u_0 \text{ in } H_{per}^1(T), \quad u_\infty^\varepsilon \to u_0 \text{ in } L^2(T), \quad u_\infty^\varepsilon \to u_0 \quad \text{a.e. in } T.$$

Going back to (4.61) we derive easily from the boundedness of u_∞^ε independently of ε that

$$\int_T \beta(u_\infty^\varepsilon - \varphi)(u_\infty^\varepsilon - \varphi^+)\, dx \leq C\varepsilon.$$

Passing to the limit we obtain

$$\int_T \beta(u_0 - \varphi)(u_0 - \varphi^+)\, dx \leq 0.$$

On the set $\{u_0 \geq \varphi\} = \{x \mid u_0(x) \geq \varphi(x)\}$ one has $\beta(u_0 - \varphi) = 0$ and thus the inequality above reads

$$\int_{\{u_0 < \varphi\}} \beta(u_0 - \varphi)(u_0 - \varphi^+)\, dx \leq 0.$$

But on $\{u_0 < \varphi\}$ one has $u_0 < \varphi \leq \varphi^+$ and the integrand above is positive. This shows that the measure of the set $\{u_0 < \varphi\}$ vanishes i.e. $u_0 \geq \varphi$ a.e.. Then taking $v = w - u_\infty^\varepsilon$ in (4.58) where $w \in K_\varphi$ we get

$$\int_T A\nabla u_\infty^\varepsilon \cdot \nabla(w - u_\infty^\varepsilon) + a u_\infty^\varepsilon(w - u_\infty^\varepsilon)$$
$$+ \frac{\beta}{\varepsilon}(u_\infty^\varepsilon - \varphi)(w - u_\infty^\varepsilon)\, dx = \int_T f(w - u_\infty^\varepsilon)\, dx.$$

Since

$$\frac{\beta}{\varepsilon}(u_\infty^\varepsilon - \varphi)(w - u_\infty^\varepsilon) = \left(\frac{\beta}{\varepsilon}(u_\infty^\varepsilon - \varphi) - \frac{\beta}{\varepsilon}(w - \varphi)\right)(w - u_\infty^\varepsilon) \leq 0$$

we have for any $w \in K_\varphi$

$$\int_T A\nabla u_\infty^\varepsilon \cdot \nabla(w - u_\infty^\varepsilon) + a u_\infty^\varepsilon(w - u_\infty^\varepsilon)\, dx \geq \int_T f(w - u_\infty^\varepsilon)\, dx$$

i.e.

$$\int_T A\nabla u_\infty^\varepsilon \cdot \nabla w + a u_\infty^\varepsilon w\, dx - \int_T f(w - u_\infty^\varepsilon)\, dx$$
$$\geq \int_T A\nabla u_\infty^\varepsilon \cdot \nabla u_\infty^\varepsilon + a(u_\infty^\varepsilon)^2\, dx. \qquad (4.62)$$

Taking the lim inf of both sides we obtain by the weak lower semicontinuity of the convex functional in the right hand side of (4.62)

$$\int_T A\nabla u_0 \cdot \nabla w + au_0 w \, dx - \int_T f(w - u_0) \, dx$$

$$\geq \int_T A\nabla u_0 \cdot \nabla u_0 + a(u_0)^2 \, dx$$

i.e. $u_0 = u_\infty$. By uniqueness of the limit the whole "sequence" u_∞^ε converges to u_0. To see that the convergence is strong in $H^1(T)$, from (4.62) written with $w = u_0$ we get

$$\int_T A\nabla u_\infty^\varepsilon \cdot \nabla u_0 + au_\infty^\varepsilon u_0 \, dx - \int_T f(u_0 - u_\infty^\varepsilon) \, dx$$

$$\geq \int_T A\nabla u_\infty^\varepsilon \cdot \nabla u_\infty^\varepsilon + a(u_\infty^\varepsilon)^2 \, dx.$$

Thus taking the lim sup from both sides when $\varepsilon \to 0$ we obtain

$$\limsup_{\varepsilon \to 0} \int_T A\nabla u_\infty^\varepsilon \cdot \nabla u_\infty^\varepsilon + a(u_\infty^\varepsilon)^2 \, dx \leq \int_T A\nabla u_0 \cdot \nabla u_0 + a(u_0)^2 \, dx.$$

Since

$$\liminf_{\varepsilon \to 0} \int_T A\nabla u_\infty^\varepsilon \cdot \nabla u_\infty^\varepsilon + a(u_\infty^\varepsilon)^2 \, dx \geq \int_T A\nabla u_0 \cdot \nabla u_0 + a(u_0)^2 \, dx$$

we have obtained

$$\lim_{\varepsilon \to 0} \int_T A\nabla u_\infty^\varepsilon \cdot \nabla u_\infty^\varepsilon + a(u_\infty^\varepsilon)^2 \, dx = \int_T A\nabla u_0 \cdot \nabla u_0 + a(u_0)^2 \, dx.$$

It follows that

$$\lambda \int_T |\nabla(u_\infty^\varepsilon - u_0)|^2 + (u_\infty^\varepsilon - u_0)^2 \, dx$$

$$\leq \int_T A\nabla(u_\infty^\varepsilon - u_0) \cdot (\nabla u_\infty^\varepsilon - u_0) + a(u_\infty^\varepsilon - u_0)^2 \, dx$$

$$= \int_T A\nabla u_\infty^\varepsilon \cdot \nabla u_\infty^\varepsilon + a(u_\infty^\varepsilon)^2 + A\nabla u_0 \cdot \nabla u_0 + a(u_0)^2 \, dx$$

$$- \int_T A\nabla u_0 \cdot \nabla u_\infty^\varepsilon + au_0 u_\infty^\varepsilon - A\nabla u_\infty^\varepsilon \cdot \nabla u_0 + au_\infty^\varepsilon u_0 \, dx \to 0$$

when $\varepsilon \to 0$. This completes the proof of the theorem. $\square$

We have also

Theorem 4.6. *Under the above assumptions there exist positive constants c, σ independent of ε such that*

$$|u_\ell^\varepsilon - u_\infty^\varepsilon|_{H^1(\Omega_{\ell/2})}^2 \leq ce^{-\sigma\ell}\{|f|_{2,\Omega_\ell'}^2 + |u_\infty^\varepsilon|_{H^1(\Omega_\ell)}^2 + |\varphi^+|_{H^1(\Omega_\ell')}^2\}. \tag{4.63}$$

Proof. This is an immediate consequence of Theorem 4.1 if we set

$$\beta(x, u) = \frac{\beta}{\varepsilon}(u - \varphi).$$

$\beta(x, u)$ defined this way satisfies all the assumptions of the theorem except for $\beta(x, 0) = 0$. This last assumption is only used to prove (4.30). To estimate u_ℓ^ε one can proceed the following way. Taking $v = u_\ell^\varepsilon - \varphi^+$ in (4.59) we get

$$\int_{\Omega_\ell'} A(x)\nabla u_\ell^\varepsilon \cdot \nabla(u_\ell^\varepsilon - \varphi^+) + a(x)u_\ell^\varepsilon(u_\ell^\varepsilon - \varphi^+)$$

$$+ \frac{\beta}{\varepsilon}(u_\ell^\varepsilon - \varphi)(u_\ell^\varepsilon - \varphi^+)\,dx = \int_{\Omega_\ell'} f(u_\ell^\varepsilon - \varphi^+)\,dx.$$

Since $\frac{\beta}{\varepsilon}(u_\ell^\varepsilon - \varphi)(u_\ell^\varepsilon - \varphi^+) \geq 0$, one gets

$$\int_{\Omega_\ell'} A(x)\nabla u_\ell^\varepsilon \cdot \nabla u_\ell^\varepsilon + a(x)(u_\ell^\varepsilon)^2\,dx$$

$$\leq \int_{\Omega_\ell'} A(x)\nabla u_\ell^\varepsilon \cdot \nabla\varphi^+ + a(x)u_\ell^\varepsilon\varphi^+\,dx + \int_{\Omega_\ell'} f(u_\ell^\varepsilon - \varphi^+)\,dx.$$

Using (4.4)-(4.6) we derive

$$\lambda \int_{\Omega_\ell'} |\nabla u_\ell^\varepsilon|^2 + (u_\ell^\varepsilon)^2\,dx$$

$$\leq \Lambda \int_{\Omega_\ell'} |\nabla u_\ell^\varepsilon||\nabla\varphi^+| + |u_\ell^\varepsilon||\varphi^+|\,dx + \int_{\Omega_\ell'} |f||u_\ell^\varepsilon|\,dx + \int_{\Omega_\ell'} |f||\varphi^+|\,dx$$

i.e.

$$|u_\ell^\varepsilon|^2_{H^1(\Omega_\ell')} \leq \frac{\Lambda}{\lambda}|u_\ell^\varepsilon|_{H^1(\Omega_\ell')}|\varphi^+|_{H^1(\Omega_\ell')}$$

$$+ \frac{1}{\lambda}|f|_{2,\Omega_\ell'}|u_\ell^\varepsilon|_{H^1(\Omega_\ell')} + \frac{1}{\lambda}|f|_{2,\Omega_\ell'}|\varphi^+|_{H^1(\Omega_\ell')}.$$

Using the Young inequality $ab \leq \frac{\delta}{2}a^2 + \frac{1}{2\delta}b^2$ for δ small enough in each of the products above, we arrive easily to

$$|u_\ell^\varepsilon|^2_{H^1(\Omega_\ell')} \leq C\{|f|^2_{2,\Omega_\ell'} + |\varphi^+|^2_{H^1(\Omega_\ell')}\}$$

as a replacement of (4.30). C is a constant depending on λ and Λ only. This completes the proof of the theorem. $\qquad\square$

Analogously to Theorem 4.3 we have

Theorem 4.7. *Under the assumptions above and if u_∞, u_ℓ are solutions to (4.57), (4.59) respectively suppose that*

$$|f|^2_{2,\Omega_\ell'} + |\varphi^+|^2_{H^1(\Omega_\ell')} \leq Ce^{(\sigma-\delta)\ell} \tag{4.64}$$

for some $0 < \delta < \sigma$ and some positive constant C. Then we have for some constant c'

$$|u_\ell - u_\infty|^2_{H^1(\Omega_{\ell/2})} \leq c' e^{-\delta\ell} \tag{4.65}$$

i.e. u_ℓ converges toward u_∞ at an exponential rate on $\Omega_{\ell/2}$.

Proof. It follows from (4.63) that

$$|u_\ell - u_\infty|_{H^1(\Omega_{\ell/2})}$$
$$\leq |u_\ell - u_\ell^\varepsilon|_{H^1(\Omega_{\ell/2})} + |u_\ell^\varepsilon - u_\infty^\varepsilon|_{H^1(\Omega_{\ell/2})} + |u_\infty^\varepsilon - u_\infty|_{H^1(\Omega_{\ell/2})}$$
$$\leq |u_\ell - u_\ell^\varepsilon|_{H^1(\Omega_{\ell/2})} + \left(ce^{-\sigma\ell}\{|f|^2_{2,\Omega'_\ell} + |u_\infty^\varepsilon|^2_{H^1(\Omega_\ell)} + |\varphi^+|^2_{H^1(\Omega'_\ell)}\}\right)^{\frac{1}{2}}$$
$$\quad + |u_\infty^\varepsilon - u_\infty|_{H^1(\Omega_{\ell/2})}.$$

It follows then by (4.60), taking the limit in ε that one has

$$|u_\ell - u_\infty|^2_{H^1(\Omega_{\ell/2})} \leq ce^{-\sigma\ell}\{|f|^2_{2,\Omega'_\ell} + |u_\infty|^2_{H^1(\Omega_\ell)} + |\varphi^+|^2_{H^1(\Omega'_\ell)}\}$$

One can proceed then as in the proof of Theorem 4.3. □

Remark 4.5. Note that (4.64) holds in particular when $\Omega'_\ell = \Omega_\ell$.

Chapter 5

Anisotropic singular perturbation problems

This theory is devoted to study problems with small diffusion velocity. One is mainly concerned with the asymptotic behaviour of the solution of such problems when the diffusion velocity approaches 0. Let us explain the situation on a classical example (see [62], [63]).

5.1 Introduction

Let Ω be a bounded open subset of $\mathbb{R}^n$. For $f \in H^{-1}(\Omega)$, $\varepsilon > 0$ consider u_ε the weak solution to

$$\begin{cases} -\varepsilon \Delta u_\varepsilon + \beta(u_\varepsilon) = f & \text{in } \Omega, \\ u_\varepsilon \in H_0^1(\Omega). \end{cases} \tag{5.1}$$

We will suppose here that

$$\beta(0) = 0, \quad \eta|u - v|^2 \le (\beta(u) - \beta(v))(u - v) \le \mu|u - v|^2 \quad \forall u, v \in \mathbb{R} \tag{5.2}$$

(η, μ are positive constants). It is easy to show under these assumptions that a solution to this problem exists and is unique (see the Appendix). The question is then to determine what happens when $\varepsilon \to 0$. Passing to the limit formally one expects that

$$\beta(u_\varepsilon) - f \to 0. \tag{5.3}$$

Indeed we have:

Theorem 5.1. *Let u_ε be the solution to (5.1). Then*

$$\beta(u_\varepsilon) - f \to 0 \quad \text{in } H^{-1}(\Omega). \tag{5.4}$$

Proof. We suppose $H_0^1(\Omega)$ equipped with the norm

$$\left\|\nabla v\right\|_{2,\Omega}. \tag{5.5}$$

If $g \in H^{-1}(\Omega)$ then, by the Riesz representation theorem, there exists a unique $u \in H_0^1(\Omega)$ such that

$$\int_\Omega \nabla u \cdot \nabla v \, dx = \langle g, v \rangle \quad \forall v \in H_0^1(\Omega)$$

and

$$|g|_{H^{-1}(\Omega)} = \left\| \nabla u \right\|_2$$

(see [16]) – we denote by $|g|_{H^{-1}(\Omega)}$ the strong dual norm of g in $H^{-1}(\Omega)$ defined as

$$|g|_{H^{-1}(\Omega)} = \mathop{\mathrm{Sup}}_{\substack{v \in H_0^1(\Omega) \\ v \neq 0}} \frac{\langle g, v \rangle}{\left\| \nabla v \right\|_{2,\Omega}}.$$

Since (5.1) is meant in a weak sense we have

$$\langle f - \beta(u_\varepsilon), v \rangle = \int_\Omega \nabla(\varepsilon u_\varepsilon) \cdot \nabla v \, dx \quad \forall v \in H_0^1(\Omega)$$

and thus

$$|f - \beta(u_\varepsilon)|_{H^{-1}(\Omega)} = \left\| \nabla(\varepsilon u_\varepsilon) \right\|_{2,\Omega}. \tag{5.6}$$

As we just saw:

$$\varepsilon \int_\Omega \nabla u_\varepsilon \cdot \nabla w + \beta(u_\varepsilon) w \, dx = \langle f, w \rangle \quad \forall w \in H_0^1(\Omega). \tag{5.7}$$

Choosing $w = \varepsilon u_\varepsilon$ we get

$$\left\| \nabla(\varepsilon u_\varepsilon) \right\|_{2,\Omega}^2 + \varepsilon \int_\Omega \beta(u_\varepsilon) u_\varepsilon \, dx = \langle f, \varepsilon u_\varepsilon \rangle \leq |f|_{H^{-1}(\Omega)} \left\| \nabla(\varepsilon u_\varepsilon) \right\|_{2,\Omega}. \tag{5.8}$$

From this we derive (since by (5.2) $\beta(u)u \geq 0$) that

$$\left\| \nabla(\varepsilon u_\varepsilon) \right\|_{2,\Omega}^2 \leq |f|_{H^{-1}(\Omega)} \left\| \nabla(\varepsilon u_\varepsilon) \right\|_{2,\Omega}$$

and thus

$$\left\| \nabla(\varepsilon u_\varepsilon) \right\|_{2,\Omega} \leq |f|_{H^{-1}(\Omega)}.$$

This shows that $\varepsilon u_\varepsilon$ is bounded in $H_0^1(\Omega)$ independently of ε. So, up to a subsequence there exists $v_0 \in H_0^1(\Omega)$ such that

$$\varepsilon u_\varepsilon \rightharpoonup v_0 \quad \text{in } H_0^1(\Omega). \tag{5.9}$$

From (5.8), (5.2) we derive then that

$$\varepsilon |u_\varepsilon|_{2,\Omega}^2 \leq \frac{\varepsilon}{\eta} \int_\Omega \beta(u_\varepsilon) u_\varepsilon \, dx \leq \frac{|f|_{H^{-1}(\Omega)}^2}{\eta}.$$

Thus

$$|\varepsilon u_\varepsilon|^2_{2,\Omega} = \varepsilon \cdot \varepsilon |u_\varepsilon|^2_{2,\Omega} \to 0.$$

This means that

$$\varepsilon u_\varepsilon \to 0 \quad in \ L^2(\Omega) \qquad (5.10)$$

and also in $\mathcal{D}'(\Omega)$. Since the convergence (5.9) implies convergence in $\mathcal{D}'(\Omega)$ we have $v_0 = 0$ and by uniqueness of the limit it follows that $\varepsilon u_\varepsilon \rightharpoonup 0$ in $H^1_0(\Omega)$. From (5.8) we derive then

$$\left\|\nabla(\varepsilon u_\varepsilon)\right\|^2_2 \leq \langle f, \varepsilon u_\varepsilon \rangle \to 0.$$

By (5.6) this completes the proof of the theorem. □

The convergence we just obtained is a very weak convergence, however it will always hold. If we wish to have a convergence of u_ε in $L^2(\Omega)$ one needs to assume $f \in L^2(\Omega)$, a convergence in $H^1_0(\Omega)$ one needs $f \in H^1_0(\Omega)$ and so on. Let us examine the convergence in $L^2(\Omega)$.

Theorem 5.2. *Suppose that $f \in L^2(\Omega)$ then if u_ε is the solution to (5.1) we have*

$$u_\varepsilon \to \beta^{-1}(f) \quad in \ L^2(\Omega). \qquad (5.11)$$

(We denote here by β^{-1} the inverse function of β which, by (5.2), is uniformly Lipschitz continuous).

Proof. Since β is monotone and uniformly Lipschitz continuous one has

$$\int_\Omega \nabla u \cdot \nabla \beta(u)\, dx \geq 0 \quad \forall u \in H^1_0(\Omega)$$

(one can see that first assuming β of class C^1 and proceed by approximation). The weak formulation of (5.1) is

$$\int_\Omega \varepsilon \nabla u_\varepsilon \cdot \nabla v + \beta(u_\varepsilon)v\, dx = \int_\Omega fv\, dx \quad \forall v \in H^1_0(\Omega). \qquad (5.12)$$

Taking $v = \beta(u_\varepsilon)$ it follows

$$\int_\Omega \beta(u_\varepsilon)^2\, dx \leq \int_\Omega f\beta(u_\varepsilon)\, dx \leq |f|_{2,\Omega}|\beta(u_\varepsilon)|_{2,\Omega} \qquad (5.13)$$

by the Cauchy–Schwarz inequality. Thus we get

$$|\beta(u_\varepsilon)|_{2,\Omega} \leq |f|_{2,\Omega}.$$

Since $\beta(u_\varepsilon)$ is bounded in $L^2(\Omega)$ independently of ε, there exists $\beta_0 \in L^2(\Omega)$ such that up to a subsequence

$$\beta(u_\varepsilon) \rightharpoonup \beta_0 \quad \text{in } L^2(\Omega). \tag{5.14}$$

Since the convergence is taking place also in $\mathcal{D}'(\Omega)$, by Theorem 5.1, we have $\beta_0 = f$ and the whole "sequence" $\beta(u_\varepsilon)$ converges weakly to f in $L^2(\Omega)$. From (5.13) one derives

$$\limsup_{\varepsilon \to 0} \int_\Omega \beta(u_\varepsilon)^2 \, dx \leq \limsup_{\varepsilon \to 0} \int_\Omega f\beta(u_\varepsilon) \, dx$$
$$= \int_\Omega f^2 \, dx \leq \liminf_{\varepsilon \to 0} \int_\Omega \beta(u_\varepsilon)^2 \, dx \tag{5.15}$$

by the weak lower semicontinuity of the norm in $L^2(\Omega)$. Thus

$$\lim_{\varepsilon \to 0} \int_\Omega \beta(u_\varepsilon)^2 \, dx = \int_\Omega f^2 \, dx$$

and $\beta(u_\varepsilon) \to f$ strongly in $L^2(\Omega)$. The result follows since by (5.2) β^{-1} is well defined and Lipschitz continuous. This completes the proof of the theorem.

$\square$

5.2   Anisotropic singular perturbation problems

Let Ω be a bounded domain in $\mathbb{R}^2$ and f a function such that for instance

$$f = f(x_1, x_2) \in L^2(\Omega). \tag{5.16}$$

Then for every $\varepsilon > 0$ there exists a unique u_ε solution to

$$\begin{cases} -\varepsilon^2 \partial_{x_1}^2 u_\varepsilon - \partial_{x_2}^2 u_\varepsilon = f & \text{in } \Omega, \\ u_\varepsilon = 0 & \text{on } \partial\Omega. \end{cases} \tag{5.17}$$

This is what we call an anisotropic singular perturbation problem since the diffusion velocity is very small in the x_1 direction but not in the other direction. Of course (5.17) is understood in a weak sense, namely u_ε satisfies

$$\begin{cases} u_\varepsilon \in H_0^1(\Omega), \\ \displaystyle\int_\Omega \varepsilon^2 \partial_{x_1} u_\varepsilon \partial_{x_1} v + \partial_{x_2} u_\varepsilon \partial_{x_2} v \, dx = \int_\Omega fv \, dx \quad \forall\, v \in H_0^1(\Omega). \end{cases} \tag{5.18}$$

We would like to study the behaviour of u_ε when ε goes to 0. Formally, at the limit, if u_0 denotes the limit of u_ε when $\varepsilon \to 0$, we have

$$\int_\Omega \partial_{x_2} u_0 \partial_{x_2} v \, dx = \int_\Omega fv \, dx \quad \forall\, v \in H_0^1(\Omega). \tag{5.19}$$

Even though we can show that $u_0 \in H_0^1(\Omega)$ the bilinear form on the left hand side of (5.19) is not coercive on $H_0^1(\Omega)$ and the problem (5.19) is not well posed.

Let us denote by $\Pi_{x_1}(\Omega)$ the projection of Ω on the x_1-axis defined by

$$\Pi_{x_1}(\Omega) = \{\, (x_1, 0) \mid \exists\, x_2 \text{ s.t. } (x_1, x_2) \in \Omega \,\}. \qquad (5.20)$$

Then for every $x_1 \in \Pi_{x_1}(\Omega) = \Pi_\Omega$ – we drop the x_1 index for simplicity – we denote by Ω_{x_1} the section of Ω above x_1 – i.e.

$$\Omega_{x_1} = \{\, x_2 \mid (x_1, x_2) \in \Omega \,\}. \qquad (5.21)$$

The closest problem to (5.19) to be well defined and suitable for the limit

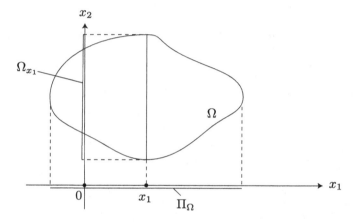

Fig. 5.1   Sectional slicing of Ω.

is the following. For almost all x_1 – or for every $x_1 \in \Pi_\Omega$ if f is a smooth function – we have

$$f(x_1, \cdot) \in L^2(\Omega_{x_1}), \qquad (5.22)$$

(this is a well known result of integration).

Then there exists a unique $u_0 = u_0(x_1, \cdot)$ solution to

$$\begin{cases} u_0 \in H_0^1(\Omega_{x_1}), \\ \displaystyle\int_{\Omega_{x_1}} \partial_{x_2} u_0 \partial_{x_2} v \, dx_2 = \int_{\Omega_{x_1}} f(x_1, \cdot) v \, dx_2 \quad \forall\, v \in H_0^1(\Omega_{x_1}). \end{cases} \qquad (5.23)$$

Note that the existence and uniqueness of a solution to (5.23) is an easy consequence of the Lax–Milgram theorem since

$$\left\{ \int_{\Omega_{x_1}} (\partial_{x_2} v)^2 \, dx \right\}^{\frac{1}{2}}$$

is a norm on $H_0^1(\Omega_{x_1})$. What we would like to show now is the convergence of u_ε toward u_0 solution of (5.23) for instance in $L^2(\Omega)$. Instead of doing that on this simple problem we show how it could be embedded relatively simply in a more general class of problems. However in a first lecture the reader can just apply the technique below to this simple example. It will help him to catch the main ideas.

Let Ω be a bounded open subset of $\mathbb{R}^n$. We denote by $x = (x_1, \ldots, x_n)$ the points in $\mathbb{R}^n$ and use the decomposition

$$x = (X_1, X_2), \quad X_1 = x_1, \ldots, x_p, \quad X_2 = x_{p+1}, \ldots, x_n. \tag{5.24}$$

With this notation we set

$$\nabla u = (\partial_{x_1} u, \ldots, \partial_{x_n} u)^T = \begin{pmatrix} \nabla_{X_1} u \\ \nabla_{X_2} u \end{pmatrix} \tag{5.25}$$

with

$$\nabla_{X_1} u = (\partial_{x_1} u, \ldots, \partial_{x_p} u)^T, \quad \nabla_{X_2} u = (\partial_{x_{p+1}} u, \ldots, \partial_{x_n} u)^T. \tag{5.26}$$

We denote by $A = (A_{i,j}(x))$ a $n \times n$-matrix such that

$$A_{i,j} \in L^\infty(\Omega) \quad \forall\, i, j = 1, \ldots, n, \tag{5.27}$$

and such that for some $\lambda > 0$ we have

$$\lambda |\xi|^2 \le A\xi \cdot \xi \quad \forall\, \xi \in \mathbb{R}^n, \text{ a.e. } x \in \Omega. \tag{5.28}$$

Note that the entries of A are written with a comma in between the indices. We decompose then A into four blocks by writing

$$A = \begin{pmatrix} A_{11} & A_{12} \\ A_{21} & A_{22} \end{pmatrix} \tag{5.29}$$

where A_{11}, A_{22} are respectively $p \times p$ and $(n-p) \times (n-p)$-matrices. We then set for $\varepsilon > 0$

$$A_\varepsilon = A_\varepsilon(x) = \begin{pmatrix} \varepsilon^2 A_{11} & \varepsilon A_{12} \\ \varepsilon A_{21} & A_{22} \end{pmatrix}. \tag{5.30}$$

For $\xi \in \mathbb{R}^n$ if we set $\xi = \begin{pmatrix} \bar{\xi}_1 \\ \bar{\xi}_2 \end{pmatrix}$ where $\bar{\xi}_1 = (\xi_1, \ldots, \xi_p)^T$, $\bar{\xi}_2 = (\xi_{p+1}, \ldots, \xi_n)^T$ we have for a.e. $x \in \Omega$ and every $\xi \in \mathbb{R}^n$

$$A_\varepsilon \xi \cdot \xi = A\xi_\varepsilon \cdot \xi_\varepsilon \ge \lambda |\xi_\varepsilon|^2 = \lambda \{ \varepsilon^2 |\bar{\xi}_1|^2 + |\bar{\xi}_2|^2 \} \tag{5.31}$$

where we have set $\xi_\varepsilon = (\varepsilon \bar{\xi}_1, \bar{\xi}_2)$. Thus A_ε satisfies

$$A_\varepsilon \xi \cdot \xi \ge \lambda(\varepsilon^2 \wedge 1)|\xi|^2 \quad \forall\, \xi \in \mathbb{R}^n, \text{ a.e. } x \in \Omega. \tag{5.32}$$

($\wedge$ denotes the minimum of two numbers). It follows that A_ε is positive definite and for $f \in H^{-1}(\Omega)$ there exists a unique u_ε solution to

$$\begin{cases} \int_\Omega A_\varepsilon \nabla u_\varepsilon \cdot \nabla v \, dx = \langle f, v \rangle \quad \forall v \in H_0^1(\Omega), \\ u_\varepsilon \in H_0^1(\Omega). \end{cases} \qquad (5.33)$$

Clearly u_ε is the solution of a problem generalizing (5.18). We would like then to study the behavior of u_ε when $\varepsilon \to 0$.

Let us denote by Π_{X_1} the orthogonal projection from $\mathbb{R}^n$ onto the space $X_2 = 0$. For any $X_1 \in \Pi_{X_1}(\Omega) := \Pi_\Omega$ we denote by Ω_{X_1} the section of Ω above X_1 defined as

$$\Omega_{X_1} = \{ X_2 \mid (X_1, X_2) \in \Omega \}.$$

To avoid unnecessary complications we will suppose that

$$f \in L^2(\Omega). \qquad (5.34)$$

Then for a.e. $X_1 \in \Pi_\Omega$ we have $f(X_1, \cdot) \in L^2(\Omega_{X_1})$ and there exists a unique $u_0 = u_0(X_1, \cdot)$ solution to

$$\begin{cases} \int_{\Omega_{X_1}} A_{22} \nabla_{X_2} u_0(X_1, X_2) \cdot \nabla_{X_2} v(X_2) \, dX_2 \\ \qquad = \int_{\Omega_{X_1}} f(X_1, X_2) v(X_2) \, dX_2 \quad \forall v \in H_0^1(\Omega_{X_1}), \quad (5.35) \\ u_0(X_1, \cdot) \in H_0^1(\Omega_{X_1}). \end{cases}$$

Clearly u_0 is the natural candidate for the limit of u_ε. Indeed we have

Theorem 5.3. *Under the assumptions above if u_ε is the solution to (5.33)*

$$u_\varepsilon \to u_0, \qquad \nabla_{X_2} u_\varepsilon \to \nabla_{X_2} u_0, \qquad \varepsilon \nabla_{X_1} u_\varepsilon \to 0 \quad \text{in } L^2(\Omega) \qquad (5.36)$$

where u_0 is the solution to (5.35).

(In these convergences the vectorial convergence in $L^2(\Omega)$ means the convergence component by component).

Proof. Let us take $v = u_\varepsilon$ in (5.33). By (5.31) we derive

$$\lambda \int_\Omega \varepsilon^2 |\nabla_{X_1} u_\varepsilon|^2 + |\nabla_{X_2} u_\varepsilon|^2 \, dx \le \langle f, u_\varepsilon \rangle \le |f|_{2,\Omega} |u_\varepsilon|_{2,\Omega}. \qquad (5.37)$$

Since Ω is bounded, by the Poincaré inequality we have for some constant C independent of ε

$$|u_\varepsilon|_{2,\Omega} \le C \big\| \nabla_{X_2} u_\varepsilon \big\|_{2,\Omega}. \qquad (5.38)$$

We then obtain

$$\lambda \int_\Omega \varepsilon^2 |\nabla_{X_1} u_\varepsilon|^2 + |\nabla_{X_2} u_\varepsilon|^2 \, dx \le C|f|_{2,\Omega} ||\nabla_{X_2} u_\varepsilon||_{2,\Omega}. \tag{5.39}$$

Dropping in this inequality the first term we get

$$\lambda ||\nabla_{X_2} u_\varepsilon||_{2,\Omega}^2 \le C|f|_{2,\Omega} ||\nabla_{X_2} u_\varepsilon||_{2,\Omega}$$

and thus

$$||\nabla_{X_2} u_\varepsilon||_{2,\Omega} \le C \frac{|f|_{2,\Omega}}{\lambda}.$$

Using this in (5.39) we are ending up with

$$\int_\Omega \varepsilon^2 |\nabla_{X_1} u_\varepsilon|^2 + |\nabla_{X_2} u_\varepsilon|^2 \, dx \le C^2 \frac{|f|_{2,\Omega}^2}{\lambda^2}. \tag{5.40}$$

Thus – due to (5.38) – we deduce that

$$u_\varepsilon, \quad |\varepsilon \nabla_{X_1} u_\varepsilon|, \quad |\nabla_{X_2} u_\varepsilon| \quad \text{are bounded in } L^2(\Omega), \tag{5.41}$$

independently of ε. It follows that there exist $u_0 \in L^2(\Omega)$, $u_1 \in (L^2(\Omega))^p$, $u_2 \in (L^2(\Omega))^{n-p}$ such that – up to a subsequence

$$u_\varepsilon \rightharpoonup u_0, \quad \varepsilon \nabla_{X_1} u_\varepsilon \rightharpoonup u_1, \quad \nabla_{X_2} u_\varepsilon \rightharpoonup u_2 \quad \text{in ``}L^2(\Omega)\text{''}.$$

(u_1, u_2 are vectors with components in $L^2(\Omega)$. The convergence is meant component by component). Of course the convergence in $L^2(\Omega)$ – weak – implies the convergence in $\mathcal{D}'(\Omega)$ and by the continuity of the derivation in $\mathcal{D}'(\Omega)$ we deduce that

$$u_\varepsilon \rightharpoonup u_0, \quad \varepsilon \nabla_{X_1} u_\varepsilon \rightharpoonup 0, \quad \nabla_{X_2} u_\varepsilon \rightharpoonup \nabla_{X_2} u_0 \quad \text{in } L^2(\Omega). \tag{5.42}$$

We then go back to the equation satisfied by u_ε that we expand using the different blocks of A. This gives

$$\int_\Omega \varepsilon^2 A_{11} \nabla_{X_1} u_\varepsilon \cdot \nabla_{X_1} v \, dx + \int_\Omega \varepsilon A_{12} \nabla_{X_2} u_\varepsilon \cdot \nabla_{X_1} v \, dx$$

$$+ \int_\Omega \varepsilon A_{21} \nabla_{X_1} u_\varepsilon \cdot \nabla_{X_2} v \, dx + \int_\Omega A_{22} \nabla_{X_2} u_\varepsilon \cdot \nabla_{X_2} v \, dx$$

$$= \int_\Omega f v \, dx \quad \forall v \in H_0^1(\Omega).$$

Passing to the limit in each term using (5.42) we get

$$\int_\Omega A_{22} \nabla_{X_2} u_0 \cdot \nabla_{X_2} v \, dx = \int_\Omega f v \, dx \quad \forall v \in H_0^1(\Omega). \tag{5.43}$$

At this point we do not know yet if for a.e. $X_1 \in \Pi_\Omega$ we have

$$u_0(X_1, \cdot) \in H_0^1(\Omega_{X_1}). \tag{5.44}$$

To see this, one remarks first that taking $v = u_\varepsilon$ in (5.43) and passing to the limit, we get

$$\int_\Omega A_{22} \nabla_{X_2} u_0 \cdot \nabla_{X_2} u_0 \, dx = \int_\Omega f u_0 \, dx. \tag{5.45}$$

Next we compute

$$I_\varepsilon = \int_\Omega A_\varepsilon \begin{pmatrix} \nabla_{X_1} u_\varepsilon \\ \nabla_{X_2}(u_\varepsilon - u_0) \end{pmatrix} \cdot \begin{pmatrix} \nabla_{X_1} u_\varepsilon \\ \nabla_{X_2}(u_\varepsilon - u_0) \end{pmatrix} \, dx. \tag{5.46}$$

We get

$$I_\varepsilon = \int_\Omega \varepsilon^2 A_{11} \nabla_{X_1} u_\varepsilon \cdot \nabla_{X_1} u_\varepsilon \, dx + \int_\Omega \varepsilon A_{12} \nabla_{X_2}(u_\varepsilon - u_0) \cdot \nabla_{X_1} u_\varepsilon \, dx$$

$$+ \int_\Omega \varepsilon A_{21} \nabla_{X_1} u_\varepsilon \cdot \nabla_{X_2}(u_\varepsilon - u_0) \, dx$$

$$+ \int_\Omega A_{22} \nabla_{X_2}(u_\varepsilon - u_0) \cdot \nabla_{X_2}(u_\varepsilon - u_0) \, dx.$$

Using (5.33) with $v = u_\varepsilon$, we obtain

$$I_\varepsilon = \int_\Omega f u_\varepsilon \, dx - \int_\Omega \varepsilon A_{12} \nabla_{X_2} u_0 \cdot \nabla_{X_1} u_\varepsilon \, dx - \int_\Omega \varepsilon A_{21} \nabla_{X_1} u_\varepsilon \cdot \nabla_{X_2} u_0 \, dx$$

$$- \int_\Omega A_{22} \nabla_{X_2} u_0 \cdot \nabla_{X_2} u_\varepsilon \, dx - \int_\Omega A_{22} \nabla_{X_2} u_\varepsilon \cdot \nabla_{X_2} u_0 \, dx$$

$$+ \int_\Omega A_{22} \nabla_{X_2} u_0 \nabla_{X_2} u_0 \, dx.$$

Passing to the limit in ε, we get

$$\lim_{\varepsilon \to 0} I_\varepsilon = \int_\Omega f u_0 \, dx - \int_\Omega A_{22} \nabla_{X_2} u_0 \cdot \nabla_{X_2} u_0 \, dx = 0.$$

Using the coerciveness assumption we have (see (5.46))

$$\lambda \int_\Omega \varepsilon^2 |\nabla_{X_1} u_\varepsilon|^2 + |\nabla_{X_2}(u_\varepsilon - u_0)|^2 \, dx \leq I_\varepsilon.$$

It follows that

$$\varepsilon \nabla_{X_1} u_\varepsilon \to 0, \qquad \nabla_{X_2} u_\varepsilon \to \nabla_{X_2} u_0 \quad \text{in } L^2(\Omega).$$

Now we also have

$$\int_{\Pi_\Omega} \int_{\Omega_{X_1}} |\nabla_{X_2}(u_\varepsilon - u_0)|^2 \, dX_2 \, dX_1 \to 0. \tag{5.47}$$

It follows that for almost every X_1

$$\int_{\Omega_{X_1}} |\nabla_{X_2}(u_\varepsilon - u_0)|^2 \, dX_2 \to 0.$$

Since

$$\left\{ \int_{\Omega_{X_1}} |\nabla_{X_2} v|^2 \, dX_2 \right\}^{\frac{1}{2}}$$

is a norm on $H_0^1(\Omega_{X_1})$ and $u_\varepsilon(X_1, \cdot) \in H_0^1(\Omega_{X_1})$ we have

$$u_0(X_1, \cdot) \in H_0^1(\Omega_{X_1})$$

and this for almost every X_1. Using then the Poincaré inequality we get

$$\int_{\Omega_{X_1}} |u_\varepsilon - u_0|^2 \, dX_2 \leq C \int_{\Omega_{X_1}} |\nabla_{X_2}(u_\varepsilon - u_0)|^2 \, dX_2$$

$$\implies \quad \int_\Omega |u_\varepsilon - u_0|^2 \, dx \leq C \int_\Omega |\nabla_{X_2}(u_\varepsilon - u_0)|^2 \, dx \to 0$$

(by (5.47)) and thus

$$u_\varepsilon \to u_0 \quad \text{in } L^2(\Omega). \tag{5.48}$$

All this is up to a "subsequence". If we can identify u_0 uniquely then all the convergences above will hold for the whole "sequence". For this purpose recall first that

$$u_0(X_1, \cdot) \in H_0^1(\Omega_{X_1}). \tag{5.49}$$

One can cover Ω by a countable family of open sets of the form

$$U_i \times V_i \subset \Omega, \quad i \in \mathbb{N}$$

where U_i, V_i are open subsets of $\mathbb{R}^p$, $\mathbb{R}^{n-p}$ respectively. One can even choose U_i, V_i hypercubes. Then choosing $\varphi \in H_0^1(V_i)$, $\eta \in \mathcal{D}(U_i)$ we derive from (5.43)

$$\int_{U_i} \eta(X_1) \int_{V_i} A_{22}(X_1, X_2) \nabla_{X_2} u_0(X_1, X_2) \cdot \nabla_{X_2} \varphi(X_2) \, dX_2 \, dX_1$$

$$= \int_{U_i} \eta(X_1) \int_{V_i} f(X_1, X_2) \varphi(X_2) \, dX_2 \, dX_1 \quad \forall \, \eta \in \mathcal{D}(U_i),$$

since $\eta \varphi \in H_0^1(\Omega)$. Thus there exists a set of measure zero, $N(\varphi)$, such that

$$\int_{V_i} A_{22}(X_1, X_2) \nabla_{X_2} u_0(X_1, X_2) \cdot \nabla_{X_2} \varphi(X_2) \, dX_2$$

$$= \int_{V_i} f(X_1, X_2) \varphi(X_2) \, dX_2 \tag{5.50}$$

for all $X_1 \in U_i \setminus N(\varphi)$. Denote by φ_n a Hilbert basis of $H_0^1(V_i)$. Then (5.50) holds (replacing φ by φ_n) for all X_1 such that

$$X_1 \in U_i \setminus N_i(\varphi_n)$$

where $N_i(\varphi_n)$ is a set of measure 0. Thus for

$$X_1 \in U_i \setminus \cup_n N_i(\varphi_n)$$

we have (5.50) for any $\varphi \in H_0^1(V_i)$. This follows easily from the density in $H_0^1(V_i)$ of the linear combinations of the φ_n. Let us then choose

$$X_1 \in \Pi_\Omega \setminus \cup_i \cup_n N_i(\varphi_n)$$

(note that $\cup_i \cup_n N_i(\varphi_n)$ is a set of measure 0). Let

$$\varphi \in \mathcal{D}(\Omega_{X_1}).$$

If K denotes the support of φ we have clearly

$$K \subset \cup_i V_i$$

and thus K can be covered by a finite number of V_i that for simplicity we will denote by $V_1, \ldots, V_k$. Using a partition of unity (see [71], [79]) there exists $\psi_i \in \mathcal{D}(V_i)$ such that

$$\sum_{i=1}^k \psi_i = 1 \quad \text{on } K.$$

By (5.50) we derive

$$\int_K A_{22} \nabla_{X_2} u_0 \cdot \nabla_{X_2} \varphi \, dX_2 = \int_K A_{22} \nabla_{X_2} u_0 \cdot \nabla_{X_2} \sum_i (\psi_i \varphi) \, dX_2$$

$$= \sum_i \int_{V_i} A_{22} \nabla_{X_2} u_0 \cdot \nabla_{X_2} (\psi_i \varphi) \, dX_2$$

$$= \sum_i \int_{V_i} f \psi_i \varphi \, dX_2$$

$$= \int_K f \varphi \, dX_2.$$

This is also

$$\int_{\Omega_{X_1}} A_{22} \nabla_{X_2} u_0 \cdot \nabla_{X_2} \varphi \, dX_2 = \int_{\Omega_{X_1}} f \varphi \, dX_2 \quad \forall \varphi \in \mathcal{D}(\Omega_{X_1})$$

and thus u_0 is the unique solution to (5.35) for a.e. $X_1 \in \Pi_\Omega$. This completes the proof of the theorem. $\qquad\square$

5.3   Estimates for the rate of convergence

We would like to apply the result of the chapter 2 to the anisotropic singular perturbation problem. Indeed the two problems can be connected to each other via a scaling as we explained in Section 1.4.

With the notation of the Chapter 2 consider

$$\Omega_1 = \omega_1 \times \omega_2 \tag{5.51}$$

where ω_1 is a bounded convex open set containing 0 (see Chapter 2). Let us denote by $A = A(x)$ a matrix like in (5.29) but such that

$$A(x) = \begin{pmatrix} A_{11}(x) & A_{12}(X_2) \\ A_{21}(x) & A_{22}(X_2) \end{pmatrix},$$

i.e. A_{12}, A_{22} are depending on X_2 only. Denote then by A_ε the matrix defined in (5.30). Due to (5.31) and (5.32) for

$$f = f(X_2) \in L^2(\omega_2), \tag{5.52}$$

there exists a unique u_ε solution to

$$\begin{cases} \displaystyle\iint_{\Omega_1} A_\varepsilon \nabla u_\varepsilon \cdot \nabla v \, dx = \int_{\Omega_1} f v \, dx \quad \forall\, v \in H_0^1(\Omega_1), \\ u_\varepsilon \in H_0^1(\Omega_1). \end{cases} \tag{5.53}$$

Due to Theorem 5.3 we know that when $\varepsilon \to 0$, $u_\varepsilon \to u_0$ in $L^2(\Omega_1)$ where u_0 is the solution to

$$\begin{cases} \displaystyle\int_{\omega_2} A_{22} \nabla_{X_2} u_0 \cdot \nabla_{X_2} v \, dX_2 = \int_{\omega_2} f(X_2) v \, dX_2 \quad \forall\, v \in H_0^1(\omega_2), \\ u_0 \in H_0^1(\omega_2). \end{cases} \tag{5.54}$$

Note that here since A_{22}, f are independent of X_1 so is u_0. Due to this special situation we can have more information on the convergence of u_ε. We have

Theorem 5.4. *If $\Omega_a = a\omega_1 \times \omega_2$, $a \in (0,1)$ there exist two positive constants c, α such that*

$$\int_{\Omega_a} |\nabla(u_\varepsilon - u_0)|^2 \, dx \le c e^{-\frac{\alpha}{\varepsilon}} |f|_{2,\omega_2}^2. \tag{5.55}$$

Proof. To prove that, we rely on a scaling argument. Indeed we set

$$\varepsilon = \frac{1}{\ell}, \qquad u_\ell(X_1, X_2) = u_\varepsilon\left(\frac{X_1}{\ell}, X_2\right). \tag{5.56}$$

With this definition it is clear that

$$u_\ell \in H_0^1(\Omega_\ell). \tag{5.57}$$

Moreover

$$\nabla_{X_1} u_\ell(X_1, X_2) = \frac{1}{\ell} \nabla_{X_1} u_\varepsilon\left(\frac{X_1}{\ell}, X_2\right) \tag{5.58}$$

$$\nabla_{X_2} u_\ell(X_1, X_2) = \nabla_{X_2} u_\varepsilon\left(\frac{X_1}{\ell}, X_2\right). \tag{5.59}$$

Making the change of variable $X_1 \to \frac{X_1}{\ell}$ in the equation (5.53), we obtain

$$\int_{\Omega_\ell} A_\varepsilon\left(\frac{X_1}{\ell}, X_2\right) \nabla u_\varepsilon\left(\frac{X_1}{\ell}, X_2\right) \cdot \nabla v\left(\frac{X_1}{\ell}, X_2\right) dx$$
$$= \int_{\Omega_\ell} f(X_2) v\left(\frac{X_1}{\ell}, X_2\right) dx \quad \forall\, v \in H_0^1(\Omega_1).$$

For $w \in H_0^1(\Omega_\ell)$, we have $v(X_1, X_2) = w(\ell X_1, X_2) \in H_0^1(\Omega_1)$ and

$$(\nabla v)\left(\frac{X_1}{\ell}, X_2\right) = (\ell \nabla_{X_1} w(X_1, X_2), \nabla_{X_2} w(X_1, X_2))^T.$$

Combining this with (5.58), (5.59) – see also (5.56), we obtain

$$A_\varepsilon\left(\frac{X_1}{\ell}, X_2\right) \nabla u_\varepsilon\left(\frac{X_1}{\ell}, X_2\right) \cdot \nabla v\left(\frac{X_1}{\ell}, X_2\right)$$

$$= \begin{pmatrix} \frac{1}{\ell^2} A_{11}\left(\frac{X_1}{\ell}, X_2\right) & \frac{1}{\ell} A_{12}(X_2) \\ \frac{1}{\ell} A_{21}\left(\frac{X_2}{\ell}, X_2\right) & A_{22}(X_2) \end{pmatrix} \begin{pmatrix} \ell \nabla_{X_1} u_\ell \\ \nabla_{X_2} u_\ell \end{pmatrix} \cdot \begin{pmatrix} \ell \nabla_{X_1} w \\ \nabla_{X_2} w \end{pmatrix}$$

$$= \begin{pmatrix} \frac{1}{\ell} A_{11}\left(\frac{X_1}{\ell}, X_2\right) \nabla_{X_1} u_\ell + \frac{1}{\ell} A_{12} \nabla_{X_2} u_\ell \\ A_{21}\left(\frac{X_1}{\ell}, X_2\right) \nabla_{X_1} u_\ell + A_{22} \nabla_{X_2} u_\ell \end{pmatrix} \cdot \begin{pmatrix} \ell \nabla_{X_1} w \\ \nabla_{X_2} w \end{pmatrix}$$

$$= \begin{pmatrix} A_{11}\left(\frac{X_1}{\ell}, X_2\right) & A_{12}(X_2) \\ A_{12}\left(\frac{X_1}{\ell}, X_2\right) & A_{22}(X_2) \end{pmatrix} \begin{pmatrix} \nabla_{X_1} u_\ell \\ \nabla_{X_2} u_\ell \end{pmatrix} \cdot \begin{pmatrix} \nabla_{X_1} w \\ \nabla_{X_2} w \end{pmatrix}.$$

Thus setting

$$\widetilde{A}_\ell(X_1, X_2) = \begin{pmatrix} A_{11}\left(\frac{X_1}{\ell}, X_2\right) & A_{12}(X_2) \\ A_{12}\left(\frac{X_1}{\ell}, X_2\right) & A_{22}(X_2) \end{pmatrix}$$

we see that u_ℓ is solution to

$$\begin{cases} \displaystyle \int\!\!\int_{\Omega_\ell} \widetilde{A}_\ell \nabla u_\ell \cdot \nabla w \, dx = \int_{\Omega_\ell} fw \, dx \quad \forall\, w \in H_0^1(\Omega_\ell), \\ u_\ell \in H_0^1(\Omega_\ell). \end{cases}$$

Applying Theorem 2.1, taking into account Remarks 2.2, we obtain the existence of some constants c_0, α_0 such that

$$\int_{\Omega_{a\ell}} |\nabla(u_\ell - u_0)|^2 \, dx \leq c_0 e^{-\alpha_0 \ell} |f|_{2,\omega_2}^2.$$

(Note that $u_0 = u_\infty$ and that in (2.6) the open set $\Omega_{\frac{\ell}{2}}$ can be easily replaced by $\Omega_{a\ell}$). Changing $X_1 \to \ell X_1$ in the integral above, we obtain

$$\ell^p \int_{\Omega_a} |\nabla(u_\ell - u_0)|^2 (\ell X_1, X_2) \, dx \leq c_0 e^{-\alpha_0 \ell} |f|_{2,\omega_2}^2$$

and by (5.58), (5.59) for $\ell > 1$ i.e. $\varepsilon < 1$

$$\frac{\ell^p}{\ell^2} \int_{\Omega_a} |\nabla(u_\varepsilon - u_0)|^2 \, dx \leq c_0 e^{-\alpha_0 \ell} |f|_{2,\omega_2}^2$$

$$\Longleftrightarrow \qquad \int_{\Omega_a} |\nabla(u_\varepsilon - u_0)|^2 \, dx \leq c_0 \ell^{2-p} e^{-\alpha_0 \ell} |f|_{2,\omega_2}^2$$

$$\leq c e^{-\frac{\alpha}{\varepsilon}} |f|_{2,\omega_2}^2$$

for some constant c and any α smaller than α_0. This completes the proof of the theorem. $\qquad \square$

Chapter 6

Eigenvalue problems

6.1 Introduction

The goal of this chapter is to analyze the asymptotic behaviour of the eigenvalues and the eigenfunctions of elliptic problems set in domains which can become unbounded in one or several directions. More precisely we will consider first the following eigenvalue problems set on cylindrical domains $\Omega_\ell = \ell\omega_1 \times \omega_2$,

$$- \operatorname{div}\left(A(x)\nabla u_\ell\right) = \lambda_\ell u_\ell \text{ in } \Omega_\ell, \qquad u_\ell = 0 \text{ on } \partial\Omega_\ell \qquad (6.1)$$

and the eigenvalue problem set on the section ω_2 of Ω_ℓ, namely

$$- \operatorname{div}\left(A_{22}(X_2)\nabla u\right) = \mu u \text{ in } \omega_2, \qquad u = 0 \text{ on } \partial\omega_2. \qquad (6.2)$$

When the positive parameter ℓ goes to plus infinity, the asymptotic behaviour of the unknowns of Problem (6.1) will be described in terms of the solutions u, μ of Problem (6.2). Indeed, under our assumptions, we will prove that the sequence of the kth eigenvalues of Problem (6.1) converges toward the first eigenvalue of (6.2) as ℓ goes to plus infinity. Moreover the convergence rate is $\frac{1}{\ell^2}$ which is, in some sense optimal (Cf. Chapter 1).

Then we will show for block diagonal matrix A that the normalized first eigenfunction of Problem (6.1) converges on any fixed subdomain of Ω_ℓ toward a well identified first eigenfunction of (6.2) as ℓ goes to plus infinity. The case of the sequence of the kth eigenfunction is considered in [33]. Finally we will see that the convergence above does not carry out for Neumann boundary conditions at the cylinder's ends.

6.2   Convergence of the eigenvalues

Let us first introduce some notation. We denote by Ω_ℓ the open subset of $\mathbb{R}^n$ defined as

$$\Omega_\ell = \ell\omega_1 \times \omega_2.$$

ℓ is a positive number, ω_1 a bounded convex subset of $\mathbb{R}^p$, $p \geq 1$, containing 0, ω_2 is a bounded open subset of $\mathbb{R}^{n-p}$. The points in $\mathbb{R}^n$ will be denoted by

$$x = (X_1, X_2),$$

with

$$X_1 = (x_1, \ldots, x_p), \qquad X_2 = (x_{p+1}, \ldots, x_n).$$

Let $A = A(X_1, X_2)$ be a $n \times n$-symmetric matrix of the type

$$A = A(X_1, X_2) = \begin{pmatrix} A_{11}(X_1, X_2) & A_{12}(X_2) \\ A_{12}^T(X_2) & A_{22}(X_2) \end{pmatrix}. \tag{6.3}$$

In (6.3), A_{11} is a $p \times p$-symmetric matrix. We will assume that A is uniformly bounded and uniformly positive definite on $\mathbb{R}^p \times \omega_2$ that is to say that

$$\|A(x)\| \leq \Lambda \qquad\qquad \text{a.e. } x \in \mathbb{R}^p \times \omega_2, \tag{6.4}$$

$$A(x)\xi \cdot \xi \geq \lambda|\xi|^2 \qquad\qquad \text{a.e. } x \in \mathbb{R}^p \times \omega_2, \quad \forall \xi \in \mathbb{R}^n. \tag{6.5}$$

In the above formula $\|\ \|$ stands for a norm on matrices, Λ, λ are some positive constants, $|\ |$ is the euclidean norm, the dot "$\cdot$" denotes the usual euclidean scalar product. If Ω is a bounded open set of $\mathbb{R}^k$ we will denote by $H^1(\Omega)$, $H_0^1(\Omega)$ the usual spaces of functions defined as

$$H^1(\Omega) = \{\, v \in L^2(\Omega) \mid \partial_{x_i} v \in L^2(\Omega) \ \forall i = 1, \ldots, k \,\},$$

$$H_0^1(\Omega) = \{\, v \in H^1(\Omega) \mid v = 0 \text{ on } \partial\Omega \,\}.$$

In the above definition ∂_{x_i} denotes the partial derivation in x_i, $\partial\Omega$ the boundary of Ω. We will always assume that $H_0^1(\Omega)$ is equipped with the Dirichlet norm

$$\|\nabla v\|_{2,\Omega} = \left\{ \int_\Omega |\nabla v(x)|^2 \, dx \right\}^{1/2},$$

($|\nabla v|$ denotes the euclidean norm of the gradient of v).

We would like to consider the eigenvalue problem

$$\begin{cases} u_\ell \in H_0^1(\Omega_\ell), \\ \displaystyle\int_{\Omega_\ell} A\nabla u_\ell \cdot \nabla v \, dx = \lambda_\ell \int_{\Omega_\ell} u_\ell v \, dx \quad \forall\, v \in H_0^1(\Omega_\ell). \end{cases} \tag{6.6}$$

To be more precise, we say that λ_ℓ is an eigenvalue to (6.6) if there exists $u_\ell \neq 0$ solution of the above system. When $\ell \to +\infty$, we expect that the limit eigenvalue problem (6.6) will be the eigenvalue problem on ω_2 defined as

$$\begin{cases} u \in H_0^1(\omega_2), \\ \displaystyle\int_{\omega_2} A_{22}(X_2)\nabla_{X_2}u \cdot \nabla_{X_2}v\,dX_2 = \mu \int_{\omega_2} uv\,dX_2 \quad \forall v \in H_0^1(\omega_2). \end{cases} \tag{6.7}$$

In the above equation $dX_2 = dx_{p+1}\ldots dx_n$, $\nabla_{X_2} = (\partial_{x_{p+1}},\ldots,\partial_{x_n})$ – i.e. ∇_{X_2} denotes the gradient in the last variables. By "limit problem" we mean that we expect that the eigenmodes of (6.6) will converge towards the eigenmodes of (6.7) when $\ell \to +\infty$.

Let us denote by λ_ℓ^1 and μ_1 the first eigenvalues of (6.6) and (6.7) respectively. It is well known that

$$\lambda_\ell^1 = \inf\left\{ \int_{\Omega_\ell} A\nabla u \cdot \nabla u\,dx \mid u \in H_0^1(\Omega_\ell), \int_{\Omega_\ell} u^2\,dx = 1 \right\}$$

$$\left(= \inf_{H_0^1(\Omega_\ell)\setminus\{0\}} \frac{\int_{\Omega_\ell} A\nabla u \cdot \nabla u\,dx}{\int_{\Omega_\ell} u^2\,dx} \right), \tag{6.8}$$

and

$$\mu_1 = \inf\left\{ \int_{\omega_2} A_{22}\nabla_{X_2}u \cdot \nabla_{X_2}u\,dX_2 \mid u \in H_0^1(\omega_2), \int_{\omega_2} u^2\,dX_2 = 1 \right\}$$

$$\left(= \inf_{H_0^1(\omega_2)\setminus\{0\}} \frac{\int_{\omega_2} A_{22}\nabla_{X_2}u \cdot \nabla_{X_2}u\,dX_2}{\int_{\omega_2} u^2\,dX_2} \right). \tag{6.9}$$

Moreover, it is well known that these eigenvalues are simple and the corresponding eigenfunctions do not change sign in Ω_ℓ or ω_2 – see [50] or [52].

Then we can show:

Theorem 6.1. *We have*

$$\mu_1 \leq \lambda_\ell^1 \leq \mu_1 + \frac{C}{\ell^2} \tag{6.10}$$

where $C = |A_{11}|_\infty \nu_1$, where ν_1 is the first eigenvalue of the Laplace operator in ω_1, $|A_{11}|_\infty = \mathrm{esssup}_{\mathbb{R}^n} |A_{11}(x)|$, $|\cdot|$ is the norm of matrices subordinated to the euclidean norm.

To prove this theorem we will need the following lemma:

Lemma 6.1. *Let Ω be a bounded open set of $\mathbb{R}^n$. Let A_ε be a family of symmetric matrices such that for λ, Λ positive independent of ε we have*

$$\lambda|\xi|^2 \leq A_\varepsilon(x)\xi \cdot \xi \qquad a.e.\ x \in \Omega, \quad \forall\,\xi \in \mathbb{R}^n,$$

$$\|A_\varepsilon(x)\| \leq \Lambda \qquad a.e.\ x \in \Omega.$$

Suppose that

$$A_\varepsilon(x) \to A(x) \quad a.e. \ x \in \Omega \ as \ \varepsilon \to 0. \tag{6.11}$$

Set

$$\lambda_\varepsilon = \inf\left\{ \int_\Omega A_\varepsilon \nabla u \cdot \nabla u \, dx \mid u \in H_0^1(\Omega), \ \int_\Omega u^2 \, dx = 1 \right\}, \tag{6.12}$$

$$\lambda_0 = \inf\left\{ \int_\Omega A \nabla u \cdot \nabla u \, dx \mid u \in H_0^1(\Omega), \ \int_\Omega u^2 \, dx = 1 \right\},$$

then we have

$$\lim_{\varepsilon \to 0} \lambda_\varepsilon = \lambda_0. \tag{6.13}$$

(i.e. the first eigenvalue of the operator $-\nabla \cdot (A_\varepsilon \nabla \cdot)$ with Dirichlet boundary conditions converges towards the first eigenvalue of $-\nabla \cdot (A\nabla \cdot)$ with the same boundary conditions).

Proof of the Lemma. Let u_ε be the first eigenfunction realizing the infimum of (6.12). (Recall that it does not change sign in Ω and is unique if we assume it to be positive – see [50]). Denote by v a fixed function in $H_0^1(\Omega)$ satisfying

$$\int_\Omega v^2 \, dx = 1.$$

By (6.4), (6.5), (6.12) we have

$$\lambda \|\nabla u_\varepsilon\|_{2,\Omega}^2 \leq \int_\Omega A_\varepsilon(x) \nabla u_\varepsilon \cdot \nabla u_\varepsilon \, dx$$
$$= \lambda_\varepsilon \leq \int_\Omega A_\varepsilon(x) \nabla v \cdot \nabla v \, dx \leq C \tag{6.14}$$

where C is a constant independent of ε. Thus λ_ε is bounded and so is u_ε in $H_0^1(\Omega)$. At the expense of extracting a "subsequence" we can assume that

$$\lambda_\varepsilon \to \lambda^0,$$
$$u_\varepsilon \rightharpoonup u_0 \quad \text{in} \quad H_0^1(\Omega), \qquad u_\varepsilon \to u_0 \quad \text{in} \quad L^2(\Omega).$$

Now, by definition of the first eigenvalue one has also since A_ε is symmetric

$$\int_\Omega A_\varepsilon(x) \nabla u_\varepsilon \cdot \nabla v \, dx = \int_\Omega \nabla u_\varepsilon \cdot A_\varepsilon(x) \nabla v \, dx$$
$$= \lambda_\varepsilon \int_\Omega u_\varepsilon v \, dx \quad \forall v \in H_0^1(\Omega). \tag{6.15}$$

Noticing that by the Lebesgue theorem we have

$$A_\varepsilon \nabla v \to A \nabla v \quad \text{in} \quad L^2(\Omega),$$

(see (6.11)) we obtain by passing to the limit in (6.15)

$$\int_\Omega A \nabla u_0 \cdot \nabla v \, dx = \lambda^0 \int_\Omega u_0 v \, dx \quad \forall \, v \in H_0^1(\Omega). \tag{6.16}$$

Thus λ^0 is an eigenvalue of the problem (6.16). Note that $u_0 \neq 0$ since

$$1 = \int_\Omega u_\varepsilon^2 \, dx \to \int_\Omega u_0^2 \, dx = 1.$$

From (6.14) we derive also

$$\lambda^0 \le \int_\Omega A(x) \nabla v \cdot \nabla v \, dx \quad \forall \, v \in H_0^1(\Omega), \; \int_\Omega v^2 \, dx = 1.$$

Thus λ^0 is the first eigenvalue of (6.16) – i.e. $\lambda^0 = \lambda_0$. By uniqueness of the possible limits this shows (6.13) and completes the proof of the Lemma. $\square$

Remark 6.1. Since when $\varepsilon \to 0$

$$\int_\Omega u_\varepsilon^2 \, dx \to \int_\Omega u_0^2 \, dx, \qquad \int_\Omega A_\varepsilon \nabla u_\varepsilon \cdot \nabla u_\varepsilon \, dx \to \int_\Omega A \nabla u_0 \cdot \nabla u_0 \, dx$$

it is easy to show that $u_\varepsilon \to u_0$ in $H^1(\Omega)$ and u_0 is the first eigenfunction (normalized and positive) of (6.16).

We then now turn to the proof of Theorem 6.1.

Proof of Theorem 6.1. Let us first establish the lower bound in (6.10). For this consider A_ε defined by

$$A_\varepsilon = \begin{cases} A \quad \text{on} \quad \Omega_{\ell-\varepsilon}, \\[2em] \begin{pmatrix} \lambda & & 0 & \vdots & \\ & \ddots & & \vdots & 0 \\ 0 & & \lambda & \vdots & \\ \cdots & \cdots & \cdots & \cdots & \cdots \\ & 0 & & \vdots & A_{22} \end{pmatrix} \quad \text{on} \quad \Omega_\ell \setminus \Omega_{\ell-\varepsilon}, \end{cases} \tag{6.17}$$

where λ is the constant in the inequality (6.5). Clearly A_ε satisfies all the assumptions of Lemma 6.1 with $\Omega = \Omega_\ell$. Define

$$\lambda_{\ell,\varepsilon}^1 = \inf \left\{ \int_{\Omega_\ell} A_\varepsilon \nabla u \cdot \nabla u \, dx \mid u \in H_0^1(\Omega_\ell), \int_{\Omega_\ell} u^2 \, dx = 1 \right\}, \tag{6.18}$$

the first eigenvalue of the operator $-\nabla \cdot (A_\varepsilon \nabla \cdot)$ with Dirichlet boundary conditions. Denote also by u_ε the function – that we can assume > 0 – where the infimum of (6.18) is achieved. We have

$$\int_{\Omega_\ell} A_\varepsilon(x) \nabla u_\varepsilon \cdot \nabla v \, dx = \lambda^1_{\ell,\varepsilon} \int_{\Omega_\ell} u_\varepsilon v \, dx \quad \forall v \in H^1_0(\Omega_\ell). \tag{6.19}$$

Denote then by ϱ_ε a smooth function on $\ell\omega_1$ such that

$$0 \leq \varrho_\varepsilon \leq 1, \quad \varrho_\varepsilon = 1 \text{ on } (\ell - \varepsilon)\omega_1, \quad \varrho_\varepsilon = 0 \text{ on } \partial(\ell\omega_1), \tag{6.20}$$

$$\varrho_\varepsilon \text{ is concave, i.e. } \partial^2_{x_i}\varrho_\varepsilon \leq 0 \quad \forall i = 1, \cdots, p. \tag{6.21}$$

Denote by w_1 the positive function realizing the infimum of (6.9) and in (6.19) choose

$$v = w_1(X_2)\varrho_\varepsilon(X_1). \tag{6.22}$$

We obtain (we decompose A_ε as A in (6.3) and put the ε above as an upper index)

$$\int_{\Omega_\ell} \{ A^\varepsilon_{11} \nabla_{X_1} u_\varepsilon \cdot \nabla_{X_1} v + A^\varepsilon_{12} \nabla_{X_2} u_\varepsilon \cdot \nabla_{X_1} v$$

$$+ A^{\varepsilon T}_{12} \nabla_{X_1} u_\varepsilon \cdot \nabla_{X_2} v + A^\varepsilon_{22} \nabla_{X_2} u_\varepsilon \cdot \nabla_{X_2} v \} \, dx = \lambda^1_{\ell,\varepsilon} \int_{\Omega_\ell} u_\varepsilon v \, dx.$$

Using (6.22) we get

$$\int_{\Omega_\ell} \{ A^\varepsilon_{11} \nabla_{X_1} u_\varepsilon \cdot \nabla_{X_1} \varrho_\varepsilon w_1 + A^\varepsilon_{12} \nabla_{X_2} u_\varepsilon \cdot \nabla_{X_1} \varrho_\varepsilon w_1$$

$$+ A^{\varepsilon T}_{12} \nabla_{X_1} u_\varepsilon \cdot \nabla_{X_2} w_1 \varrho_\varepsilon + A^\varepsilon_{22} \nabla_{X_2} u_\varepsilon \cdot \nabla_{X_2} w_1 \varrho_\varepsilon \} \, dx \tag{6.23}$$

$$= \lambda^1_{\ell,\varepsilon} \int_{\Omega_\ell} u_\varepsilon w_1 \varrho_\varepsilon \, dx.$$

First remark that

$$\nabla_{X_1} \varrho_\varepsilon = 0 \quad \text{on} \quad \Omega_{\ell-\varepsilon}$$

so that by (6.17) the second integral in (6.23) vanishes. Next, for the first integral we have

$$\int_{\Omega_\ell} A^\varepsilon_{11} \nabla_{X_1} u_\varepsilon \cdot \nabla_{X_1} \varrho_\varepsilon w_1 \, dx$$

$$= \int_{\Omega_\ell \setminus \Omega_{\ell-\varepsilon}} A^\varepsilon_{11} \nabla_{X_1} u_\varepsilon \cdot \nabla_{X_1} \varrho_\varepsilon \, w_1 \, dx$$

$$= \lambda \int_{\Omega_\ell \setminus \Omega_{\ell-\varepsilon}} \nabla_{X_1} u_\varepsilon \cdot \nabla_{X_1} \varrho_\varepsilon \, w_1 \, dx$$

$$= \lambda \int_{\Omega_\ell \setminus \Omega_{\ell-\varepsilon}} \nabla_{X_1} \cdot \{ u_\varepsilon \nabla_{X_1} \varrho_\varepsilon \, w_1 \} - u_\varepsilon \sum_{i=1}^{p} \partial^2_{x_i} \varrho_\varepsilon \, dx$$

$$\geq \lambda \int_{\Omega_\ell \setminus \Omega_{\ell-\varepsilon}} \nabla_{X_1} \cdot \{ u_\varepsilon \nabla_{X_1} \varrho_\varepsilon w_1 \} \, dx$$

by (6.20), (6.21). But since $u_\varepsilon \nabla_{X_1}\varrho_\varepsilon$ vanishes on the boundary of $\Omega_\ell \setminus \Omega_{\ell-\varepsilon}$ the above integral is identically equal to 0. Thus, we derive from (6.23)

$$\int_{\Omega_\ell} A_{12}^{\varepsilon T} \nabla_{X_1} u_\varepsilon \cdot \nabla_{X_2} w_1 \varrho_\varepsilon \, dx + \int_{\Omega_\ell} A_{22}^\varepsilon \nabla_{X_2} u_\varepsilon \cdot \nabla_{X_2} w_1 \varrho_\varepsilon \, dx$$

$$\leq \lambda_{\ell,\varepsilon}^1 \int_{\Omega_\ell} u_\varepsilon w_1 \varrho_\varepsilon \, dx$$

which is also by (6.17)

$$\int_{\Omega_{\ell-\varepsilon}} A_{12}^T \nabla_{X_1} u_\varepsilon \cdot \nabla_{X_2} w_1 \varrho_\varepsilon \, dx + \int_{\Omega_\ell} A_{22} \nabla_{X_2} u_\varepsilon \cdot \nabla_{X_2} w_1 \varrho_\varepsilon \, dx \tag{6.24}$$

$$\leq \lambda_{\ell,\varepsilon}^1 \int_{\Omega_\ell} u_\varepsilon w_1 \varrho_\varepsilon \, dx.$$

The second integral above can be written as

$$\int_{\Omega_\ell} A_{22} \nabla_{X_2} u_\varepsilon \cdot \nabla_{X_2} w_1 \varrho_\varepsilon \, dx$$

$$= \int_{\ell\omega_1} \varrho_\varepsilon \int_{\omega_2} A_{22} \nabla_{X_2} u_\varepsilon(X_1, X_2) \cdot \nabla_{X_2} w_1(X_2) \, dX_2 \, dX_1$$

$$= \int_{\ell\omega_1} \varrho_\varepsilon \int_{\omega_2} \mu_1 u_\varepsilon w_1 \, dX_2 \, dX_1$$

$$= \mu_1 \int_{\Omega_\ell} u_\varepsilon w_1 \varrho_\varepsilon \, dx$$

by definition of w_1 and since for a.e. X_1, $u_\varepsilon(X_1, \cdot) \in H_0^1(\omega_2)$. It follows that (6.24) becomes

$$\int_{\Omega_{\ell-\varepsilon}} A_{12}^T \nabla_{X_1} u_\varepsilon \cdot \nabla_{X_2} w_1 \varrho_\varepsilon \, dx + \mu_1 \int_{\Omega_\ell} u_\varepsilon w_1 \varrho_\varepsilon \, dx$$

$$\leq \lambda_{\ell,\varepsilon}^1 \int_{\Omega_\ell} u_\varepsilon w_1 \varrho_\varepsilon, \, dx.$$

Passing to the limit in ε – see also Remark 6.1 – we get (see Lemma 6.1)

$$\int_{\Omega_\ell} A_{12}^T \nabla_{X_1} u_0 \cdot \nabla_{X_2} w_1 \, dx + \mu_1 \int_{\Omega_\ell} u_0 w_1 \, dx \leq \lambda_\ell^1 \int_{\Omega_\ell} u_0 w_1 \, dx$$

($\varrho_\varepsilon \to 1$ as $\varepsilon \to 0$). But since $u_0 \in H_0^1(\Omega_\ell)$ and A_{12}^T, w_1 depend on X_2 only we have

$$\int_{\Omega_\ell} A_{12}^T \nabla_{X_1} u_0 \cdot \nabla_{X_2} w_1 \, dx = \int_{\Omega_\ell} \nabla_{X_1} \cdot (u_0 A_{12} \nabla_{X_2} w_1) \, dx = 0.$$

We derive then

$$\mu_1 \int_{\Omega_\ell} u_0 w_1 \, dx \leq \lambda_\ell^1 \int_{\Omega_\ell} u_0 w_1 \, dx$$

which implies the first inequality of (6.10) since u_0, w_1 are positive (see Remark 6.1). It remains to show the second inequality of (6.10). For that let v be a smooth function in $H_0^1(\ell\omega_1)$ such that

$$\int_{\ell\omega_1} v^2 \, dX_1 = 1 \qquad (6.25)$$

and w_1 as above. It is clear that

$$u = vw_1 = v(X_1)w_1(X_2) \in H_0^1(\Omega_\ell) \quad \text{and} \quad \int_{\Omega_\ell} u^2 \, dx = 1.$$

Thus, by (6.8) we have

$$\lambda_\ell^1 \le \int_{\Omega_\ell} A\nabla(vw_1) \cdot \nabla(vw_1) \, dx$$

$$= \int_{\Omega_\ell} A_{11}\nabla_{X_1}v \cdot \nabla_{X_1}vw_1^2 \, dx + \int_{\Omega_\ell} A_{12}\nabla_{X_2}w_1 \cdot \nabla_{X_1}v(vw_1) \, dx$$

$$+ \int_{\Omega_\ell} A_{12}^T\nabla_{X_1}v \cdot \nabla_{X_2}w_1(vw_1) \, dx$$

$$+ \int_{\Omega_\ell} A_{22}\nabla_{X_2}w_1 \cdot \nabla_{X_2}w_1v^2 \, dx \qquad (6.26)$$

$$= \int_{\Omega_\ell} A_{11}\nabla_{X_1}v \cdot \nabla_{X_1}vw_1^2 \, dx + 2\int_{\Omega_\ell} A_{12}\nabla_{X_2}w_1 \cdot \nabla_{X_1}v(vw_1) \, dx$$

$$+ \int_{\Omega_\ell} A_{22}\nabla_{X_2}w_1 \cdot \nabla_{X_2}w_1v^2 \, dx.$$

One remarks that the second integral is equal to 0 since $v \in H_0^1(\ell\omega_1)$. Indeed – recall that A_{12}, w_1 are depending on X_2 only – and thus we have

$$2\int_{\Omega_\ell} A_{12}\nabla_{X_2}w_1 \cdot \nabla_{X_1}v \, vw_1 \, dx$$

$$= \int_{\omega_2} \int_{\ell\omega_1} \nabla_{X_1} \cdot \{v^2 A_{12}\nabla_{X_2}w_1w_1\} \, dX_1 \, dX_2 = 0.$$

Next, due to the definition of w_1 we have – see (6.25)

$$\int_{\Omega_\ell} A_{22}\nabla_{X_2}w_1 \cdot \nabla_{X_2}w_1v^2 \, dx$$

$$= \int_{\ell\omega_1} v^2 \int_{\omega_2} A_{22}\nabla_{X_2}w_1 \cdot \nabla_{X_2}w_1 \, dX_2 \, dX_1$$

$$= \int_{\ell\omega_1} v^2\mu_1 dX_1 = \mu_1.$$

Thus, from (6.26) we derive

$$\lambda_\ell^1 \leq \mu_1 + \int_{\Omega_\ell} A_{11} \nabla_{X_1} v \cdot \nabla_{X_1} v w_1^2 \, dx$$

$$\forall v \in H_0^1(\ell\omega_1), \quad \int_{\ell\omega_1} v^2 \, dX_1 = 1.$$

It follows that

$$\lambda_\ell^1 \leq \mu_1 + \int_{\ell\omega_1} |A_{11}| |\nabla_{X_1} v|^2 \, dX_1 \int_{\omega_2} w_1^2 \, dX_2$$

$$\leq \mu_1 + |A_{11}|_\infty \int_{\ell\omega_1} |\nabla_{X_1} v|^2 \, dX_1 \tag{6.27}$$

$$\forall v \in H_0^1(\ell\omega_1), \quad \int_{\ell\omega_1} v^2 \, dX_1 = 1.$$

It is easy to see by an elementary change of variable (Cf. Proposition 6.1 below) that

$$\inf_{v \in H_0^1(\ell\omega_1)} \frac{\int_{\ell\omega_1} |\nabla_{X_1} v(X_1)|^2 \, dX_1}{\int_{\ell\omega_1} v(X_1)^2 \, dX_1} = \frac{1}{\ell^2} \inf_{v \in H_0^1(\omega_1)} \frac{\int_{\omega_1} |\nabla_{X_1} v(X_1)|^2 \, dX_1}{\int_{\omega_1} v(X_1)^2 \, dX_1}.$$

It follows from (6.27) taking the infimum in v that

$$\lambda_\ell^1 \leq \mu_1 + \frac{|A_{11}|_\infty}{\ell^2} \nu_1.$$

Note that ν_1 is the first eigenvalue of the Laplacian in ω_1. This completes the proof of Theorem 6.1. □

As a corollary we have

Theorem 6.2 (Convergence of the first eigenvalue). *Let Ω_ℓ' denote a domain in $\mathbb{R}^n$ such that*

$$\Omega_\ell \subset \Omega_\ell' \subset \Omega_{\ell'} \tag{6.28}$$

for some ℓ'. Let $\lambda_\ell'^1$ be the first eigenvalue of $-\nabla \cdot (A\nabla \cdot)$ with Dirichlet boundary conditions on Ω_ℓ' – i.e.

$$\lambda_\ell'^1 = \inf\left\{ \int_{\Omega_\ell'} A\nabla u \cdot \nabla u \, dx \mid u \in H_0^1(\Omega_\ell'), \int_{\Omega_\ell'} u^2 \, dx = 1 \right\} \tag{6.29}$$

then it holds

$$\lim_{\ell \to +\infty} \lambda_\ell'^1 = \mu_1.$$

Proof. It is easy to see from the definition – see for instance (6.29) – that the first eigenvalue of the operator $-\nabla \cdot (A\nabla\cdot)$ with Dirichlet boundary conditions on Ω is nonincreasing when the size of Ω is increasing. Thus we have by (6.28)

$$\mu_1 \le \lambda_{\ell'}^1 \le \lambda_\ell'^1 \le \lambda_\ell^1 \le \mu_1 + \frac{c}{\ell^2}$$

and the result follows. $\square$

As we explained in our introduction (Chapter 1) it is reasonable to expect that the λ_ℓ^k – the kth eigenvalue of the operator $-\nabla \cdot (A\nabla\cdot)$ with Dirichlet boundary conditions on Ω_ℓ converges towards μ_1 as ℓ goes to plus infinity. We will restrict ourselves to the case when $\omega_1 = (-1,1)^p$ i.e. $\Omega_\ell = (-\ell,\ell)^p \times \omega_2$. We have

Theorem 6.3 (Convergence of the kth eigenvalue λ_ℓ^k). *With the notation above there exists a constant $C_k = C(k, p, |A_{11}|_\infty)$ such that*

$$\mu_1 \le \lambda_\ell^k \le \mu_1 + \frac{C_k}{\ell^2}.$$

Proof. Of course we can assume without loss of generality that $k \ge 2$. Let us split the domain Ω_ℓ in k subdomains Q_i in the x_1-direction – i.e. let us set

$$Q_i = \left(-\ell + (i-1)\frac{2\ell}{k}, -\ell + i\frac{2\ell}{k}\right) \times (-\ell,\ell)^{p-1} \times \omega_2 \quad i = 1, \ldots, k.$$

Moreover let us denote by $\lambda_{Q_i}^1$ the first eigenvalue defined by

$$\begin{aligned}
\lambda_{Q_i}^1 &= \lambda_{Q_i}^1(A) \\
&= \inf\left\{\int_{Q_i} A\nabla u \cdot \nabla u \, dx \mid u \in H_0^1(Q_i), \int_{Q_i} u^2 \, dx = 1\right\}
\end{aligned} \tag{6.30}$$

and by u_i the first eigenfunction – i.e. the only positive function achieving the infimum above. Set now

$$Q = \left(-\frac{\ell}{k}, \frac{\ell}{k}\right) \times (-\ell,\ell)^{p-1} \times \omega.$$

Denote by $\lambda_Q^1 = \lambda_Q^1(A)$ the first eigenvalue defined by (6.30) where Q_i is replaced by Q. Performing a translation in x_1 in the integral

$$\int_{Q_i} A\nabla u \cdot \nabla u \, dx$$

it is easy to see that

$$\lambda_{Q_i}^1(A) = \lambda_Q^1(\tilde{A})$$

where $\tilde{A}(x) = A(x_1 + (2i-1)\frac{\ell}{k} - \ell, x_2, \ldots, x_n)$. Since

$$\Omega_{\ell/k} \subset Q \subset \Omega_\ell$$

it follows from Theorem 6.1 that

$$\mu_1 \le \lambda^1_{Q_i}(A) = \lambda^1_Q(\tilde{A}) \le \lambda^1_{\ell/k} \le \mu_1 + \frac{k^2|A_{11}|_\infty \nu_1}{\ell^2}. \tag{6.31}$$

Suppose now that u_i is extended by 0 in $\Omega_\ell \setminus Q_i$. Denote by u_ℓ^j the jth eigenfunction corresponding to λ_ℓ^j (our numbering of the eigenvalues is such that the λ_ℓ^j are not necessarily pairwise distinct). Since one has more unknowns than equations it is possible to find $\alpha_1, \ldots, \alpha_k$ not all equal to zero such that

$$u = \sum_{i=1}^k \alpha_i u_i$$

satisfies

$$(u, u_\ell^j) = \sum_{i=1}^k \alpha_i(u_i, u_\ell^j) = 0 \quad \forall j = 1, \ldots, k-1. \tag{6.32}$$

$(\cdot, \cdot)$ denotes the scalar product in $L^2(\Omega_\ell)$. It is well known – see [67] – that

$$\lambda_\ell^1 \le \lambda_\ell^k = \inf_{\substack{v \in H_0^1(\Omega_\ell) \setminus \{0\} \\ v \perp u_\ell^1, \ldots, u_\ell^{k-1}}} \left\{ \int_{\Omega_\ell} A\nabla v \cdot \nabla v \, dx \Big/ \int_{\Omega_\ell} v^2 \, dx \right\}.$$

Thus from (6.32) we get

$$\mu_1 \le \lambda_\ell^k \le \int_{\Omega_\ell} A\nabla u \cdot \nabla u \, dx \Big/ \int_{\Omega_\ell} u^2 \, dx.$$

Since the u_i have disjoint supports it follows

$$\mu_1 \le \lambda_\ell^k \le \sum_{i=1}^k \alpha_i^2 \int_{Q_i} A\nabla u_i \cdot \nabla u_i \, dx \Big/ \sum_{i=1}^k \alpha_i^2 \int_{Q_i} u_i^2 \, dx$$

$$= \sum_{i=1}^k \alpha_i^2 \lambda^1_{Q_i}(A) \Big/ \sum_{i=1}^k \alpha_i^2$$

(recall that u_i is such that $\int_{Q_i} u_i^2 \, dx = 1$). Going back to (6.31) we obtain

$$\mu_1 \le \lambda_\ell^k \le \mu_1 + \frac{k^2|A_{11}|_\infty \nu_1}{\ell^2}$$

which completes the proof of the theorem. $\qquad\square$

6.3   Convergence of the eigenfunctions

We are coming back here to the situation and the notation of Section 6.1. As we have seen in the introduction (Section 1.5) the convergence of the eigenfunctions depends strongly on the scaling of them – recall that they are defined up to a multiplicative constant. To stress further this point we consider the case where the eigenfunctions are from separated variables i.e. the case when

$$A = \begin{pmatrix} A_{11}(X_1) & 0 \\ 0 & A_{22}(X_2) \end{pmatrix}. \tag{6.33}$$

A satisfying the assumptions of the preceding sections, in particular for some constants $\lambda, \Lambda > 0$ we have

$$\lambda |\xi|^2 \le A(x)\xi \cdot \xi \le \Lambda |\xi|^2 \quad \text{a.e. } x \in \mathbb{R}^n, \ \forall \xi \in \mathbb{R}^n. \tag{6.34}$$

Let us denote by λ_ℓ the first eigenvalue of the operator $-\nabla_{X_1} \cdot (A_{11}(X_1)\nabla_{X_1} \cdot)$ on $\ell\omega_1$ with Dirichlet boundary conditions – i.e.

$$\lambda_\ell = \inf \left\{ \int_{\ell\omega_1} A_{11}\nabla_{X_1} u \cdot \nabla_{X_1} u \, dX_1 \mid u \in H_0^1(\ell\omega_1), \int_{\ell\omega_1} u^2 \, dX_1 = 1 \right\}.$$

We denote by u_ℓ the first eigenfunction corresponding to λ_ℓ, positive and such that

$$\int_{\ell\omega_1} u_\ell^2 \, dX_1 = \ell^p.$$

We introduce v_ℓ as

$$v_\ell(y) = u_\ell(\ell y) \quad \text{a.e.} \quad y \in \omega_1. \tag{6.35}$$

Then we have:

Proposition 6.1. *$\ell^2 \lambda_\ell$ is the first eigenvalue of the operator*

$$-\nabla \cdot (A_{11}(\ell y)\nabla \cdot)$$

on ω_1 with Dirichlet boundary conditions and v_ℓ is the first positive normalized eigenfunction, i.e. we have in particular

$$\begin{cases} \int_{\omega_1} A_{11}(\ell y)\nabla v_\ell \cdot \nabla v \, dy = \ell^2 \lambda_\ell \int_{\omega_1} v_\ell v \, dy \quad \forall v \in H_0^1(\omega_1), \\ v_\ell \in H_0^1(\omega_1), \int_{\omega_1} v_\ell^2 \, dy = 1. \end{cases} \tag{6.36}$$

(∇ is the gradient in the y-variable).

Proof. Clearly $v_\ell \in H_0^1(\omega_1)$. Moreover we have

$$\int_{\omega_1} v_\ell^2(y)\, dy = \int_{\omega_1} u_\ell^2(\ell y)\, dy = \int_{\ell\omega_1} u_\ell^2(X_1)\frac{dX_1}{\ell^p} = 1.$$

By definition of λ_ℓ and u_ℓ we have

$$\lambda_\ell = \frac{\int_{\ell\omega_1} A_{11}(X_1)\nabla_{X_1} u_\ell \cdot \nabla_{X_1} u_\ell\, dX_1}{\int_{\ell\omega_1} u_\ell^2\, dX_1}$$

$$= \int_{\ell\omega_1} A_{11}(X_1)\nabla_{X_1} u_\ell \cdot \nabla_{X_1} u_\ell \frac{dX_1}{\ell^p}$$

$$\leq \int_{\ell\omega_1} A_{11}(X_1)\nabla v \cdot \nabla v \frac{dX_1}{\ell^p} \qquad \forall\, v \in H_0^1(\ell\omega_1),\ \int_{\ell\omega_1} v^2\, dX_1 = \ell^p.$$

One notices that by (6.35)

$$\nabla v_\ell(y) = \ell \nabla_{X_1} u_\ell(\ell y).$$

Moreover for every $w \in H_0^1(\omega_1)$ with $\int_{\omega_1} w^2\, dy = 1$ we have

$$v(X_1) = w\left(\frac{X_1}{\ell}\right) \in H_0^1(\ell\omega_1), \qquad \nabla_{X_1} v(X_1) = \frac{1}{\ell}\nabla w\left(\frac{X_1}{\ell}\right)$$

and

$$\int_{\ell\omega_1} v^2(X_1)\, dX_1 = \ell^p.$$

From these remarks we derive for every $w \in H_0^1(\omega_1)$, $\int_{\omega_1} w^2\, dy = 1$:

$$\lambda_\ell = \frac{1}{\ell^2}\int_{\omega_1} A_{11}(\ell y)\nabla v_\ell \cdot \nabla v_\ell\, dy \leq \frac{1}{\ell^2}\int_{\omega_1} A_{11}(\ell y)\nabla w \cdot \nabla w\, dy. \qquad (6.37)$$

This shows that

$$\ell^2 \lambda_\ell \text{ is the first eigenvalue of } -\nabla \cdot (A_{11}(\ell y)\nabla\cdot) \text{ on } \omega_1,$$

$$v_\ell \text{ is a first eigenfunction of } -\nabla \cdot (A_{11}(\ell y)\nabla\cdot) \text{ on } \omega_1,$$

for the Dirichlet boundary conditions and this completes the proof of the proposition. $\qquad\square$

Remark 6.2. Applying this result for $A_{11} = Id$ where Id denotes the $p \times p$-Identity matrix we get

$$\inf_{v \in H_0^1(\ell\omega_1)} \frac{\int_{\ell\omega_1} |\nabla_{X_1} v|^2\, dX_1}{\int_{\ell\omega_1} v^2\, dX_1} = \frac{1}{\ell^2} \inf_{v \in H_0^1(\omega_1)} \frac{\int_{\omega_1} |\nabla_{X_1} v|^2\, dX_1}{\int_{\omega_1} v^2\, dX_1} = \frac{\nu_1}{\ell^2} \qquad (6.38)$$

where ν_1 denotes the first eigenvalue of the Laplace operator in ω_1 with Dirichlet boundary conditions.

We now study the asymptotic behaviour of v_ℓ. For this we suppose that

$$\lim_{|X_1| \to +\infty} A_{11}(X_1) = A_\infty \tag{6.39}$$

where A_∞ is a constant $p \times p$ matrix. Then we have

Proposition 6.2. *Under the assumption* (6.39)

$$\ell^2 \lambda_\ell \to \lambda_\infty, \qquad v_\ell \to v_\infty \text{ in } H_0^1(\omega_1)$$

where λ_∞ is the first eigenvalue, v_∞ the first positive normalized eigenfunction of the operator $-\nabla \cdot (A_\infty \nabla \cdot)$ with Dirichlet boundary conditions on ω_1 – i.e. in particular

$$\begin{cases} \displaystyle\int_{\omega_1} A_\infty \nabla v_\infty \cdot \nabla v \, dy = \lambda_\infty \int_{\omega_1} v_\infty v \, dy \quad \forall v \in H_0^1(\omega_1), \\ v_\infty \in H_0^1(\omega_1), \int_{\omega_1} v_\infty^2 \, dy = 1. \end{cases}$$

Proof. One has

$$\int_{\omega_1} A_{11}(\ell y) \nabla v_\ell \cdot \nabla v_\ell \, dy \le \int_{\omega_1} A_{11}(\ell y) \nabla_{X_1} w \cdot \nabla_{X_1} w \, dy$$

$$\le \Lambda \int_{\omega_1} |\nabla_{X_1} w|^2 \, dy \quad \forall w \in H_0^1(\omega_1), \int_{\omega_1} w^2 = 1.$$

Taking the infimum in w, we derive

$$\ell^2 \lambda_\ell = \int_{\omega_1} A_{11}(\ell y) \nabla v_\ell \cdot \nabla v_\ell \, dy \le \Lambda \nu_1$$

therefore $\ell^2 \lambda_\ell$ is bounded. Moreover we have

$$\lambda \int_{\omega_1} |\nabla v_\ell|^2 \, dy \le \int_{\omega_1} A_{11}(\ell y) \nabla v_\ell \cdot \nabla v_\ell \, dy = \ell^2 \lambda_\ell \le \Lambda \nu_1.$$

Thus v_ℓ is also bounded in $H_0^1(\omega_1)$ and up to a "subsequence" we can assume that for some λ_∞, v_∞ we have that

$$\ell^2 \lambda_\ell \to \lambda_\infty, \tag{6.40}$$

$$v_\ell \rightharpoonup v_\infty \text{ in } H_0^1(\omega_1), \qquad v_\ell \to v_\infty \text{ in } L^2(\omega_1).$$

We would like to identify these limits. First by (6.39) we have

$$A_{11}(\ell y) \nabla v \to A_\infty \nabla v \quad \text{in } L^2(\omega_1)$$

for any $v \in H_0^1(\omega_1)$. Thus passing to the limit in (6.36) – recall that A_{11} is symmetric – we obtain

$$\begin{cases} \displaystyle\int_{\omega_1} A_\infty \nabla v_\infty \cdot \nabla v \, dy = \lambda_\infty \int_{\omega_1} v_\infty v \, dy \quad \forall v \in H_0^1(\omega_1), \\ v_\infty \in H_0^1(\omega_1), \int_{\omega_1} v_\infty^2 \, dy = 1. \end{cases} \tag{6.41}$$

Thus v_∞ is an eigenfunction of $-\nabla \cdot (A_\infty \nabla \cdot)$. We also have by (6.36)

$$\int_{\omega_1} A_{11}(\ell y) \nabla v_\ell \cdot \nabla v_\ell \, dy = \ell^2 \lambda_\ell \to \lambda_\infty.$$

It follows that

$$\int_{\omega_1} A_{11}(\ell y) \nabla(v_\ell - v_\infty) \cdot \nabla(v_\ell - v_\infty) \, dy$$

$$= \int_{\omega_1} A_{11}(\ell y) \nabla v_\ell \cdot \nabla v_\ell \, dy - 2 \int_{\omega_1} A_{11}(\ell y) \nabla v_\infty \cdot \nabla v_\ell$$

$$+ \int_{\omega_1} A_{11}(\ell y) \nabla v_\infty \cdot \nabla v_\infty \, dy$$

$$\to \lambda_\infty - 2 \int_{\omega_1} A_\infty \nabla v_\infty \cdot \nabla v_\infty \, dy + \int_{\omega_1} A_\infty \nabla v_\infty \cdot \nabla v_\infty \, dy$$

$$= \lambda_\infty - \int_{\omega_1} A_\infty \nabla v_\infty \cdot \nabla v_\infty.$$

It results from (6.41) – with $v = v_\infty$ – that this last quantity is 0 and therefore

$$v_\ell \to v_\infty \quad \text{in} \quad H_0^1(\omega_1). \tag{6.42}$$

Going back to (6.37) we have

$$\ell^2 \lambda_\ell \leq \int_{\omega_1} A_{11}(\ell y) \nabla w \cdot \nabla w \, dy \quad \forall \, w \in H_0^1(\omega_1), \int_{\omega_1} w^2 \, dy = 1.$$

Passing to the limit we get

$$\lambda_\infty = \int_{\omega_1} A_\infty \nabla v_\infty \cdot \nabla v_\infty \, dy$$

$$\leq \int_{\omega_1} A_\infty \nabla w \cdot \nabla w \, dy \quad \forall \, w \in H_0^1(\omega_1), \int_{\omega_1} w^2 \, dy = 1.$$

This shows that λ_∞ is the first eigenvalue of the operator $-\nabla \cdot (A_\infty \nabla \cdot)$ and v_∞ its first normalized (positive) eigenfunction. Since λ_∞, v_∞ are uniquely determined we have obtained that (6.40), (6.42) hold for every "subsequence" and thus the whole "sequences" converge. This completes the proof of the proposition. □

In addition we can show:

Proposition 6.3. *Under the assumptions of Proposition 6.2 it holds that*

$$|v_\ell - v_\infty|_{\infty,\omega_1} \to 0$$

when $\ell \to +\infty$. *$| \cdot |_{\infty,\omega_1}$ denotes the usual $L^\infty(\omega_1)$-norm.*

Proof. First it is clear from the usual regularity results for elliptic equations that v_∞ is a smooth function. By definition of v_ℓ and v_∞ we have in a weak sense

$$-\nabla \cdot (A_{11}(\ell y)\nabla v_\ell) - \ell^2 \lambda_\ell v_\ell = 0 = -\nabla \cdot (A_\infty \nabla v_\infty) - \lambda_\infty v_\infty.$$

This implies

$$\begin{aligned}- \nabla \cdot (A_{11}(\ell y)\nabla(v_\ell - v_\infty)) &- \ell^2 \lambda_\ell(v_\ell - v_\infty) \\ &= \nabla \cdot ((A_{11}(\ell y) - A_\infty)\nabla v_\infty) - (\lambda_\infty - \ell^2 \lambda_\ell)v_\infty.\end{aligned}$$

Since $\ell^2 \lambda_\ell$ is bounded away from 0 independently of ℓ, using the well known L^∞-estimates of Stampacchia (see [60] or [72]), we obtain

$$\begin{aligned}|v_\ell - v_\infty|_{\infty,\omega_1} \leq C\big\{ &\big||(A_{11}(\ell y) - A_\infty)|\nabla v_\infty|\big|_{p,\omega_1} + \\ &|\lambda_\infty - \ell^2\lambda_\ell||v_\infty|_{p,\omega_1} + |v_\ell - v_\infty|_{2,\omega_1}\big\}\end{aligned}$$

where C is a constant independent of ℓ, $|\cdot|_{p,\omega_1}$ denotes the usual $L^p(\omega_1)$-norm, $p > n$. The results follow then from the Lebesgue theorem and Proposition 6.2. $\qquad\square$

We can now state a convergence theorem for the eigenfunctions. We have

Theorem 6.4 (Convergence of the eigenfunction). *Let A be a symmetric matrix satisfying* (6.33), (6.34), (6.39). *Let $u_{1,\ell}$ be the first eigenfunction (positive) and normalized by*

$$\int_{\Omega_\ell} u_{1,\ell}^2 \, dx = \ell^p, \tag{6.43}$$

corresponding to λ_ℓ^1 (see (6.8)). *Let w_1 be the first positive eigenfunction realizing the infimum of* (6.9). *Then for any $\ell_0 > 0$ we have*

$$|u_{1,\ell} - v_\infty(0)w_1|_{\infty,\Omega_{\ell_0}} \to 0$$

when $\ell \to +\infty$. $|\cdot|_{\infty,\Omega_{\ell_0}}$ is the usual $L^\infty(\Omega_{\ell_0})$-norm, v_∞ has been defined in (6.41).

Proof. We claim that

$$u_{1,\ell} = u_\ell w_1 \tag{6.44}$$

where u_ℓ has been defined above (6.35). It is indeed clear that $u_\ell w_1 \in H_0^1(\Omega_\ell)$ and that

$$\int_{\Omega_\ell} (u_\ell w_1)^2 \, dx = \ell^p.$$

Moreover for $v \in H_0^1(\Omega_\ell)$ it holds

$$\int_{\Omega_\ell} A\nabla v \cdot \nabla v \, dx = \int_{\Omega_\ell} A_{11}\nabla_{X_1} v \cdot \nabla_{X_1} v \, dx + \int_{\Omega_\ell} A_{22}\nabla_{X_2} v \cdot \nabla_{X_2} v \, dx$$

$$= \int_{\omega_2} \int_{\ell\omega_1} A_{11}(X_1)\nabla_{X_1} v \cdot \nabla_{X_1} v \, dX_1 \, dX_2$$

$$+ \int_{\ell\omega_1} \int_{\omega_2} A_{22}(X_2)\nabla_{X_2} v \cdot \nabla_{X_2} v \, dX_2 \, dX_1.$$

Using the definitions of λ_ℓ, μ_1 respectively (and the fact that $v(\cdot, X_2) \in H_0^1(\ell\omega_1)$ a.e. X_2) we get

$$\int_{\Omega_\ell} A\nabla v \cdot \nabla v \, dx \geq \int_{\omega_2} \lambda_\ell \int_{\ell\omega_1} v^2 \, dX_1 \, dX_2 + \int_{\ell\omega_1} \mu_1 \int_{\omega_2} v^2 \, dX_2 \, dX_1$$

$$= (\lambda_\ell + \mu_1) \int_{\Omega_\ell} v^2 \, dX_1 \, dX_2.$$

Now one has also

$$\int_{\Omega_\ell} A\nabla(u_\ell w_1) \cdot \nabla(u_\ell w_1) \, dx = \int_{\omega_2} \int_{\ell\omega_1} A_{11}(X_1)\nabla_{X_1} u_\ell \cdot \nabla_{X_1} u_\ell \cdot w_1^2 \, dx$$

$$+ \int_{\ell\omega_1} \int_{\omega_2} A_{22}(X_2)\nabla_{X_2} w_1 \cdot \nabla_{X_2} w_1 u_\ell^2 \, dx$$

$$= (\lambda_\ell + \mu_1) \int_{\Omega_\ell} (u_\ell w_1)^2 \, dx.$$

This shows that

$$\lambda_\ell^1 = \lambda_\ell + \mu_1, \qquad u_{1,\ell} = u_\ell w_1.$$

Using now (6.35) we have

$$u_{1,\ell}(x) - v_\infty(0)w_1 = u_\ell(X_1)w_1(X_2) - v_\infty(0)w_1(X_2)$$

$$= \left\{ v_\ell\left(\frac{X_1}{\ell}\right) - v_\infty(0) \right\} w_1(X_2)$$

$$= \left\{ v_\ell\left(\frac{X_1}{\ell}\right) - v_\infty\left(\frac{X_1}{\ell}\right) + v_\infty\left(\frac{X_1}{\ell}\right) - v_\infty(0) \right\} w_1(X_2)$$

and thus

$$|u_{1,\ell}(x) - v_\infty(0)w_1|_{\infty,\Omega_{\ell_0}}$$

$$\leq |w_1|_{\infty,\omega_2} \left\{ |v_\ell - v_\infty|_{\infty,\omega_1} + \left| v_\infty\left(\frac{X_1}{\ell}\right) - v_\infty(0) \right|_{\infty,\ell_0\omega_1} \right\}.$$

The result follows then from Proposition 6.3 and from the continuity of v_∞. $\qquad\square$

Remark 6.3. The scaling of the eigenfunctions is a very delicate issue. For instance (6.43) is not enough to insure the convergence of $u_{1,\ell}$ toward w_1, one has to further scale by $v_\infty(0)$ a constant depending on the behaviour of the operator at ∞. From (6.44) we always have

$$u_{1,\ell} = v_\ell\left(\frac{X_1}{\ell}\right)w_1(X_2).$$

Now the behaviour of v_ℓ is not so easy to control. For instance (see (6.36)) if A_{11} is periodic then v_ℓ is the solution of a homogenization problem for which only weak convergence is known (see [47]). We refer the reader to [33] for results regarding the convergence of the k-eigenfunctions.

6.4   An application

We consider here a symmetric matrix defined on $\mathbb{R}^p \times \omega_2$ by (6.3) i.e. such that

$$A = A(X_1, X_2) = \begin{pmatrix} A_{11}(X_1, X_2) & A_{12}(X_2) \\ A_{12}^T(X_2) & A_{22}(X_2) \end{pmatrix}.$$

We will assume that (6.4), (6.5) hold. Let $\beta(X_2, u)$ be a Carathéodory function (i.e. measurable in X_2 for every u and continuous in u for a.e. X_2) defined on $\omega_2 \times \mathbb{R}$ and satisfying

$$\beta(X_2, 0) = 0 \quad \text{a.e. } X_2 \in \omega_2, \tag{6.45}$$

and

$$-\mu(u - v)^2 \le (\beta(X_2, u) - \beta(X_2, v))(u - v)$$
$$\le \mu'(u - v)^2 \quad \text{a.e. } X_2 \in \omega_2, \forall u, v, \in \mathbb{R} \tag{6.46}$$

where μ, μ' are two positive constants and μ is such that

$$\mu < \mu_1 \tag{6.47}$$

where μ_1 is the first eigenvalue of the operator

$$-\nabla_{X_2} \cdot (A_{22}\nabla_{X_2} \cdot)$$

for the Dirichlet problem in ω_2.

Then we have

Theorem 6.5. *Let* $f = f(X_2) \in L^2(\omega_2)$. *There exist* u_ℓ, u_∞ *unique weak solutions respectively to*

$$-\,-\nabla \cdot (A\nabla u_\ell) + \beta(X_2, u_\ell) = f \quad in \ \Omega_\ell, \quad u_\ell = 0 \quad on \ \partial\Omega_\ell \tag{6.48}$$

and

$$-\nabla_{X_2} \cdot (A_{22}\nabla_{X_2} u_\infty) + \beta(X_2, u_\infty) = f \quad in \ \omega_2, \quad u_\infty = 0 \quad on \ \partial\omega_2. \tag{6.49}$$

$(\Omega_\ell = \ell\omega_1 \times \omega_2).$

Proof. We are giving the proof for u_ℓ, the proof for u_∞ being similar. By a weak solution to (6.48) we mean a function u_ℓ satisfying

$$\begin{cases} u_\ell \in H_0^1(\Omega_\ell), \\ \int_{\Omega_\ell} A\nabla u_\ell \cdot \nabla v + \beta(X_2, u_\ell)v dx = \int_{\Omega_\ell} fv dx \ \forall v \in H_0^1(\Omega_\ell). \end{cases} \quad (6.50)$$

First notice that by (6.46) written for $v = 0$ one has

$$-\mu u^2 \le \beta(X_2, u)u \le \mu' u^2 \quad \text{a.e. } X_2 \in \omega_2, \forall u \in \mathbb{R}$$

i.e.

$$|\beta(X_2, u)| \le (\mu' \vee \mu)|u| \quad \text{a.e. } X_2 \in \omega_2, \forall u \in \mathbb{R},$$

($\vee$ the maximum of two numbers). Thus for $u \in L^2(\Omega_\ell)$ we have $\beta(X_2, u) \in L^2(\Omega_\ell)$. This implies that the first integral in (6.50) makes sense and one can define an operator $\mathcal{A} : H_0^1(\Omega_\ell) \to H^{-1}(\Omega_\ell)$ by setting

$$\langle \mathcal{A}(u), v \rangle = \int_{\Omega_\ell} A\nabla u \cdot \nabla v + \beta(X_2, u)v dx.$$

One has for $u_1, u_2 \in H_0^1(\Omega_\ell)$

$$\langle \mathcal{A}(u_1) - \mathcal{A}(u_2), u_1 - u_2 \rangle$$

$$= \int_{\Omega_\ell} A\nabla(u_1 - u_2) \cdot \nabla(u_1 - u_2) + (\beta(X_2, u_1) - \beta(X_2, u_2))(u_1 - u_2)dx$$

$$\ge \int_{\Omega_\ell} A\nabla(u_1 - u_2) \cdot \nabla(u_1 - u_2)dx - \mu \int_{\Omega_\ell} (u_1 - u_2)^2 dx.$$

By definition of λ_ℓ^1 (see (6.8)) one has

$$\int_{\Omega_\ell} (u_1 - u_2)^2 dx \le \frac{1}{\lambda_\ell^1} \int_{\Omega_\ell} A\nabla(u_1 - u_2) \cdot \nabla(u_1 - u_2)dx$$

which leads to

$$\langle \mathcal{A}(u_1) - \mathcal{A}(u_2), u_1 - u_2 \rangle \ge (1 - \frac{\mu}{\lambda_\ell^1}) \int_{\Omega_\ell} A\nabla(u_1 - u_2) \cdot \nabla(u_1 - u_2)dx$$

$$\ge (1 - \frac{\mu}{\mu_1})\lambda \int_{\Omega_\ell} |\nabla(u_1 - u_2)|^2 dx.$$

It is clear that $\mathcal{A}$ is monotone, hemicontinuous and coercive and the existence and uniqueness of u_ℓ follows from classical results (Cf. the Appendix). The existence and uniqueness of u_∞ proceeds of the same arguments.

Remark 6.4. The assumption $\beta(X_2, 0) = 0$ can easily be removed by replacing $\beta(X_2, u)$ by $\beta(X_2, u) - \beta(X_2, 0)$ and $f(X_2)$ by $f(X_2) - \beta(X_2, 0)$. Note also that a suitable choice of $\beta(X_2, u)$ is for instance

$$\beta(X_2, u) = a(X_2)u$$

where a is a measurable function satisfying

$$-\mu_1 < -\mu \le a(X_2) \le \mu' \quad \text{a.e. } X_2 \in \omega_2$$

i.e. one can add to the problem studied in Chapter 2 a lower order term.

$\square$

We can now prove:

Theorem 6.6. *Under the assumptions of the previous theorem there exist positive constants C, β such that*

$$\left|\left|\nabla(u_\ell - u_\infty)\right|\right|_{2,\Omega_{\frac{\ell}{2}}} \leq Ce^{-\beta\ell} \tag{6.51}$$

i.e. $u_\ell \to u_\infty$ with an exponential rate of convergence.

Proof. Let $v \in H_0^1(\Omega_\ell)$. For a.e. $X_1 \in \ell\omega_1$ one has

$$v(X_1, \cdot) \in H_0^1(\omega_2).$$

Thus, by definition of u_∞, one has for a.e. $X_1 \in \ell\omega_1$

$$\int_{\omega_2} A_{22}\nabla_{X_2}u_\infty \cdot \nabla_{X_2}v(X_1, \cdot) + \beta(X_2, u_\infty)v(X_1, \cdot)dX_2 = \int_{\omega_2} fv(X_1, \cdot)dX_2.$$

Integrating this inequality in X_1 on $\ell\omega_1$ leads to

$$\int_{\Omega_\ell} A_{22}\nabla_{X_2}u_\infty \cdot \nabla_{X_2}v + \beta(X_2, u_\infty)v\,dx = \int_{\Omega_\ell} fv\,dx. \tag{6.52}$$

Since u_∞ is independent of X_1, one has

$$\begin{aligned}
A\nabla u_\infty \cdot \nabla v &= \begin{pmatrix} A_{11} & A_{12} \\ A_{12}^T & A_{22} \end{pmatrix} \begin{pmatrix} 0 \\ \nabla_{X_2}u_\infty \end{pmatrix} \cdot \begin{pmatrix} \nabla_{X_1}v \\ \nabla_{X_2}v \end{pmatrix} \\
&= \begin{pmatrix} A_{12}\nabla_{X_2}u_\infty \\ A_{22}\nabla_{X_2}u_\infty \end{pmatrix} \cdot \begin{pmatrix} \nabla_{X_1}v \\ \nabla_{X_2}v \end{pmatrix} \\
&= A_{12}\nabla_{X_2}u_\infty \cdot \nabla_{X_1}v + A_{22}\nabla_{X_2}u_\infty \cdot \nabla_{X_2}v.
\end{aligned}$$

Since $A_{12}\nabla_{X_2}u_\infty$ is independent of X_1 and v vanishes on the boundary of Ω_ℓ, one has

$$\begin{aligned}
\int_{\Omega_\ell} A\nabla u_\infty \cdot \nabla v\,dx &= \int_{\Omega_\ell} A_{12}\nabla_{X_2}u_\infty \cdot \nabla_{X_1}v\,dx + \int_{\Omega_\ell} A_{22}\nabla_{X_2}u_\infty \cdot \nabla_{X_2}v\,dx \\
&= \int_{\Omega_\ell} \nabla_{X_1} \cdot (A_{12}\nabla_{X_2}u_\infty v)dx + \int_{\Omega_\ell} A_{22}\nabla_{X_2}u_\infty \cdot \nabla_{X_2}v\,dx \\
&= \int_{\Omega_\ell} A_{22}\nabla_{X_2}u_\infty \cdot \nabla_{X_2}v\,dx.
\end{aligned}$$

(One can first consider v smooth to justify the integration). Thus (6.52) can be written

$$\int_{\Omega_\ell} A\nabla u_\infty \cdot \nabla v + \beta(X_2, u_\infty)v\,dx = \int_{\Omega_\ell} fv\,dx \quad \forall v \in H_0^1(\Omega_\ell). \tag{6.53}$$

Subtracting this from (6.50), one arrives to

$$\int_{\Omega_\ell} A\nabla(u_\ell - u_\infty) \cdot \nabla v + (\beta(X_2, u_\ell) - \beta(X_2, u_\infty))v dx = 0 \quad \forall v \in H_0^1(\Omega_\ell).$$

For $\ell_1 \leq \ell - 1$, let us denote by $\rho = \rho_{\ell_1}(X_1)$ the function introduced in (2.7). Since

$$\rho^2(u_\ell - u_\infty) \in H_0^1(\Omega_\ell)$$

one has

$$\int_{\Omega_\ell} A\nabla(u_\ell - u_\infty) \cdot \nabla(\rho^2(u_\ell - u_\infty)) dx$$

$$+ \int_{\Omega_\ell} (\beta(X_2, u_\ell) - \beta(X_2, u_\infty))\rho^2(u_\ell - u_\infty) dx = 0.$$

(6.54)

For a function v, one has by simple computations

$$A\nabla v \cdot \nabla(\rho^2 v) = A\nabla v \cdot (\rho^2 \nabla v + 2\rho v \nabla \rho)$$

$$= \rho^2 A\nabla v \cdot \nabla v + 2\rho v A\nabla v \cdot \nabla \rho$$

and –since A is symmetric

$$A\nabla(\rho v) \cdot \nabla(\rho v) = A(\rho\nabla v + v\nabla\rho) \cdot (\rho\nabla v + v\nabla\rho)$$

$$= \rho^2 A\nabla v \cdot \nabla v + 2\rho v A\nabla v \cdot \nabla\rho + v^2 A\nabla\rho \cdot \nabla\rho$$

and thus

$$A\nabla v \cdot \nabla(\rho^2 v) = A\nabla(\rho v) \cdot \nabla(\rho v) - v^2 A\nabla\rho \cdot \nabla\rho.$$

Going back to (6.54), we deduce

$$\int_{\Omega_\ell} A\nabla(\rho(u_\ell - u_\infty)) \cdot \nabla(\rho(u_\ell - u_\infty)) dx$$

$$+ \int_{\Omega_\ell} (\beta(X_2, u_\ell) - \beta(X_2, u_\infty))\rho^2(u_\ell - u_\infty) dx$$

$$= \int_{\Omega_\ell} (u_\ell - u_\infty)^2 A\nabla\rho \cdot \nabla\rho dx.$$

Using (6.46) it follows that

$$\int_{\Omega_\ell} A\nabla(\rho(u_\ell - u_\infty)) \cdot \nabla(\rho(u_\ell - u_\infty)) dx$$

$$- \mu \int_{\Omega_\ell} \rho^2(u_\ell - u_\infty)^2 dx$$

$$\leq \int_{\Omega_\ell} (u_\ell - u_\infty)^2 A\nabla\rho \cdot \nabla\rho dx.$$

Since
$$\mu_1 \leq \lambda_\ell^1$$
and
$$\int_{\Omega_\ell} \rho^2 (u_\ell - u_\infty)^2 dx \leq \frac{1}{\lambda_\ell^1} \int_{\Omega_\ell} A\nabla(\rho(u_\ell - u_\infty)) \cdot \nabla(\rho(u_\ell - u_\infty)) dx$$

one deduces
$$(1 - \frac{\mu}{\mu_1}) \int_{\Omega_\ell} A\nabla(\rho(u_\ell - u_\infty)) \cdot \nabla(\rho(u_\ell - u_\infty)) dx \leq \int_{\Omega_\ell} (u_\ell - u_\infty)^2 A\nabla\rho \cdot \nabla\rho \, dx.$$

Since $\rho = \rho_{\ell_1}$ vanishes outside Ω_{ℓ_1+1}, it is enough in these integrals to integrate on Ω_{ℓ_1+1}. Moreover $\nabla\rho = 0$ on Ω_{ℓ_1}, thus we easily obtain by (6.4), (6.5) for different constants C

$$\lambda(1 - \frac{\mu}{\mu_1}) \int_{\Omega_{\ell_1+1}} |\nabla(\rho(u_\ell - u_\infty))|^2 dx \leq C \int_{\Omega_{\ell_1+1} \backslash \Omega_{\ell_1}} (u_\ell - u_\infty)^2$$

$$\leq C \int_{\Omega_{\ell_1+1} \backslash \Omega_{\ell_1}} |\nabla_{X_2}(u_\ell - u_\infty)|^2 dx$$

(in the last inequality we applied the Poincaré inequality on ω_2). Integrating on a smaller domain –namely Ω_{ℓ_1}– in the first integral we obtain for some constant C

$$\int_{\Omega_{\ell_1}} |\nabla(u_\ell - u_\infty)|^2 dx \leq C \int_{\Omega_{\ell_1+1} \backslash \Omega_{\ell_1}} |\nabla(u_\ell - u_\infty)|^2 dx$$

i.e.
$$\int_{\Omega_{\ell_1}} |\nabla(u_\ell - u_\infty)|^2 dx \leq \gamma \int_{\Omega_{\ell_1+1}} |\nabla(u_\ell - u_\infty)|^2 dx$$

where $\gamma = \frac{C}{C+1} < 1$. Proceeding as before –i.e. iterating this formula from $\ell_1 = \frac{\ell}{2}$, $[\frac{\ell}{2}]$ times– one derives easily that

$$\int_{\Omega_{\frac{\ell}{2}}} |\nabla(u_\ell - u_\infty)|^2 dx \leq \frac{1}{\gamma} \gamma^{\frac{\ell}{2}} \int_{\Omega_\ell} |\nabla(u_\ell - u_\infty)|^2 dx. \qquad (6.55)$$

Then one is reduced to estimate u_ℓ and u_∞. For that, taking $v = u_\ell$ in (6.50) one deduces

$$\int_{\Omega_\ell} A\nabla u_\ell \cdot \nabla u_\ell + \beta(X_2, u_\ell) u_\ell dx = \int_{\Omega_\ell} f u_\ell dx.$$

Using (6.45) and (6.46) with $v = 0$ it comes

$$\int_{\Omega_\ell} A\nabla u_\ell \cdot \nabla u_\ell dx - \mu \int_{\Omega_\ell} u_\ell^2 dx \leq \int_{\Omega_\ell} f u_\ell dx.$$

Next using as before the estimate

$$\int_{\Omega_\ell} u_\ell^2 dx \leq \frac{1}{\lambda_\ell^1} \int_{\Omega_\ell} A\nabla u_\ell \cdot \nabla u_\ell dx \leq \frac{1}{\mu_1} \int_{\Omega_\ell} A\nabla u_\ell \cdot \nabla u_\ell dx$$

one derives

$$(1 - \frac{\mu}{\mu_1}) \int_{\Omega_\ell} A\nabla u_\ell \cdot \nabla u_\ell dx \leq \int_{\Omega_\ell} f u_\ell dx.$$

Using then (6.4) and the Poincaré inequality on ω_2, one gets

$$\lambda(1 - \frac{\mu}{\mu_1}) \int_{\Omega_\ell} |\nabla u_\ell|^2 dx \leq C|f|_{2,\Omega_\ell} (\int_{\Omega_\ell} |\nabla u_\ell|^2 dx)^{\frac{1}{2}}$$

i.e. for some other constant C

$$\int_{\Omega_\ell} |\nabla u_\ell|^2 dx \leq C|f|_{2,\Omega_\ell}^2 = C \int_{\ell\omega_1} \int_{\omega_2} f^2 dX_1 dX_2$$

$$= C\ell^p |\omega_1| |f|_{2,\omega_2}^2$$

where $|\omega_1|$ denotes the measure of ω_1. The same estimate holds for u_∞ –i.e. one can show following the steps above –see (6.52)

$$\int_{\Omega_\ell} |\nabla u_\infty|^2 dx \leq C\ell^p |\omega_1| |f|_{2,\omega_2}^2.$$

(The dependence in ℓ^p is clear since $|\nabla u_\infty|^2$ is independent of X_1). Going back to (6.55) one deduces then that for some constant C

$$\int_{\Omega_{\frac{\ell}{2}}} |\nabla(u_\ell - u_\infty)|^2 dx \leq C\ell^p \gamma^{\frac{\ell}{2}} = C\ell^p e^{-\frac{\ell}{2} \ln \frac{1}{\gamma}}.$$

the result follows by choosing $\beta < \frac{\ell}{4} \ln \frac{1}{\gamma}$. □

6.5 The case of Neumann boundary conditions

As mentioned in the remark 2.3 the boundary conditions on $\ell\partial\omega_1 \times \omega_2$ are playing no role when passing to the limit in ℓ. Thus in the particular case of Neumann boundary conditions, one expects for the eigenvalues of the operator the same behaviour as for Dirichlet boundary conditions. As we are about to see this is not true. Let us start by a lemma of linear algebra. We consider a $n \times n$-matrix A

$$A = \begin{pmatrix} A_{11} & A_{12} \\ A_{12}^T & A_{22} \end{pmatrix}$$

where A_{11} is a $p \times p$-matrix and A_{22} a $(n-p) \times (n-p)$, T denotes the transposition. One has:

Lemma 6.2. *Suppose that A is symmetric positive definite then*

$$A\xi \cdot \xi \geq (A_{22} - A_{12}^T A_{11}^{-1} A_{12})\xi_2 \cdot \xi_2 \quad \forall \xi = (\xi_1, \xi_2)^T \tag{6.56}$$

($\xi_1 \in \mathbb{R}^p$, $\xi_2 \in \mathbb{R}^{(n-p)}$) moreover the matrix

$$A_{22} - A_{12}^T A_{11}^{-1} A_{12}$$

is symmetric positive definite.

Proof. One has

$$A\xi \cdot \xi = \begin{pmatrix} A_{11} & A_{12} \\ A_{12}^T & A_{22} \end{pmatrix} \begin{pmatrix} \xi_1 \\ \xi_2 \end{pmatrix} \cdot \begin{pmatrix} \xi_1 \\ \xi_2 \end{pmatrix} = \begin{pmatrix} A_{11}\xi_1 + A_{12}\xi_2 \\ A_{12}^T\xi_1 + A_{22}\xi_2 \end{pmatrix} \cdot \begin{pmatrix} \xi_1 \\ \xi_2 \end{pmatrix}$$

$$= A_{11}\xi_1 \cdot \xi_1 + 2A_{12}\xi_2 \cdot \xi_1 + A_{22}\xi_2 \cdot \xi_2.$$

Thus the inequality (6.56) reads

$$A_{11}\xi_1 \cdot \xi_1 + 2A_{12}\xi_2 \cdot \xi_1 + A_{12}^T A_{11}^{-1} A_{12}\xi_2 \cdot \xi_2 \geq 0$$

or equivalently

$$S := A_{11}\xi_1 \cdot \xi_1 + 2A_{12}\xi_2 \cdot \xi_1 + A_{11}^{-1} A_{12}\xi_2 \cdot A_{12}\xi_2 \geq 0.$$

Since A is positive definite so is A_{11} and its square root $A_{11}' = (A_{11})^{\frac{1}{2}}$ is well defined and symmetric positive definite. Thus it comes

$$S := A_{11}'\xi_1 \cdot A_{11}'\xi_1 + 2A_{11}'(A_{11}')^{-1}A_{12}\xi_2 \cdot \xi_1 + ((A_{11}')^{-1})^2 A_{12}\xi_2 \cdot A_{12}\xi_2$$

$$= A_{11}'\xi_1 \cdot A_{11}'\xi_1 + 2(A_{11}')^{-1}A_{12}\xi_2 \cdot A_{11}'\xi_1$$

$$+ (A_{11}')^{-1}A_{12}\xi_2 \cdot (A_{11}')^{-1}A_{12}\xi_2$$

$$= (A_{11}'\xi_1 + (A_{11}')^{-1}A_{12}\xi_2) \cdot (A_{11}'\xi_1 + (A_{11}')^{-1}A_{12}\xi_2) \geq 0.$$

This completes the proof of (6.56).

Next we notice that for $\xi_2 \in \mathbb{R}^{(n-p)}$ one has $-A_{11}^{-1}A_{12}\xi_2 \in \mathbb{R}^p$. Thus if λ denotes the ellipticity constant of A we have then

$$A \begin{pmatrix} -A_{11}^{-1}A_{12}\xi_2 \\ \xi_2 \end{pmatrix} \cdot \begin{pmatrix} -A_{11}^{-1}A_{12}\xi_2 \\ \xi_2 \end{pmatrix} \geq \lambda \{ |-A_{11}^{-1}A_{12}\xi_2|^2 + |\xi_2|^2 \} \tag{6.57}$$

$$\geq \lambda |\xi_2|^2 \quad \forall \xi_2 \in \mathbb{R}^{(n-p)}.$$

Computing now the left hand side of this inequality it comes

$$\begin{pmatrix} A_{11} & A_{12} \\ A_{12}^T & A_{22} \end{pmatrix} \begin{pmatrix} -A_{11}^{-1}A_{12}\xi_2 \\ \xi_2 \end{pmatrix} \cdot \begin{pmatrix} -A_{11}^{-1}A_{12}\xi_2 \\ \xi_2 \end{pmatrix}$$

$$= \begin{pmatrix} 0 \\ -A_{12}^T A_{11}^{-1} A_{12}\xi_2 + A_{22}\xi_2 \end{pmatrix} \cdot \begin{pmatrix} -A_{11}^{-1}A_{12}\xi_2 \\ \xi_2 \end{pmatrix}$$

$$= (A_{22} - A_{12}^T A_{11}^{-1} A_{12})\xi_2 \cdot \xi_2.$$

Combining this with (6.57) we get

$$(A_{22} - A_{12}^T A_{11}^{-1} A_{12})\xi_2 \cdot \xi_2 \geq \lambda |\xi_2|^2 \quad \forall \xi_2 \in \mathbb{R}^{(n-p)}.$$

This completes the proof of the lemma. $\qquad\square$

Considering $\Omega_\ell = \ell\omega_1 \times \omega_2$ we define

$$V_\ell = \{v \in H^1(\Omega_\ell) \mid v = 0 \text{ on } \ell\omega_1 \times \partial\omega_2\}$$

and for

$$A = A(X_2) = \begin{pmatrix} A_{11}(X_2) & A_{12}(X_2) \\ A_{12}^T(X_2) & A_{22}(X_2) \end{pmatrix}$$

bounded and uniformly elliptic i.e. satisfying (6.4), (6.5) we define

$$\lambda_\ell^1 = \inf\{\int_{\Omega_\ell} A\nabla v \cdot \nabla v dx \mid v \in V_\ell, \int_{\Omega_\ell} v^2 dx = 1\}$$

the first eigenvalue of the mixed boundary value problem associated to A and Ω_ℓ. We denote by u_ℓ the corresponding first positive normalized eigenfunction. We introduce also (see Lemma 6.2)

$$\mu_1' = \inf\{\int_{\omega_2} (A_{22} - A_{12}^T A_{11}^{-1} A_{12})\nabla_{X_2} v \cdot \nabla_{X_2} v dX_2$$

$$\mid v \in H_0^1(\omega_2), \int_{\omega_2} v^2 dX_2 = 1\}. \tag{6.58}$$

We are going to consider the behaviour of λ_ℓ^1 when $\ell \to 0$. We have

Theorem 6.7. *Under the above assumptions one has*

$$\lambda_\ell^1 \geq \mu_1' \tag{6.59}$$

and if ω_1 is symmetric with respect to 0

$$\lim_{\ell \to 0} \lambda_\ell^1 = \mu_1'. \tag{6.60}$$

Proof. From (6.56) one has

$$\lambda_\ell^1 = \int_{\Omega_\ell} A\nabla u_\ell \cdot \nabla u_\ell dx \geq \int_{\Omega_\ell} (A_{22} - A_{12}^T A_{11}^{-1} A_{12})\nabla_{X_2} u_\ell \cdot \nabla_{X_2} u_\ell dx$$

$$= \int_{\ell\omega_1} \int_{\omega_2} (A_{22} - A_{12}^T A_{11}^{-1} A_{12})\nabla_{X_2} u_\ell \cdot \nabla_{X_2} u_\ell dX_2 dX_1.$$

From the definition of μ_1' and since $u_\ell(X_1, \cdot) \in H_0^1(\omega_2)$ it comes

$$\lambda_\ell^1 \geq \int_{\ell\omega_1} \mu_1' \int_{\omega_2} u_\ell^2 dX_2 dX_1 = \mu_1' \int_{\Omega_\ell} u_\ell^2 dx = \mu_1'. \tag{6.61}$$

This proves (6.59).

Next we denote by u_∞' the first eigenfunction associated to (6.58) i.e. the positive function achieving the infimum (6.58). We set

$$v_\ell = u_\infty' - (A_{12}^T A_{11}^{-1} X_1 \cdot \nabla_{X_2} u_\infty')\rho_\ell(X_2),$$

$$L(X_2)(X_1) = (A_{12}^T A_{11}^{-1} X_1 \cdot \nabla_{X_2} u'_\infty)$$

i.e.

$$v_\ell = u'_\infty - L(X_2)(X_1)\rho_\ell(X_2),$$

where ρ_ℓ is the function vanishing on $\partial \omega_2$ and defined as

$$\rho_\ell = \frac{\text{dist}(X_2, \partial \omega_2)}{\ell^\alpha} \wedge 1, \quad 0 < \alpha < 1.$$

(dist$(X_2, \partial \omega_2)$ denotes the distance from X_2 to the boundary of ω_2, $\wedge$ the minimum of two numbers. Since A is depending on X_2 only note that the function $L(X_2)(X_1)$ is linear in X_1).

Assuming A smooth it is clear that $v_\ell \in V_\ell$. However if we assume the coefficients of A being only essentially bounded this is no more the case and some approximation is in order (see [34]). For that purpose assuming that

$$L(X_2)(X_1) = \sum_{k=1}^p g_k(X_2) x_k$$

we will denote by g_k^ε smooth functions with compact support such that

$$g_k^\varepsilon \to g_k \quad \text{in } L^2(\omega_2) \text{ and a.e. in } \omega_2.$$

We set

$$L^\varepsilon(X_2)(X_1) = \sum_{k=1}^p g_k^\varepsilon(X_2) x_k$$

and will work with

$$v_\ell^\varepsilon = u'_\infty - L^\varepsilon(X_2)(X_1)\rho_\ell(X_2) \in V_\ell.$$

First note that

$$\int_{\Omega_\ell} (v_\ell^\varepsilon)^2 dx = \int_{\Omega_\ell} (u'_\infty)^2 - 2u'_\infty L^\varepsilon(X_2)(X_1))\rho_\ell + (L^\varepsilon(X_2)(X_1))\rho_\ell)^2 dx.$$

Since we assumed ω_1 to be symmetric with respect to the origin one has

$$\int_{\Omega_\ell} u'_\infty L^\varepsilon(X_2)(X_1) dx = \int_{\omega_2} u'_\infty L^\varepsilon(X_2)(\int_{\ell\omega_1} X_1 dX_1))\rho_\ell dX_2 = 0$$

and thus

$$\int_{\Omega_\ell} (v_\ell^\varepsilon)^2 dx \geq \int_{\Omega_\ell} (u'_\infty)^2 dx = \int_{\ell\omega_1}\int_{\omega_2} (u'_\infty)^2 dX_2 dX_1 = |\ell\omega_1| \qquad (6.62)$$

where $|\ |$ denotes here the Lebesgue measure in $\mathbb{R}^p$.

Next we compute the gradient of v_ℓ^ε with respect to X_1. One has clearly

$$\nabla_{X_1} v_\ell^\varepsilon = -\rho_\ell(g_1^\varepsilon, \cdots, g_p^\varepsilon) = -\rho_\ell g^\varepsilon \tag{6.63}$$

where we have denoted by g^ε the vector of components g_k^ε.

Thus we have

$$\int_{\Omega_\ell} A\nabla v_\ell^\varepsilon \cdot \nabla v_\ell^\varepsilon dx$$

$$= \int_{\Omega_\ell} A_{11} \nabla_{X_1} v_\ell^\varepsilon \cdot \nabla_{X_1} v_\ell^\varepsilon dx + 2 \int_{\Omega_\ell} A_{12} \nabla_{X_2} v_\ell^\varepsilon \cdot \nabla_{X_1} v_\ell^\varepsilon dx$$

$$+ \int_{\Omega_\ell} A_{22} \nabla_{X_2} v_\ell^\varepsilon \cdot \nabla_{X_2} v_\ell^\varepsilon dx$$

$$= I_1 + I_2 + I_3.$$

Computing each integral separately we get (see (6.63))

$$I_1 = \int_{\Omega_\ell} A_{11} g^\varepsilon \cdot g^\varepsilon \rho_\ell^2 dx = |\ell\omega_1| \int_{\omega_2} A_{11} g^\varepsilon \cdot g^\varepsilon \rho_\ell^2 dX_2.$$

For the second integral we get

$$I_2 = 2 \int_{\Omega_\ell} (A_{12} \nabla_{X_2} u_\infty' - A_{12} \nabla_{X_2}(L(X_2)(X_1)\rho_\ell)) \cdot (-\rho_\ell g^\varepsilon) dx$$

$$= -2 \int_{\Omega_\ell} A_{12} \nabla_{X_2} u_\infty' \cdot g^\varepsilon \rho_\ell dx = -2|\ell\omega_1| \int_{\omega_2} A_{12} \nabla_{X_2} u_\infty' \cdot g^\varepsilon \rho_\ell dX_2.$$

(As before the integral of linear term in X_1 vanishes).

Finally for the third integral we have denoting $L^\varepsilon(X_2)(X_1)$ by L^ε

$$I_3 = \int_{\Omega_\ell} A_{22}(\nabla_{X_2} u_\infty' - \nabla_{X_2}(L^\varepsilon \rho_\ell)) \cdot (\nabla_{X_2} u_\infty' - \nabla_{X_2}(L^\varepsilon \rho_\ell)) dx$$

$$= \int_{\Omega_\ell} A_{22} \nabla_{X_2} u_\infty' \cdot \nabla_{X_2} u_\infty' + A_{22} \nabla_{X_2}(L^\varepsilon \rho_\ell) \cdot \nabla_{X_2}(L^\varepsilon \rho_\ell) dx$$

since the integral of the linear term in X_1 vanishes. Thus

$$I_3 = |\ell\omega_1| \int_{\omega_2} A_{22} \nabla_{X_2} u_\infty' \cdot \nabla_{X_2} u_\infty' dX_2$$

$$+ \int_{\Omega_\ell} A_{22} \nabla_{X_2}(L^\varepsilon \rho_\ell) \cdot \nabla_{X_2}(L^\varepsilon \rho_\ell) dx.$$

Recalling (6.62) we obtain

$$\lambda_\ell^1 \le \frac{\int_{\Omega_\ell} A\nabla v_\ell^\varepsilon \cdot \nabla v_\ell^\varepsilon dx}{\int_{\Omega_\ell} (v_\ell^\varepsilon)^2 dx} \le \frac{I_1 + I_2 + I_3}{|\ell\omega_1|}$$

$$= \int_{\omega_2} A_{11} g^\varepsilon \cdot g^\varepsilon \rho_\ell^2 dX_2 - 2 \int_{\omega_2} A_{12} \nabla_{X_2} u_\infty' \cdot g^\varepsilon \rho_\ell dX_2$$

$$+ \int_{\omega_2} A_{22} \nabla_{X_2} u_\infty' \cdot \nabla_{X_2} u_\infty' dX_2$$

$$+ \frac{1}{|\ell\omega_1|} \int_{\Omega_\ell} A_{22} \nabla_{X_2}(L^\varepsilon \rho_\ell) \cdot \nabla_{X_2}(L^\varepsilon \rho_\ell) dx.$$

We claim that

$$\lim_{\ell \to 0} \frac{1}{|\ell \omega_1|} \int_{\Omega_\ell} A_{22} \nabla_{X_2}(L^\varepsilon \rho_\ell) \cdot \nabla_{X_2}(L^\varepsilon \rho_\ell) dx = 0. \qquad (6.64)$$

Thus, since $\rho_\ell \to 1$ a.e. in ω_2 we get

$$\limsup_{\ell \to 0} \lambda_\ell^1$$

$$\leq \int_{\omega_2} A_{22} \nabla_{X_2} u'_\infty \cdot \nabla_{X_2} u'_\infty + A_{11} g^\varepsilon \cdot g^\varepsilon - 2A_{12} \nabla_{X_2} u'_\infty \cdot g^\varepsilon dX_2.$$

Since this inequality is true for every ε, passing to the limit in ε we obtain

$$\limsup_{\ell \to 0} \lambda_\ell^1 \leq \int_{\omega_2} A_{22} \nabla_{X_2} u'_\infty \cdot \nabla_{X_2} u'_\infty + A_{11} g \cdot g - 2A_{12} \nabla_{X_2} u'_\infty \cdot g dX_2 \quad (6.65)$$

where g denotes the vector with components g_k. If $\gamma_{i,k}$, $i = p+1, \cdots, n$, $k = 1, \cdots, p$ denote the entries of the matrix $A_{12}^T A_{11}^{-1}$. One has

$$L(X_2)(X_1) = \sum_k g_k x_k = \sum_i \sum_k \gamma_{i,k} x_k \partial_{x_i} u'_\infty$$

and thus

$$g_k = \sum_i \gamma_{i,k} \partial_{x_i} u'_\infty$$

i.e. since A_{11} is symmetric

$$g = (A_{12}^T A_{11}^{-1})^T \nabla_{X_2} u'_\infty = A_{11}^{-1} A_{12} \nabla_{X_2} u'_\infty.$$

Going back to (6.65) we obtain

$$\lim_{\ell \to 0} \lambda_\ell^1 \leq \int_{\omega_2} A_{22} \nabla_{X_2} u'_\infty \cdot \nabla_{X_2} u'_\infty dX_2$$

$$- \int_{\omega_2} A_{12} \nabla_{X_2} u'_\infty \cdot A_{11}^{-1} A_{12} \nabla_{X_2} u'_\infty dX_2$$

$$= \int_{\omega_2} (A_{22} - A_{12}^T A_{11}^{-1} A_{12}) \nabla_{X_2} u'_\infty \cdot \nabla_{X_2} u'_\infty dX_2 = \mu'_1.$$

This combined with (6.61) would complete the proof of the theorem. Thus remains just to prove (6.64). Since the matrix A is bounded one has

$$I = \frac{1}{|\ell \omega_1|} \int_{\Omega_\ell} A_{22} \nabla_{X_2}(L^\varepsilon \rho_\ell)) \cdot \nabla_{X_2}(L^\varepsilon \rho_\ell) dx$$

$$\leq \frac{C}{\ell^p |\omega_1|} \int_{\Omega_\ell} |\nabla_{X_2}(L^\varepsilon \rho_\ell)|^2 dx.$$

One has

$$\nabla_{X_2}(L^\varepsilon \rho_\ell) = \rho_\ell \nabla_{X_2} L^\varepsilon + L^\varepsilon \nabla_{X_2} \rho_\ell,$$

and we get for some constant depending on ε

$$|\nabla_{X_2}(L^\varepsilon \rho_\ell)|^2 \leq C(1 + |\nabla_{X_2}\rho_\ell|^2)|X_1|^2.$$

The estimate for the integral I becomes (for possibly some other constant C, see the definition of ρ_ℓ)

$$I \leq \frac{C}{\ell^p |\omega_1|} \int_{\ell\omega_1} \int_{\omega_2} (1 + \frac{1}{\ell^{2\alpha}})|X_1|^2 dX_2 dX_1.$$

Making the change of variable

$$X_1 = \ell X_1', \quad X_1' \in \omega_1$$

it comes since $dX_1 = \ell^p dX_1'$

$$I \leq \frac{C}{|\omega_1|}(\ell^2 + \ell^{2-2\alpha}) \int_{\omega_1} \int_{\omega_2} |X_1'|^2 dX_2 dX_1'.$$

Since the right hand side of this inequality goes to 0 when $\ell \to 0$ this completes the proof of the theorem. $\qquad\square$

If w_1 denotes the function achieving the infimum (6.9), i.e. the eigenfunction corresponding to μ_1, one has:

Theorem 6.8. *One has $\mu_1' \leq \mu_1$ and $\mu_1' < \mu_1$ if and only if*

$$A_{12}\nabla_{X_2}w_1 \not\equiv 0 \ a.e. \ X_2 \in \omega_2. \tag{6.66}$$

Proof. Since A_{11}, A_{11}^{-1} are positive definite, for every $v \in H_0^1(\omega_2)$ one has

$$\int_{\omega_2} A_{12}^T A_{11}^{-1} A_{12} \nabla_{X_2}v \cdot \nabla_{X_2}v dX_2 \geq 0$$

and thus

$$\int_{\omega_2} (A_{22} - A_{12}^T A_{11}^{-1} A_{12})\nabla_{X_2}v \cdot \nabla_{X_2}v dX_2 \leq \int_{\omega_2} A_{22}\nabla_{X_2}v \cdot \nabla_{X_2}v dX_2.$$

The inequality $\mu_1' \leq \mu_1$ follows by taking the infimum in v. If (6.66) holds one has

$$\mu_1' \leq \int_{\omega_2} (A_{22} - A_{12}^T A_{11}^{-1} A_{12})\nabla_{X_2}w_1 \cdot \nabla_{X_2}w_1 dX_2$$

$$< \int_{\omega_2} A_{22}\nabla_{X_2}w_1 \cdot \nabla_{X_2}w_1 dX_2 = \mu_1$$

and the inequality $\mu_1' < \mu_1$ follows. Else if $A_{12}\nabla_{X_2}w_1 \equiv 0$ it is clear that u_∞' and w_1 are both eigenfunctions of the operator

$$-\nabla_{X_2} \cdot (A_{22} - A_{12}^T A_{11}^{-1} A_{12})\nabla_{X_2}$$

on ω_2, but only the first eigenfunction can be positive and thus $\mu_1' = \mu_1$. This completes the proof of the theorem. $\qquad\square$

Remark 6.5. It is easy to have the condition (6.66) satisfied. For instance if the rank of A_{12} is equal to $n - p$ then (6.66) holds. Else one would have $\nabla_{X_2} w_1 = 0$ and thus $w_1 = 0$ which is impossible.

When we were considering the eigenvalues for Dirichlet boundary conditions we had from (6.10) $\mu_1 \leq \lambda_\ell^1$ for every ℓ. If $\mu_1' < \mu_1$ (6.60) shows that the situation for Neumann boundary condition is different since λ_ℓ^1 –the first eigenvalue for Neumann boundary conditions– is allowed to pass below μ_1 for ℓ small. We are going to precise this in the next theorem. We will restrict ourselves to the case where

$$\Omega_\ell = (-\ell, \ell) \times \omega_2$$

and the matrix A to

$$A = \begin{pmatrix} A_{1,1}(X_2) & A_{12}(X_2) \\ A_{12}(X_2)^T & A_{22}(X_2) \end{pmatrix}$$

with $A_{1,1}(X_2)$ a positive number. Then we have

Theorem 6.9. *Under the assumptions above if*

$$A_{12} \nabla_{X_2} w_1 \not\equiv 0 \ a.e. \ X_2 \in \omega_2$$

one has

$$\limsup_{\ell \to \infty} \lambda_\ell^1 < \mu_1.$$

Proof. First notice that in the proof of Theorem 6.7 in definition of v_ℓ^ε, one can replace u_∞' by w_1 and arrive with the same arguments to

$$\limsup_{\ell \to 0} \lim_{\varepsilon \to 0} \frac{\int_{\Omega_\ell} A \nabla v_\ell^\varepsilon \cdot \nabla v_\ell^\varepsilon dx}{\int_{\Omega_\ell} (v_\ell^\varepsilon)^2 dx}$$

$$= \int_{\omega_2} (A_{22} - A_{12}^T A_{11}^{-1} A_{12}) \nabla_{X_2} w_1 \cdot \nabla_{X_2} w_1 dX_2 < \mu_1.$$

Thus we can select ε_0 and ℓ_0 such that

$$\frac{\int_{\Omega_\ell} A \nabla v_{\ell_0}^{\varepsilon_0} \cdot \nabla v_{\ell_0}^{\varepsilon_0} dx}{\int_{\Omega_\ell} (v_{\ell_0}^{\varepsilon_0})^2 dx} = \mu_0 < \mu_1. \tag{6.67}$$

We define ψ_ℓ as

$$\psi_\ell = \begin{cases} v_{\ell_0}^{\varepsilon_0}(x_1 - \ell + \ell_0, X_2) & \text{on } (\ell - \ell_0, \ell) \times \omega_2, \\ \frac{1}{\eta}(x_1 - \ell + \ell_0 + \eta)w_1 & \text{on } (\ell - \ell_0 - \eta, \ell - \ell_0) \times \omega_2, \\ 0 & \text{on } \Omega_{\ell - \ell_0 - \eta}, \\ \frac{1}{\eta}(-x_1 - \ell + \ell_0 + \eta)w_1 & \text{on } (-\ell + \ell_0, -\ell + \ell_0 + \eta) \times \omega_2, \\ v_{\ell_0}^{\varepsilon_0}(x_1 + \ell - \ell_0, X_2) & \text{on } (-\ell, -\ell + \ell_0) \times \omega_2. \end{cases} \tag{6.68}$$

Note that at $x_1 = \pm(\ell - \ell_0)$ each of the functions above are taking the value w_1, η will be chosen later on. We have

$$\int_{\Omega_\ell} \psi_\ell^2 dx = \int_{\Omega_\ell \setminus \Omega_{\ell-\ell_0}} \psi_\ell^2 dx + \int_{\Omega_{\ell-\ell_0}} \psi_\ell^2 dx = J_1 + J_2.$$

Due to the definition of ψ_ℓ it comes

$$J_1 = \int_{\Omega_{\ell_0}} (v_{\ell_0}^{\varepsilon_0})^2 dx,$$

and

$$J_2 = 2 \int_{\ell-\ell_0-\eta}^{\ell-\ell_0} \frac{1}{\eta^2} (x_1 - \ell + \ell_0 + \eta)^2 dx_1 \int_{\omega_2} w_1^2 dX_2 = \frac{2}{3}\eta,$$

and thus

$$\int_{\Omega_\ell} \psi_\ell^2 dx = \int_{\Omega_{\ell_0}} (v_{\ell_0}^{\varepsilon_0})^2 dx + \frac{2}{3}\eta. \qquad (6.69)$$

Similarly

$$\int_{\Omega_\ell} A\nabla\psi_\ell \cdot \nabla\psi_\ell dx = \int_{\Omega_{\ell_0}} A\nabla v_{\ell_0}^{\varepsilon_0} \cdot \nabla v_{\ell_0}^{\varepsilon_0} dx + \int_{\Omega_{\ell-\ell_0}} A\nabla\psi_\ell \cdot \nabla\psi_\ell dx.$$

Setting $D = \Omega_{\ell-\ell_0} \setminus \Omega_{\ell-\ell_0-\eta}$ the last integral above can be written as

$$\int_{\Omega_{\ell-\ell_0}} A\nabla\psi_\ell \cdot \nabla\psi_\ell dx = \int_D A_{1,1}\frac{1}{\eta^2}w_1^2 dx + 2\int_D A_{12}\nabla_{X_2}\psi_\ell \partial_{x_1}\psi_\ell dx$$

$$+ 2\int_{\ell-\ell_0-\eta}^{\ell-\ell_0} \int_{\omega_2} A_{22}\nabla_{X_2}w_1 \cdot \nabla_{X_2}w_1 \frac{1}{\eta^2}(x_1 - \ell + \ell_0 + \eta)^2 dX_2 dx_1.$$

Since $\partial_{x_1}\psi_\ell$ is odd on D the second integral vanishes and it comes

$$\int_{\Omega_{\ell-\ell_0}} A\nabla\psi_\ell \cdot \nabla\psi_\ell dx = \frac{2}{\eta}\int_{\omega_2} A_{1,1}w_1^2 dX_2 + \frac{2}{3}\eta\int_{\omega_2} A_{22}\nabla_{X_2}w_1 \cdot \nabla_{X_2}w_1 dX_2.$$

Combining this with (6.69) we obtain dropping some measures of integration

$$\lambda_\ell^1 \leq \frac{\int_{\Omega_\ell} \nabla\psi_\ell \cdot \nabla\psi_\ell dx}{\int_{\Omega_\ell} \psi_\ell^2 dx}$$

$$\leq \frac{\int_{\Omega_{\ell_0}} \nabla v_{\ell_0}^{\varepsilon_0} \cdot \nabla v_{\ell_0}^{\varepsilon_0} + \frac{2}{\eta}\int_{\omega_2} A_{1,1}w_1^2 + \frac{2}{3}\eta\int_{\omega_2} A_{22}\nabla_{X_2}w_1 \cdot \nabla_{X_2}w_1}{\int_{\Omega_{\ell_0}} (v_{\ell_0}^{\varepsilon_0})^2 + \frac{2}{3}\eta}.$$

Going back to (6.67) we derive

$$\lambda_\ell^1 \leq \frac{(\mu_1 + \mu_0 - \mu_1)\int_{\Omega_{\ell_0}} (v_{\ell_0}^{\varepsilon_0})^2 + \frac{2}{\eta}\int_{\omega_2} A_{1,1}w_1^2 + \frac{2}{3}\eta\mu_1}{\int_{\Omega_{\ell_0}} (v_{\ell_0}^{\varepsilon_0})^2 + \frac{2}{3}\eta}$$

$$= \mu_1 + \frac{(\mu_0 - \mu_1)\int_{\Omega_{\ell_0}} (v_{\ell_0}^{\varepsilon_0})^2 + \frac{2}{\eta}\int_{\omega_2} A_{1,1}w_1^2}{\int_{\Omega_{\ell_0}} (v_{\ell_0}^{\varepsilon_0})^2 + \frac{2}{3}\eta}.$$

Since $\mu_0 < \mu_1$ it is clear that one can choose η large enough in such a way that the right hand side of the inequality above becomes less than μ_1. This completes the proof of the theorem. □

Chapter 7

Elliptic systems

7.1 Abstract formulation

We are going to consider vector valued functions. We will denote such a function with bold face letters for instance as

$$\mathbf{u} = (u^1, u^2, \cdots, u^N).$$

There are of course various meaning for such functions. It could represent a velocity for a fluid, a displacement for a material... To see how our previous analysis can be extended in a vectorial setting, let us adopt the following notation. If V denotes some space we will denote by $\mathbb{V}$ the space built with N copies of this space, i.e.

$$\mathbb{V} = V^N = \{\mathbf{u} = (u^1, u^2, \cdots, u^N) \mid u^i \in V \ \ \forall i = 1, \cdots, N\}.$$

For instance if O is some open subset of $\mathbb{R}^k$, $\mathbb{L}^2(O)$, $\mathbb{H}^1(O)$, $\mathbb{H}^1_0(O)$, $\mathbb{H}^{-1}(O)$ will stand for N copies of $L^2(O)$, $H^1(O)$, $H^1_0(O)$, $H^{-1}(O)$ respectively. N might vary from a problem to another.

Let $\Omega \subset \mathbb{R}^p \times \omega_2$ be an open subset of $\mathbb{R}^n$. Define

$$\Omega_\ell = (\ell \omega_1 \times \omega_2) \cap \Omega,$$

with our usual assumptions for ω_1 and ω_2. Suppose now that a is a bilinear form on $\mathbb{H}^1_0(\Omega_\ell)$ such that

$$\text{(Coerciveness)} \quad \lambda ||\nabla \mathbf{u}||^2_{2,\Omega_\ell} \leq a(\mathbf{u}, \mathbf{u}) \ \ \forall \mathbf{u} \in \mathbb{H}^1_0(\Omega_\ell), \tag{7.1}$$

$$\text{(Continuity)} \quad |a(u, v)| \leq \Lambda ||\nabla \mathbf{u}||_{2,\Omega_\ell} ||\nabla \mathbf{v}||_{2,\Omega_\ell} \ \ \forall \mathbf{u}, \mathbf{v} \in \mathbb{H}^1_0(\Omega_\ell). \tag{7.2}$$

We suppose here that $\mathbb{H}^1_0(O)$ is metrerized with the Dirichlet norm defined as

$$||\nabla \mathbf{v}||^2_{2,O} = \int_O |\nabla \mathbf{v}|^2 dx$$

where $|\nabla \mathbf{v}|$ denotes the euclidean norm of the jacobian matrix $\nabla \mathbf{v}$ considered as a vector in $\mathbb{R}^{n \times N}$, i.e.

$$|\nabla \mathbf{v}|^2 = \sum_{i=1}^{n} \sum_{j=1}^{N} (\partial_{x_i} v^j)^2.$$

Suppose that $\mathbf{f} \in \mathbb{H}^{-1}(\Omega_\ell)$, then, by the Lax-Milgram theorem, there exists a unique $\mathbf{u}_\ell$ solution to

$$\mathbf{u}_\ell \in \mathbb{H}_0^1(\Omega_\ell), \quad a(\mathbf{u}_\ell, \mathbf{v}) = \langle \mathbf{f}, \mathbf{v} \rangle \quad \forall \mathbf{v} \in \mathbb{H}_0^1(\Omega_\ell). \tag{7.3}$$

$\langle \, , \, \rangle$ denotes the duality bracket between $\mathbb{H}^{-1}(\Omega_\ell)$ and $\mathbb{H}_0^1(\Omega_\ell)$ defined as

$$\langle \mathbf{f}, \mathbf{v} \rangle = \sum_{i=1}^{N} \langle f^i, v^i \rangle \quad \forall \, \mathbf{f} = (f^1, \cdots, f^N) \in \mathbb{H}^{-1}(\Omega_\ell),$$

$$\forall \mathbf{v} = (v^1, \cdots, v^N) \in \mathbb{H}_0^1(\Omega_\ell),$$

where $\langle \, , \, \rangle$ denotes also the duality bracket between $H^{-1}(\Omega_\ell)$ and $H_0^1(\Omega_\ell)$. Moreover one has

Theorem 7.1. *Suppose that for some nonnegative constant* γ

$$|\mathbf{f}|_{\mathbb{H}^{-1}(\Omega_\ell)} = O(\ell^\gamma) \tag{7.4}$$

then there exists a unique solution to

$$\begin{cases} \mathbf{u}_\infty \in \mathbb{H}_{loc}^1(\overline{\Omega}), \quad \mathbf{u}_\infty = 0 \text{ on } \partial\Omega, \\ a(\mathbf{u}_\infty, \mathbf{v}) = \langle \mathbf{f}, \mathbf{v} \rangle \quad \forall \mathbf{v} \in \mathbb{H}_0^1(\Omega_\ell), \quad \forall \ell > 0, \\ ||\nabla \mathbf{u}_\infty||_{2,\Omega_\ell} = O(\ell^\gamma), \end{cases} \tag{7.5}$$

and one has for some positive constants C, α

$$||\nabla(\mathbf{u}_\ell - \mathbf{u}_\infty)||_{2,\Omega_{\frac{\ell}{2}}} \leq Ce^{-\alpha\ell}. \tag{7.6}$$

Here $|\mathbf{f}|_{\mathbb{H}^{-1}(\Omega_\ell)}$ *denotes the strong dual norm of* $\mathbf{f}$ *defined as*

$$|\mathbf{f}|_{\mathbb{H}^{-1}(\Omega_\ell)} = \sup_{\mathbf{v} \in \mathbb{H}_0^1(\Omega_\ell), \mathbf{v} \neq 0} \frac{|\langle \mathbf{f}, \mathbf{v} \rangle|}{||\nabla \mathbf{v}||_{2,\Omega_\ell}^2}.$$

Proof. We are going to show that $\mathbf{u}_\ell$ is a Cauchy sequence and for this purpose estimate first $\mathbf{u}_\ell - \mathbf{u}_{\ell+r}$ when $r \in [0,1]$. First from (7.3) one has

$$a(\mathbf{u}_\ell - \mathbf{u}_{\ell+r}, \mathbf{v}) = 0 \quad \forall \mathbf{v} \in \mathbb{H}_0^1(\Omega_\ell) \tag{7.7}$$

(one supposes $\mathbf{v}$ extended by 0 outside of Ω_ℓ). Then if $\rho = \rho_{\ell_1}(X_1)$ is the function defined in (2.7) for $\ell_1 \leq \ell - 1$ one has

$$\rho_{\ell_1}(\mathbf{u}_\ell - \mathbf{u}_{\ell+r}) \in \mathbb{H}_0^1(\Omega_\ell) \tag{7.8}$$

and thus

$$a(\mathbf{u}_\ell - \mathbf{u}_{\ell+r}, \rho_{\ell_1}(\mathbf{u}_\ell - \mathbf{u}_{\ell+r})) = 0. \tag{7.9}$$

This can be written as

$$a(\rho_{\ell_1}(\mathbf{u}_\ell - \mathbf{u}_{\ell+r}), \rho_{\ell_1}(\mathbf{u}_\ell - \mathbf{u}_{\ell+r})) = a((\rho_{\ell_1} - 1)(\mathbf{u}_\ell - \mathbf{u}_{\ell+r}), \rho_{\ell_1}(\mathbf{u}_\ell - \mathbf{u}_{\ell+r})).$$

Using (7.1), (7.2) we derive, noting that $\rho_{\ell_1}(\mathbf{u}_\ell - \mathbf{u}_{\ell+r})$ vanishes outside of Ω_{ℓ_1+1},

$$\lambda \int_{\Omega_{\ell_1+1}} |\nabla\{\rho_{\ell_1}(\mathbf{u}_\ell - \mathbf{u}_{\ell+r})\}|^2 dx$$

$$\leq \Lambda\Big(\int_{\Omega_{\ell_1+1}} |\nabla\{(\rho_{\ell_1} - 1)(\mathbf{u}_\ell - \mathbf{u}_{\ell+r})\}|^2 dx\Big)^{\frac{1}{2}}$$

$$\Big(\int_{\Omega_{\ell_1+1}} |\nabla\{\rho_{\ell_1}(\mathbf{u}_\ell - \mathbf{u}_{\ell+r})\}|^2 dx\Big)^{\frac{1}{2}}$$

and thus

$$\int_{\Omega_{\ell_1+1}} |\nabla\{\rho_{\ell_1}(\mathbf{u}_\ell - \mathbf{u}_{\ell+r})\}|^2 dx \leq \Big(\frac{\Lambda}{\lambda}\Big)^2 \int_{\Omega_{\ell_1+1}} |\nabla\{(\rho_{\ell_1} - 1)(\mathbf{u}_\ell - \mathbf{u}_{\ell+r})\}|^2 dx.$$

Since $\rho_{\ell_1} = 1$ on Ω_{ℓ_1} it comes

$$\int_{\Omega_{\ell_1}} |\nabla(\mathbf{u}_\ell - \mathbf{u}_{\ell+r})|^2 dx \leq \Big(\frac{\Lambda}{\lambda}\Big)^2 \int_{\Omega_{\ell_1+1}\setminus\Omega_{\ell_1}} |(\rho_{\ell_1} - 1)\nabla(\mathbf{u}_\ell - \mathbf{u}_{\ell+r})|^2 dx$$

$$+ \Big(\frac{\Lambda}{\lambda}\Big)^2 \int_{\Omega_{\ell_1+1}\setminus\Omega_{\ell_1}} |\nabla\rho_{\ell_1}(\mathbf{u}_\ell - \mathbf{u}_{\ell+r})|^2 dx$$

$$\leq \Big(\frac{\Lambda}{\lambda}\Big)^2 \int_{\Omega_{\ell_1+1}\setminus\Omega_{\ell_1}} |\nabla(\mathbf{u}_\ell - \mathbf{u}_{\ell+r})|^2 dx$$

$$+ C\Big(\frac{\Lambda}{\lambda}\Big)^2 \int_{\Omega_{\ell_1+1}\setminus\Omega_{\ell_1}} |\mathbf{u}_\ell - \mathbf{u}_{\ell+r}|^2 dx$$

$$\leq \kappa \int_{\Omega_{\ell_1+1}\setminus\Omega_{\ell_1}} |\nabla(\mathbf{u}_\ell - \mathbf{u}_{\ell+r})|^2 dx$$

for some positive constant κ (we applied the Poincaré inequality in the last integral, $\nabla\rho_{\ell_1}(\mathbf{u}_\ell - \mathbf{u}_{\ell+r})$ denotes the matrix with entries $\partial_{x_i}\rho_{\ell_1}(u_\ell^j - u_{\ell+r}^j)$). Thus it comes

$$\int_{\Omega_{\ell_1}} |\nabla(\mathbf{u}_\ell - \mathbf{u}_{\ell+r})|^2 dx \leq a \int_{\Omega_{\ell_1+1}} |\nabla(\mathbf{u}_\ell - \mathbf{u}_{\ell+r})|^2 dx, \quad a = \frac{\kappa}{\kappa+1} < 1.$$

Starting from $\frac{\ell}{2}$ and making $[\frac{\ell}{2}]$ iterations one derives easily (see (2.9))

$$\int_{\Omega_{\frac{\ell}{2}}} |\nabla(\mathbf{u}_\ell - \mathbf{u}_{\ell+r})|^2 dx \leq C e^{-\alpha\ell} \int_{\Omega_\ell} |\nabla(\mathbf{u}_\ell - \mathbf{u}_{\ell+r})|^2 dx. \tag{7.10}$$

It remains then to estimate $\mathbf{u}_\ell$. For that taking $\mathbf{v} = \mathbf{u}_\ell$ in (7.3), one obtains using (7.1)

$$\lambda ||\nabla \mathbf{u}_\ell||^2_{2,\Omega_\ell} \leq a(\mathbf{u}_\ell, \mathbf{u}_\ell) = \langle \mathbf{f}, \mathbf{u}_\ell \rangle \leq |\mathbf{f}|_{\mathbb{H}^{-1}(\Omega_\ell)} ||\nabla \mathbf{u}_\ell||_{2,\Omega_\ell}$$

i.e.

$$||\nabla \mathbf{u}_\ell||^2_{2,\Omega_\ell} \leq (\frac{|\mathbf{f}|_{\mathbb{H}^{-1}(\Omega_\ell)}}{\lambda})^2 \leq C\ell^{2\gamma}.$$

Going back to (7.10), one gets easily

$$||\nabla(\mathbf{u}_\ell - \mathbf{u}_{\ell+r})||_{2,\Omega_{\frac{\ell}{2}}} \leq C e^{-\beta\ell},$$

for any constant $\beta < \frac{\alpha}{2}$. The triangular inequality (see for instance (2.11)) allows us to obtain for every t

$$||\nabla(\mathbf{u}_\ell - \mathbf{u}_{\ell+t})||_{2,\Omega_{\frac{\ell}{2}}} \leq C e^{-\beta\ell}$$

and thus $\mathbf{u}_\ell$ is a Cauchy sequence in $\mathbb{H}^1_0(\Omega_{\ell_0})$ for any $\ell_0 \leq \frac{\ell}{2}$. The rest of the proof follows the standard steps that we have seen several times before (see for instance Theorems 2.1, 3.3). $\qquad\square$

Remark 7.1. As a variant one could consider a space $\mathbb{V}_\ell$ such that

$$\mathbb{H}^1_0(\Omega_\ell) \subset \mathbb{V}_\ell \subset \{\mathbf{v} \in \mathbb{H}^1(\Omega_\ell) \mid \mathbf{v} = 0 \text{ on } \ell\omega_1 \times \partial\omega_2\}.$$

If $\mathbb{V}_\ell$ is a Hilbert space for the Dirichlet norm $||\nabla \mathbf{u}_\ell||_{2,\Omega_\ell}$, $\forall \ell > 0$, then for $\mathbf{f} \in \mathbb{V}^*_\ell$ where $\mathbb{V}^*_\ell$ is the strong dual of $\mathbb{V}_\ell$ there exists a unique $\mathbf{u}_\ell$ solution to

$$\begin{cases} \mathbf{u}_\ell \in \mathbb{V}_\ell, \\ a(\mathbf{u}_\ell, \mathbf{v}) = \langle \mathbf{f}, \mathbf{v} \rangle \quad \forall \mathbf{v} \in \mathbb{V}_\ell. \end{cases}$$

$\langle \ , \ \rangle$ denotes here the duality bracket between $\mathbb{V}^*_\ell$ and $\mathbb{V}_\ell$. Then, if $| \ |_{\mathbb{V}^*_\ell}$ denotes the strong dual norm on $\mathbb{V}^*_\ell$ and if

$$|\mathbf{f}|_{\mathbb{V}^*_\ell} = O(\ell^\gamma) \text{ or } O(e^{\gamma'\ell})$$

for $\gamma \geq 0$ arbitrary or $\gamma' \geq 0$ small enough, then $\mathbf{u}_\ell$ converges toward $\mathbf{u}_\infty$ the unique solution to (7.5). Moreover for some positive constants C, α one has

$$||\nabla(\mathbf{u}_\ell - \mathbf{u}_\infty)||_{2,\Omega_{\frac{\ell}{2}}} \leq C e^{-\alpha\ell}.$$

The reader will easily be able to establish the proof noting that (7.8), (7.9) remain valid.

7.2 Some applications

Consider a bounded domain Ω of $\mathbb{R}^n$ and functions

$$A_{\alpha\beta}^{ij} = A_{\alpha\beta}^{ij}(x), \quad i, j = 1, \cdots, n, \ \alpha, \beta = 1, \cdots, N. \quad (7.11)$$

In what follows the roman indices will run from 1 to n, the greek ones from 1 to N. An elliptic system is a set of equations of the type

$$-\partial_{x_i}(A_{\alpha\beta}^{ij}\partial_{x_j}u^\beta) = f^\alpha, \quad \alpha = 1, \cdots, N, \quad (7.12)$$

where the coefficients $A_{\alpha\beta}^{ij}$ are "elliptic" in the sense that they define a positive quadratic form on some set of matrices. Note that in (7.12) we make the summation convention on repeated indices in such a way that (7.12) has to be interpreted as

$$-\sum_{i,j=1}^{n}\sum_{\beta=1}^{N}\partial_{x_i}(A_{\alpha\beta}^{ij}\partial_{x_j}u^\beta) = f^\alpha, \quad \alpha = 1, \cdots, N.$$

$\mathbf{f} = (f^1, \cdots, f^N)$ is a given data, $\mathbf{u} = (u^1, \cdots, u^N)$ is the unknown of the system. Note that we are summing over the indices sitting at different levels. To respect this way of summation we are going to denote the $n \times N$ matrices as $\xi = (\xi_i^\alpha)$ $(i = 1, \cdots, n, \ \alpha = 1, \cdots, N)$. Then

$$Q(\xi) = A_{\alpha\beta}^{ij}\xi_i^\alpha\xi_j^\beta = \sum_{i,j,\alpha,\beta} A_{\alpha\beta}^{ij}\xi_i^\alpha\xi_j^\beta$$

is a quadratic form on the $n \times N$ matrices. When "x" is not necessary we will drop it. Denoting by $|\xi| = (\xi_i^\alpha\xi_i^\alpha)^{\frac{1}{2}} = (\sum_{i,\alpha}(\xi_i^\alpha)^2)^{\frac{1}{2}}$ the euclidean norm of matrices the most common ellipticity conditions are for some $\lambda > 0$

$$Q(\xi) \geq \lambda|\xi|^2 \ \ \forall\xi \ \ \text{(Legendre condition)}, \quad (7.13)$$

$$Q(\xi) \geq \lambda|\xi|^2 \ \ \forall\xi = (\xi_i^\alpha) = (\varphi_i\eta^\alpha) \ \ \text{(Legendre-Hadamard condition)}, \quad (7.14)$$

$$Q(\xi) \geq \lambda|\xi|^2 \ \ \forall\xi \ \ \text{such that} \ \ \xi = \xi^T \ \ \text{(Elasticity ellipticity)}. \quad (7.15)$$

Note that in the second condition $(\varphi_i\eta^\alpha)$ is a rank one matrix. The third ellipticity condition imposes obviously that $n = N$. Of course the Legendre condition is the strongest of the conditions above in the sense that if it holds then so do the two others. The converse is not true. For instance if $n = N$ and if

$$Q(\xi) = |\xi|^2 + k \det(\xi)$$

for some constant k, det denoting the determinant of a matrix, then obviously the Legendre-Hadamard condition holds with $\lambda = 1$ since the determinant of a rank one matrix is vanishing. However, choosing a matrix ξ such that $\det(\xi) < 0$, it is clear that for $k > 0$ large enough one has

$$Q(\xi) = |\xi|^2 + k \det(\xi) < 0$$

and thus the Legendre condition fails. We can show

Lemma 7.1. *Let Ω be a bounded open subset of $\mathbb{R}^n$. Denote by $A_{\alpha\beta}^{ij}$, $i, j = 1, \cdots, n$, $\alpha, \beta = 1, \cdots, N$ a set of functions such that*

$$A_{\alpha\beta}^{ij} \in L^\infty(\Omega).$$

Define for $\mathbf{u}, \mathbf{v} \in \mathbb{H}_0^1(\Omega)$ the bilinear form a by

$$a(\mathbf{u}, \mathbf{v}) = \int_\Omega A_{\alpha\beta}^{ij} \partial_{x_i} u^\alpha \partial_{x_j} v^\beta \, dx,$$

with the usual summation convention of repeated indices. Then a is well defined and for some constant Λ one has

$$|a(\mathbf{u}, \mathbf{v})| \leq \Lambda ||\nabla \mathbf{u}||_{2,\Omega} ||\nabla \mathbf{v}||_{2,\Omega} \quad \forall \mathbf{u}, \mathbf{v} \in \mathbb{H}_0^1(\Omega). \tag{7.16}$$

Moreover if

$-A_{\alpha\beta}^{ij}$ *satisfies the Legendre condition* (7.13),

or

$-A_{\alpha\beta}^{ij}$ *is independent of x and satisfies* (7.14),

or

$-\, n = N$ *and $A_{\alpha\beta}^{ij}$ satisfies* (7.15) *and is such that $A_{\alpha\beta}^{ij} = A_{\alpha j}^{i\beta} = A_{i\beta}^{\alpha j}$,*

one has

$$a(\mathbf{u}, \mathbf{u}) \geq \lambda' ||\nabla \mathbf{u}||_{2,\Omega}^2 \quad \forall \mathbf{u} \in \mathbb{H}_0^1(\Omega) \tag{7.17}$$

for some positive constant $\lambda' = \lambda$ in the cases of (7.13) *and* (7.14) *and $\lambda' = \frac{\lambda}{2}$ in the case of* (7.15).

Proof. Since the coefficients $A_{\alpha\beta}^{ij}$ are bounded, one has clearly for some constant Λ

$$|A_{\alpha\beta}^{ij} \partial_{x_i} u^\alpha| \leq \Lambda |\nabla \mathbf{u}|$$

(recall that $|\;|$ is the euclidean norm on the $n \times N$-matrices considered as vectors of $\mathbb{R}^{n \times N}$). It follows by the Cauchy-Schwarz inequality in $\mathbb{R}^{n \times N}$ that

$$
\begin{aligned}
|a(\mathbf{u}, \mathbf{v})| &\leq \int_\Omega |A_{\alpha\beta}^{ij} \partial_{x_i} u^\alpha \partial_{x_j} v^\beta| \, dx \\
&\leq \int_\Omega |A_{\alpha\beta}^{ij} \partial_{x_i} u^\alpha| |\partial_{x_j} v^\beta| \, dx \leq \Lambda \int_\Omega |\nabla \mathbf{u}| |\nabla \mathbf{v}| \, dx \\
&\leq \Lambda ||\nabla \mathbf{u}||_{2,\Omega} ||\nabla \mathbf{v}||_{2,\Omega}
\end{aligned}
$$

by application of the Cauchy-Schwarz inequality again. Of course if (7.13) holds one has

$$a(\mathbf{u}, \mathbf{u}) = \int_\Omega Q(\nabla\mathbf{u})dx \geq \lambda \int_\Omega |\nabla\mathbf{u}|^2 dx = \lambda ||\nabla\mathbf{u}||^2_{2,\Omega}.$$

In the case where the $A^{ij}_{\alpha\beta}$ are constant and satisfy (7.14) it is well known (see [44], [13]) that

$$a(\mathbf{u}, \mathbf{u}) = \int_\Omega Q(\nabla\mathbf{u})dx \geq \lambda \int_\Omega |\nabla\mathbf{u}|^2 dx = \lambda ||\nabla\mathbf{u}||^2_{2,\Omega}.$$

Finally in the third case one has due to the symmetry assumption on the coefficients

$$\int_\Omega A^{ij}_{\alpha\beta}\{\frac{\partial_{x_i}u^\alpha + \partial_{x_\alpha}u^i}{2}\}\{\frac{\partial_{x_j}u^\beta + \partial_{x_\beta}u^j}{2}\}dx$$

$$= \frac{1}{4}\int_\Omega A^{ij}_{\alpha\beta}\partial_{x_i}u^\alpha\partial_{x_j}u^\beta + A^{ij}_{\alpha\beta}\partial_{x_i}u^\alpha\partial_{x_\beta}u^j$$

$$+ A^{ij}_{\alpha\beta}\partial_{x_\alpha}u^i\partial_{x_j}u^\beta + A^{ij}_{\alpha\beta}\partial_{x_\alpha}u^i\partial_{x_\beta}u^j \, dx$$

$$= \frac{1}{4}\int_\Omega A^{ij}_{\alpha\beta}\partial_{x_i}u^\alpha\partial_{x_j}u^\beta + A^{i\beta}_{\alpha j}\partial_{x_i}u^\alpha\partial_{x_\beta}u^j$$

$$+ A^{\alpha j}_{i\beta}\partial_{x_\alpha}u^i\partial_{x_j}u^\beta + A^{\alpha\beta}_{ij}\partial_{x_\alpha}u^i\partial_{x_\beta}u^j \, dx$$

$$= \int_\Omega A^{ij}_{\alpha\beta}\partial_{x_i}u^\alpha\partial_{x_j}u^\beta dx = a(\mathbf{u}, \mathbf{u}).$$

Thus we have

$$a(\mathbf{u}, \mathbf{u}) \geq \lambda ||\frac{\nabla\mathbf{u} + \nabla\mathbf{u}^T}{2}||^2_{2,\Omega}.$$

The result follows then from the so-called first Korn inequality, namely – with the summation convention in i, α

$$\int_\Omega (\partial_{x_i}u^\alpha)^2 dx \leq 2\int_\Omega (\frac{\partial_{x_i}u^\alpha + \partial_{x_\alpha}u^i}{2})^2 dx.$$

To prove this last inequality one notices that – again with the summation convention

$$\int_\Omega (\frac{\partial_{x_i}u^\alpha + \partial_{x_\alpha}u^i}{2})^2 dx = \frac{1}{4}\int_\Omega (\partial_{x_i}u^\alpha)^2 + 2\partial_{x_i}u^\alpha\partial_{x_\alpha}u^i + (\partial_{x_\alpha}u^i)^2 dx$$

$$= \frac{1}{2}\int_\Omega |\nabla\mathbf{u}|^2 dx + \frac{1}{2}\int_\Omega \partial_{x_i}u^\alpha\partial_{x_\alpha}u^i dx$$

$$= \frac{1}{2}\int_\Omega |\nabla\mathbf{u}|^2 dx + \frac{1}{2}\int_\Omega \partial_{x_\alpha}u^\alpha\partial_{x_i}u^i$$

after two integrations by parts in the last integral assuming first $\mathbf{u} \in \mathcal{D}(\Omega)^n$. Thus we have

$$\int_\Omega (\frac{\partial_{x_i} u^\alpha + \partial_{x_\alpha} u^i}{2})^2 dx = \frac{1}{2} \int_\Omega |\nabla \mathbf{u}|^2 dx + \frac{1}{2} \int_\Omega (\operatorname{div} \mathbf{u})^2 dx$$
$$\geq \frac{1}{2} \int_\Omega |\nabla \mathbf{u}|^2 dx \tag{7.18}$$

$(\operatorname{div} \mathbf{u} = \sum_i \partial_{x_i} u^i)$. Then the result follows. This completes the proof of the lemma. □

Let us now suppose that

$$\Omega = \mathbb{R}^p \times \omega_2, \quad \Omega_\ell = \ell\omega_1 \times \omega_2$$

and consider uniformly bounded measurable coefficients $A^{ij}_{\alpha\beta}$ such that

$$A^{ij}_{\alpha\beta} = A^{ij}_{\alpha\beta}(X_2) \text{ and } (7.13) \text{ holds},$$
$$\text{or} \tag{7.19}$$
$$A^{ij}_{\alpha\beta} \text{ are constants and } (7.14) \text{ holds}.$$

(By uniformly bounded we mean that there exists a constant $\Lambda > 0$ such that $|A^{ij}_{\alpha\beta}(X_2)| \leq \Lambda$ a.e. $X_2 \in \omega_2$ for all i, j, α, β).

Let $\mathbf{f} = \mathbf{f}(X_2)$ be such that

$$\mathbf{f} \in \mathbb{L}^2(\omega_2).$$

It is then clear by the Lax-Milgram theorem that there exists $\mathbf{u}_\ell$ and $\mathbf{u}_\infty$ solutions respectively to

$$\begin{cases} \mathbf{u}_\ell \in \mathbb{H}^1_0(\Omega_\ell), \\ \int_{\Omega_\ell} A^{ij}_{\alpha\beta} \partial_{x_i} u^\alpha_\ell \partial_{x_j} v^\beta dx = \int_{\Omega_\ell} \mathbf{f} \cdot \mathbf{v} dx \quad \forall \mathbf{v} \in \mathbb{H}^1_0(\Omega_\ell), \end{cases} \tag{7.20}$$

and

$$\begin{cases} \mathbf{u}_\infty \in \mathbb{H}^1_0(\omega_2), \\ \int_{\omega_2} A^{ij}_{\alpha\beta} \partial_{x_i} u^\alpha_\infty \partial_{x_j} v^\beta dX_2 = \int_{\omega_2} \mathbf{f} \cdot \mathbf{v} dX_2 \quad \forall \mathbf{v} \in \mathbb{H}^1_0(\omega_2). \end{cases} \tag{7.21}$$

Some explanations are in order regarding the solution of this last problem. Indeed since $\mathbf{u}_\infty$ is independent of X_1 the summation in i, j in the first integral of (7.21) is performed on $i, j = p + 1, \cdots, n$.

Then we have

Theorem 7.2. *There exist positive constants C, α such that*

$$\big\|\nabla(\mathbf{u}_\ell - \mathbf{u}_\infty)\big\|_{2,\Omega_{\frac{\ell}{2}}} \leq Ce^{-\alpha\ell}.$$

Proof. It is enough to apply Theorem 7.1 setting

$$a(\mathbf{u}, \mathbf{v}) = \int_{\Omega_\ell} A^{ij}_{\alpha\beta} \partial_{x_i} u^\alpha \partial_{x_j} v^\beta dx.$$

Indeed a satisfies clearly (7.1), (7.2) thanks to (7.19) and Lemma 7.1. Moreover since

$$\langle \mathbf{f}, \mathbf{v} \rangle = \int_{\Omega_\ell} \mathbf{f} \cdot \mathbf{v} dx$$

one has by the Cauchy-Schwarz and Poincaré inequalities for some constant $C(\omega_2)$

$$|\langle \mathbf{f}, \mathbf{v} \rangle| \leq ||\mathbf{f}||_{2,\Omega_\ell} ||\mathbf{v}||_{2,\Omega_\ell} \leq C(\omega_2) ||\mathbf{f}||_{2,\Omega_\ell} ||\nabla \mathbf{v}||_{2,\Omega_\ell}.$$

Thus

$$|\mathbf{f}|_{\mathbb{H}^{-1}(\Omega_\ell)} \leq C(\omega_2) ||\mathbf{f}||_{2,\Omega_\ell} = C(\omega_2) \left(\int_{\ell\omega_1} \int_{\omega_2} |\mathbf{f}(X_2)|^2 dx \right)^{\frac{1}{2}}$$

$$= C(\omega_2) |\ell\omega_1|^{\frac{1}{2}} ||\mathbf{f}||_{2,\omega_2} = C(\omega_2) |\omega_1|^{\frac{1}{2}} ||\mathbf{f}||_{2,\omega_2} \ell^{\frac{p}{2}}$$

$$= O(\ell^{\frac{p}{2}}),$$

$| \ |$ denotes here the euclidean norm and the measure of sets as well. Now it is easy to see that $\mathbf{u}_\infty$ defined as the solution to (7.21) satisfies all the properties of (7.5) and is thus the solution to (7.5). Indeed since $\mathbf{u}_\infty$ is independent of X_1 the summation in (7.21) can run for $i, j = 1, \cdots, n$. Moreover for $\mathbf{v} \in \mathbb{H}^1_0(\Omega_\ell)$ one has $\mathbf{v}(X_1, \cdot) \in \mathbb{H}^1_0(\omega_2)$ and $\mathbf{v}(X_1, \cdot)$ is for a.e. X_1 a test function for (7.21). One obtains then for a.e X_1

$$\int_{\omega_2} A^{ij}_{\alpha\beta} \partial_{x_i} u^\alpha \partial_{x_j} v^\beta (X_1, \cdot) dX_2 = \int_{\omega_2} \mathbf{f} \cdot \mathbf{v}(X_1, \cdot) dX_2.$$

Integrating in X_1, it comes

$$a(\mathbf{u}_\infty, \mathbf{v}) = \int_{\Omega_\ell} \mathbf{f} \cdot \mathbf{v} dx = \langle \mathbf{f}, \mathbf{v} \rangle \quad \forall \mathbf{v} \in \mathbb{H}^1_0(\Omega_\ell) \quad \forall \ell > 0.$$

The fact that

$$||\nabla \mathbf{u}_\infty||_{2,\Omega_\ell} = O(\ell^{\frac{p}{2}})$$

results from the independence of $\mathbf{u}_\infty$ with respect to X_1. The result follows then from Theorem 7.1. This completes the proof of the theorem. $\square$

More interesting is the application to elasticity. We consider here the physical situation i.e. the case n=3 and a cylinder $\Omega_\ell = (-\ell, \ell) \times \omega_2$ as in Figure 7.1.

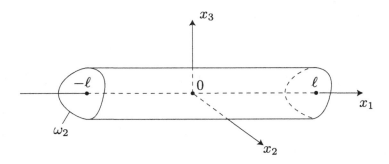

Fig. 7.1   Description of Ω_ℓ.

We suppose that this cylindrical body is subjected to the action of a density of forces

$$\mathbf{f} = (0, f^2(X_2), f^3(X_2)) \in \mathbb{L}^2(\omega_2), \quad X_2 = (x_2, x_3) \tag{7.22}$$

i.e. the density of forces is depending only on the section of the cylinder.

Assuming (7.22) the displacement of the above body supposed to be clamped (see [44], [13]) is given by $\mathbf{u}_\ell$ solution to

$$\begin{cases} \mathbf{u}_\ell \in \mathbb{H}_0^1(\Omega_\ell), \\ \lambda \int_{\Omega_\ell} \operatorname{div} \mathbf{u}_\ell \operatorname{div} \mathbf{v} dx + 2\mu \int_{\Omega_\ell} \sum_{i,j=1}^3 \epsilon_{i,j}(\mathbf{u}_\ell)\epsilon_{i,j}(\mathbf{v}) dx \\ \qquad\qquad = \int_{\Omega_\ell} \mathbf{f} \cdot \mathbf{v} dx \quad \forall \mathbf{v} \in \mathbb{H}_0^1(\Omega_\ell). \end{cases} \tag{7.23}$$

In the above system

$$\epsilon_{i,j}(\mathbf{v}) = \frac{1}{2}(\partial_{x_i} v^j + \partial_{x_j} v^i),$$

λ, μ are two constants. Setting

$$a(\mathbf{u}, \mathbf{v}) = \lambda \int_{\Omega_\ell} \operatorname{div} \mathbf{u} \operatorname{div} \mathbf{v} dx + 2\mu \int_{\Omega_\ell} \sum_{i,j=1}^3 \epsilon_{i,j}(\mathbf{u})\epsilon_{i,j}(\mathbf{v}) dx$$

it is clear from (7.18) that

$$a(\mathbf{u}, \mathbf{u}) \geq \mu \int_{\Omega_\ell} |\nabla \mathbf{u}|^2 dx + \mu \int_{\Omega_\ell} (\operatorname{div} \mathbf{u})^2 dx + \lambda \int_{\Omega_\ell} (\operatorname{div} \mathbf{u})^2 dx$$

$$\geq \mu \int_{\Omega_\ell} |\nabla \mathbf{u}|^2 dx$$

when $\lambda + \mu \geq 0$. Thus by the Lax-Milgram theorem (7.23) possesses a unique solution provided

$$\mu > 0, \lambda + \mu \geq 0. \tag{7.24}$$

Under these conditions and if
$$\mathbf{f}' = (f^2, f^3), \quad \operatorname{div}{}'\mathbf{v} = \partial_{x_2}v^2 + \partial_{x_3}v^3$$
there exists a unique $\mathbf{u}'_\infty = (u^2_\infty, u^3_\infty)$ solution to

$$\begin{cases} \mathbf{u}'_\infty = (u^2_\infty, u^3_\infty) \in \mathbb{H}^1_0(\omega_2) = (H^1_0(\omega_2))^2, \\ \lambda \int_{\omega_2} \operatorname{div}{}'\mathbf{u}'_\infty \ \operatorname{div}{}'\mathbf{v}dX_2 + 2\mu \int_{\omega_2} \sum_{i,j=2}^{3} \epsilon_{i,j}(\mathbf{u}'_\infty)\epsilon_{i,j}(\mathbf{v})dX_2 \qquad (7.25) \\ \qquad = \int_{\omega_2} \mathbf{f}' \cdot \mathbf{v}dX_2, \quad \forall \mathbf{v} = (v^2, v^3) \in \mathbb{H}^1_0(\omega_2). \end{cases}$$

We then set

$$\mathbf{u}_\infty = (0, u^2_\infty, u^3_\infty). \qquad (7.26)$$

Then we have

Theorem 7.3. *There exist positive constants* C, α *such that*
$$\left|\left|\nabla(\mathbf{u}_\ell - \mathbf{u}_\infty)\right|\right|_{2,\frac{\ell}{2}} \le Ce^{-\alpha\ell}.$$

Proof. Notice that we are about to prove that $\mathbf{u}_\ell$ converges toward the solution of the elasticity problem set on the section. In order to show that it is enough to apply Theorem 7.1, i.e. to show that $\mathbf{u}_\infty$ defined by (7.25), (7.26) satisfies (7.5) (the condition (7.4) can be established exactly as in the preceding theorem). Due to (7.26) and the fact that $\mathbf{u}_\infty$ is independent of x_1 one has

$$\operatorname{div}\mathbf{u}_\infty = \operatorname{div}{}'\mathbf{u}'_\infty, \quad \epsilon_{1,j}(\mathbf{u}_\infty) = \frac{1}{2}(\partial_{x_1}u^j_\infty + \partial_{x_j}u^1_\infty) = 0 \ \ \forall j = 1, 2, 3,$$

and for $\mathbf{v} = (v^1, v^2, v^3)$

$$\epsilon_{i,j}(\mathbf{u}_\infty)\epsilon_{i,j}(\mathbf{v}) = \epsilon_{i,j}(\mathbf{u}'_\infty)\epsilon_{i,j}(\mathbf{v}) \ \ \forall i, j = 1, 2, 3.$$

Thus for $\mathbf{v} \in \mathbb{H}^1_0(\Omega_\ell)$ since $\mathbf{v}(x_1, \cdot) \in \mathbb{H}^1_0(\omega_2)$ for a.e. $x_1 \in (-\ell, \ell)$, one gets from (7.25)

$$\lambda \int_{\omega_2} \operatorname{div}\mathbf{u}_\infty \ \operatorname{div}{}'\mathbf{v}dX_2 + 2\mu \int_{\omega_2} \sum_{i,j=1}^{3} \epsilon_{i,j}(\mathbf{u}_\infty)\epsilon_{i,j}(\mathbf{v})dX_2$$

$$= \int_{\omega_2} \mathbf{f} \cdot \mathbf{v}dX_2, \quad \forall \mathbf{v} \in \mathbb{H}^1_0(\Omega_\ell), \quad \text{a.e. } x_1 \in (-\ell, \ell).$$

Integrating in x_1 gives

$$\lambda \int_{\Omega_\ell} \operatorname{div}\mathbf{u}_\infty \ \operatorname{div}{}'\mathbf{v}dx + 2\mu \int_{\Omega_\ell} \sum_{i,j=1}^{3} \epsilon_{i,j}(\mathbf{u}_\infty)\epsilon_{i,j}(\mathbf{v})dx$$

$$\qquad (7.27)$$

$$= \int_{\Omega_\ell} \mathbf{f} \cdot \mathbf{v}dx \ \ \forall \mathbf{v} \in \mathbb{H}^1_0(\Omega_\ell).$$

Now by the divergence theorem one has

$$\int_{\Omega_\ell} \operatorname{div} \mathbf{u}_\infty \partial_{x_1} v^1 dx = \int_{\Omega_\ell} \partial_{x_1}(\operatorname{div} \mathbf{u}_\infty v^1) dx = 0$$

and thus

$$\int_{\Omega_\ell} \operatorname{div} \mathbf{u}_\infty \operatorname{div}' v dx = \int_{\Omega_\ell} \operatorname{div} \mathbf{u}_\infty \operatorname{div} \mathbf{v} dx$$

and (7.27) becomes thanks to the definition of a given above

$$a(\mathbf{u}_\infty, \mathbf{v}) = \int_{\Omega_\ell} \mathbf{f} \cdot \mathbf{v} dx, \quad \forall \mathbf{v} \in \mathbb{H}_0^1(\Omega_\ell)$$

i.e. $\mathbf{u}_\infty$ fulfills all the properties of (7.5) and the result follows from Theorem 7.1. This completes the proof of the theorem. $\qquad\square$

Remark 7.2. We refer the reader to [13] for more details about elliptic systems. He will be able after some light work to improve the rate of the convergence results given there.

Chapter 8

The Stokes problem

8.1 Introduction and notation

In this chapter we consider the Stokes problem in a cylinder of the type

$$\Omega_\ell = (-\ell, \ell) \times \omega',$$

where $\ell > 0$ is a parameter that goes to infinity (see Figure 8.1).

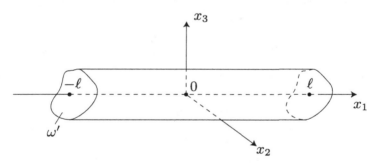

Fig. 8.1   Description of Ω_ℓ.

The section ω' of the cylinder is a bounded, connected, open subset of $\mathbb{R}^{n-1}$ with Lipschitz boundary $\partial\omega'$. We will need for the limit problem the infinite cylinder

$$\Omega_\infty = \mathbb{R} \times \omega'.$$

We will use the notation

$$x = (x_1, x'), \quad x' = x_2, \cdots, x_n$$

to denote the points in $\mathbb{R}^n$. We would like to consider the asymptotic behaviour of the Stokes problem set in Ω_ℓ. The solution of the Stokes

177

problem is a couple $(\mathbf{u}_\ell, p_\ell)$ where

$$\mathbf{u}_\ell = (u_\ell^1, u_\ell^2, \cdots, u_\ell^n)$$

satisfying

$$\begin{cases} -\mu \Delta \mathbf{u}_\ell + \nabla p_\ell = \mathbf{f} \text{ in } \Omega_\ell, \\ \operatorname{div} \mathbf{u}_\ell = 0 \text{ in } \Omega_\ell, \\ \mathbf{u}_\ell = 0 \text{ on } \partial \Omega_\ell. \end{cases} \tag{8.1}$$

This system describes the stationary motion of an incompressible fluid of viscosity μ in Ω_ℓ under the action of the force $\mathbf{f}$. We refer the reader to the huge literature [51], [73], [61]. We recall that the divergence operator is given by

$$\operatorname{div} \mathbf{u} = \sum_{i=1}^n \partial_{x_i} u^i = \partial_{x_i} u^i \ \forall \mathbf{u} = (u^1, u^2, \cdots, u^n)$$

where ∂_{x_i} denotes the partial derivative in x_i and in the second equality we used the summation convention in i. $\mathbf{u}_\ell$ stands for the velocity of the fluid and p_ℓ for its pressure. It is clear that this function is defined up to a constant and to fix this one we will introduce the space

$$\hat{L}^2(O) = \{ v \in L^2(O) \mid \int_O v dx = 0 \}$$

i.e. $L^2(O)$ is the Hilbert space of $L^2(O)$-functions of average 0. We will use also the space

$$\mathbb{H}_0^1(O) = (H_0^1(O))^n$$

introduced in the previous chapter and

$$\hat{\mathbb{H}}_0^1(O) = \{ \mathbf{v} \in \mathbb{H}_0^1(O) \mid \operatorname{div} \mathbf{v} = 0 \text{ in } O \}$$

the divergence equality taking place in $L^2(O)$ or in the distributional sense. With this notation a weak solution $(\mathbf{u}_\ell, p_\ell)$ to (8.1) is a couple satisfying

$$\begin{cases} (\mathbf{u}_\ell, p_\ell) \in \hat{\mathbb{H}}_0^1(\Omega_\ell) \times \hat{L}^2(\Omega_\ell), \\ \mu \int_{\Omega_\ell} \nabla \mathbf{u}_\ell \cdot \nabla \mathbf{v} - p_\ell \operatorname{div} \mathbf{v} dx = \int_{\Omega_\ell} \mathbf{f} \cdot \mathbf{v} dx \ \forall \mathbf{v} \in \mathbb{H}_0^1(\Omega_\ell). \end{cases} \tag{8.2}$$

In the integral above $\nabla \mathbf{v}$ denotes the Jacobian matrix of $\mathbf{v}$ given by

$$\nabla \mathbf{v} = (\partial_{x_j} v^i)$$

and the expression $\nabla \mathbf{u}_\ell \cdot \nabla \mathbf{v}$ stands for the scalar product of the two matrices involved when considered as vector in $\mathbb{R}^{n^2}$ namely

$$\nabla \mathbf{u}_\ell \cdot \nabla \mathbf{v} = \partial_{x_j} u_\ell^i \partial_{x_j} v^i$$

with the summation convention in i, j. The existence and uniqueness of the solution to (8.2) are well known for any positive number μ. Let us notice that by the Lax-Milgram theorem $\mathbf{u}_\ell$ is the unique solution to

$$\begin{cases} \mathbf{u}_\ell \in \hat{\mathbb{H}}_0^1(\Omega_\ell), \\ \mu \int_{\Omega_\ell} \nabla \mathbf{u}_\ell \cdot \nabla \mathbf{v} dx = \int_{\Omega_\ell} \mathbf{f} \cdot \mathbf{v} dx \quad \forall \mathbf{v} \in \hat{\mathbb{H}}_0^1(\Omega_\ell). \end{cases} \tag{8.3}$$

We did not precise our assumption on $\mathbf{f}$. For simplicity we assume here that

$$\mathbf{f} \in \mathbb{L}^2(\Omega_\ell) = (L^2(\Omega_\ell))^n.$$

Then it follows from (8.3) that

$$\langle \mu \Delta \mathbf{u}_\ell + \mathbf{f}, \mathbf{v} \rangle = 0 \quad \forall \mathbf{v} \in \hat{\mathbb{H}}_0^1(\Omega_\ell) \tag{8.4}$$

where $\langle \, , \, \rangle$ denotes the duality between $\mathbb{H}^{-1}(\Omega_\ell)$ and $\mathbb{H}_0^1(\Omega_\ell)$. It follows (see [51], [73], [1]) that there exists $p_\ell \in L^2(\Omega_\ell)$ such that

$$-\mu \Delta \mathbf{u}_\ell + \nabla p_\ell = \mathbf{f} \quad \text{in } \mathbb{H}^{-1}(\Omega_\ell).$$

For unbounded domains we will use the notation

$$L_{\text{loc}}^2(\overline{O}) = \{v \mid v \in L^2(U) \; \forall U \text{ bounded } \subset O\}, \quad \mathbb{L}_{\text{loc}}^2(\overline{O}) = (L_{loc}^2(\overline{O}))^n,$$

$$\hat{L}_{\text{loc}}^2(\overline{O}) = L_{\text{loc}}^2(\overline{O})/\mathbb{R},$$

$$H_{\text{loc}}^1(\overline{O}) = \{v \mid v \in H^1(U) \; \forall U \text{ bounded } \subset O\}, \quad \mathbb{H}_{\text{loc}}^1(\overline{O}) = (H_{\text{loc}}^1(\overline{O}))^n.$$

We will need some lemmas.

8.2 Auxiliary lemmas

Lemma 8.1. *Let O be a bounded domain of $\mathbb{R}^n$ and $\mathbf{u} \in \hat{\mathbb{H}}_0^1(\Omega_\ell)$. We suppose that $\mathbf{u}$ is extended by 0 outside of O. Then one has*

$$\int_{\mathbb{R}^{n-1}} u^1(x_1, x')dx' = 0 \quad a.e. \; x_1 \in \mathbb{R}.$$

Proof. Suppose that $O \subset [-A, A]^n$ and that $\mathbf{u}$ is extended by 0 outside O. Applying the divergence formula on the hyperrectangle $R_{x_1} = (-A, x_1) \times (-A, A)^{(n-1)}$ gives

$$0 = \int_{R_{x_1}} \text{div } \mathbf{u} dx = \int_{\mathbb{R}^{n-1}} u^1(x_1, x')dx'.$$

This completes the proof of the lemma. $\square$

We have also

Lemma 8.2. *Let $\ell \geq 1$ and $g \in L^2(\Omega_\ell)$ such that*

$$\int_{\Omega_\ell} g \, dx = 0.$$

Then there exists $\mathbf{u} \in \mathbb{H}_0^1(\Omega_\ell)$ such that

$$\begin{cases} \operatorname{div} \mathbf{u} = g & in \ \Omega_\ell, \\ ||\nabla \mathbf{u}||_{2,\Omega_\ell} \leq C\ell|g|_{2,\Omega_\ell}. \end{cases} \tag{8.5}$$

Proof. Let $\Omega_1 = (-1,1) \times \omega'$, $\tilde{g} \in L^2(\Omega_1)$ with

$$\int_{\Omega_1} \tilde{g} \, dx = 0.$$

It is well known (see [51], [4], [1]) that there exists $\tilde{\mathbf{u}} \in \mathbb{H}_0^1(\Omega_1)$ such that

$$\begin{cases} \operatorname{div} \tilde{\mathbf{u}} = \tilde{g} & in \ \Omega_1, \\ ||\nabla \tilde{\mathbf{u}}||_{2,\Omega_1} \leq C|\tilde{g}|_{2,\Omega_1} \end{cases} \tag{8.6}$$

where the constant C depends only on Ω_1.

Note that $(\mathbf{u}, g) \in \mathbb{H}_0^1(\Omega_\ell) \times L^2(\Omega_\ell)$ satisfies

$$\int_{\Omega_\ell} g \, dx = 0, \quad div \ \mathbf{u} = g \quad in \ \Omega_\ell$$

if and only if

$$\tilde{\mathbf{u}}(x_1, x') = (\frac{1}{\ell}u^1(\ell x_1, x'), \mathbf{u}'(\ell x_1, x')), \quad \tilde{g}(x_1, x') = g(\ell x_1, x') \tag{8.7}$$

satisfies

$$\int_{\Omega_1} \tilde{g} \, dx = 0, \quad div \ \tilde{\mathbf{u}} = \tilde{g} \quad in \ \Omega_1.$$

(We denote by $\mathbf{u}'$ the $n-1$ last components of $\mathbf{u}$, similarly we will denote below by ∇' the $n-1$ last components of ∇). By a simple change of variable we have

$$\int_{\Omega_1} \tilde{g}^2 dx = \int_{\Omega_1} g(\ell x_1, x')^2 \, dx = \frac{1}{\ell} \int_{\Omega_\ell} g^2(x) \, dx$$

i.e.

$$|\tilde{g}|_{2,\Omega_1}^2 = \frac{1}{\ell}|g|_{2,\Omega_\ell}^2.$$

Also with similar arguments we have

$$||\nabla'\tilde{\mathbf{u}}'||^2_{2,\Omega_1} = \frac{1}{\ell}||\nabla'\mathbf{u}'||^2_{2,\Omega_\ell}, \quad ||\partial_{x_1}\tilde{\mathbf{u}}'||^2_{2,\Omega_1} = \ell||\partial_{x_1}\mathbf{u}||^2_{2,\Omega_\ell}$$

$$||\nabla'\tilde{u}^1||^2_{2,\Omega_1} = \frac{1}{\ell^3}||\nabla'u^1||^2_{2,\Omega_\ell}, \quad |\partial_{x_1}\tilde{u}^1|^2_{2,\Omega_1} = \frac{1}{\ell}|\partial_{x_1}u^1|^2_{2,\Omega_\ell}.$$

It follows easily that for $\ell \geq 1$

$$||\nabla\tilde{\mathbf{u}}||^2_{2,\Omega_1} \geq \frac{1}{\ell^3}||\nabla\mathbf{u}||^2_{2,\Omega_\ell}.$$

Then starting with g of average 0 we construct $\tilde{\mathbf{u}}$ solution to (8.6). Then we construct $\mathbf{u}$ using the first equality of (8.7) to obtain a solution to the first equation of (8.5) which satisfies

$$||\nabla\mathbf{u}||^2_{2,\Omega_\ell} \leq \ell^3||\nabla\tilde{\mathbf{u}}||^2_{2,\Omega_1} \leq C\ell^2|g|^2_{2,\Omega_\ell}$$

which completes the proof of the lemma. □

8.3   Main results

We have

Theorem 8.1. *Let* $\mathbf{f} \in \mathbb{L}^2_{\text{loc}}(\overline{\Omega}_\infty)$. *Suppose that for some* $\gamma \geq 0$ *we have*

$$||\mathbf{f}||_{2,\Omega_\ell} = O(\ell^\gamma). \tag{8.8}$$

Then when $\ell \to \infty$, $(\mathbf{u}_\ell, p_\ell)$ *converges toward the unique solution* $(\mathbf{u}_\infty, p_\infty)$ *to*

$$\begin{cases} (\mathbf{u}_\infty, p_\infty) \in \mathbb{H}^1_{\text{loc}}(\overline{\Omega}_\infty) \times \hat{L}^2_{\text{loc}}(\overline{\Omega}_\infty), \\ -\mu\Delta\mathbf{u}_\infty + \nabla p_\infty = \mathbf{f}, \quad \text{div}\,\mathbf{u}_\infty = 0 \quad in \quad \mathcal{D}'(\Omega_\infty), \\ \mathbf{u}_\infty = 0 \quad on \quad \mathbb{R} \times \partial\omega', \\ \int_{\omega'} u^1_\infty(x_1, x')\,dx' = 0 \quad for \quad a.e. \ x_1 \in \mathbb{R}, \\ ||\nabla\mathbf{u}_\infty||_{2,\Omega_\ell} = O(\ell^\gamma). \end{cases} \tag{8.9}$$

(In the last equality $|\ \ |$ *denotes the euclidean norm for the matrices considered as vectors in* $\mathbb{R}^{n^2}$ *).*

  Moreover, for some positive constants C, α *one has,*

$$||\nabla(\mathbf{u}_\ell - \mathbf{u}_\infty)||_{2,\Omega_{\ell/2}} + |p_\ell - p_\infty|_{2,\Omega_{\ell/2}} \leq Ce^{-\alpha\ell}. \tag{8.10}$$

Remark 8.1. *If* $\mathbf{f} \in \mathbb{L}^2(\overline{\Omega}_\infty)$ *then the existence of a solution to (8.9) follows from the Lax-Milgram theorem but when* $\gamma > 0$ *this is part of the result.*

Proof. As usual we are going to show that $\mathbf{u}_\ell$ is a Cauchy "sequence". For that we start to estimate $\mathbf{u}_\ell - \mathbf{u}_{\ell+r}$ when $r \in [0,1]$. So we notice that by (8.3) one has

$$\mu \int_{\Omega_\ell} \nabla(\mathbf{u}_\ell - \mathbf{u}_{\ell+r}) \cdot \nabla \mathbf{v} dx = 0 \quad \forall \mathbf{v} \in \hat{\mathbb{H}}_0^1(\Omega_\ell). \qquad (8.11)$$

(We assume $\mathbf{v}$ extended by 0 outside Ω_ℓ). For $\ell_1 \leq \ell - 1$ we denote by $\rho = \rho_{\ell_1}$ the function depicted in Figure 1.4. One has clearly

$$\rho(\mathbf{u}_\ell - \mathbf{u}_{\ell+r}) \in \mathbb{H}_0^1(\Omega_\ell).$$

Moreover, since $\mathbf{u}_\ell$, $\mathbf{u}_{\ell+r}$ are divergence free, one has

$$\operatorname{div} \rho(\mathbf{u}_\ell - \mathbf{u}_{\ell+r}) = \partial_{x_1}\rho(u_\ell^1 - u_{\ell+r}^1).$$

Due to Lemma 8.1, one notices that

$$\int_{\ell_1}^{\ell_1+1} \int_{\omega'} \partial_{x_1}\rho(u_\ell^1 - u_{\ell+r}^1)dx'dx_1 = -\int_{\ell_1}^{\ell_1+1} \int_{\omega'} (u_\ell^1 - u_{\ell+r}^1)dx'dx_1 = 0.$$

Similarly

$$\int_{-\ell_1-1}^{-\ell_1} \int_{\omega'} \partial_{x_1}\rho(u_\ell^1 - u_{\ell+r}^1)dx'dx_1 = 0.$$

Then (see [51], [4]) if $D_{\ell_1} = \Omega_{\ell_1+1} \backslash \Omega_{\ell_1}$, $D_{\ell_1}^\pm = D_{\ell_1} \cap (\mathbb{R}^\pm \times \omega')$ one can find in each domain $D_{\ell_1}^+$ and $D_{\ell_1}^-$ functions β^+ and β^- such that

$$\begin{cases} \beta^\pm \in \mathbb{H}_0^1(D_{\ell_1}^\pm) \\ \operatorname{div} \beta^\pm = \partial_{x_1}\rho(u_\ell^1 - u_{\ell+r}^1) \text{ in } D_{\ell_1}^\pm \\ ||\nabla\beta^\pm||_{2,D_{\ell_1}^\pm} \leq C|u_\ell^1 - u_{\ell+r}^1|_{2,D_{\ell_1}^\pm} \end{cases} \qquad (8.12)$$

(C is independent of ℓ since by translation one can transport the problem on $(0,1) \times \omega'$). We denote by β the function given by β^+ on $D_{\ell_1}^+$ and β^- on $D_{\ell_1}^-$. It is clear that

$$\rho(\mathbf{u}_\ell - \mathbf{u}_{\ell+r}) - \beta \in \hat{\mathbb{H}}_0^1(\Omega_\ell)$$

and moreover

$$||\nabla\beta||_{2,D_{\ell_1}} \leq C|u_\ell^1 - u_{\ell+r}^1|_{2,D_{\ell_1}}. \qquad (8.13)$$

Thus by (8.11) one deduces - since ρ and β vanish outside of Ω_{ℓ_1+1}

$$\int_{\Omega_{\ell_1+1}} \nabla(\mathbf{u}_\ell - \mathbf{u}_{\ell+r}) \cdot \nabla(\rho(\mathbf{u}_\ell - \mathbf{u}_{\ell+r}) - \beta)dx = 0.$$

This implies easily

$$\int_{\Omega_{\ell_1+1}} |\nabla(\mathbf{u}_\ell - \mathbf{u}_{\ell+r})|^2 \rho dx = -\int_{\Omega_{\ell_1+1}} \partial_{x_1}(\mathbf{u}_\ell - \mathbf{u}_{\ell+r}) \cdot (\mathbf{u}_\ell - \mathbf{u}_{\ell+r})\partial_{x_1}\rho dx$$

$$+\int_{\Omega_{\ell_1+1}} \nabla(\mathbf{u}_\ell - \mathbf{u}_{\ell+r}) \cdot \nabla\beta dx.$$

In these two last integrals one integrates only on D_{ℓ_1}. Moreover one will lower the integral on the left hand side by integrating on Ω_{ℓ_1} where $\rho = 1$. Thus it comes

$$\int_{\Omega_{\ell_1}} |\nabla(\mathbf{u}_\ell - \mathbf{u}_{\ell+r})|^2 dx \leq \int_{D_{\ell_1}} |\nabla(\mathbf{u}_\ell - \mathbf{u}_{\ell+r})||\mathbf{u}_\ell - \mathbf{u}_{\ell+r}|dx$$

$$+\int_{D_{\ell_1}} |\nabla(\mathbf{u}_\ell - \mathbf{u}_{\ell+r})||\nabla\beta|dx.$$

Using the Cauchy-Young inequality $ab \leq \frac{1}{2}(a^2 + b^2)$, one obtains

$$\int_{\Omega_{\ell_1}} |\nabla(\mathbf{u}_\ell - \mathbf{u}_{\ell+r})|^2 dx \leq \int_{D_{\ell_1}} |\nabla(\mathbf{u}_\ell - \mathbf{u}_{\ell+r})|^2 dx$$

$$+\frac{1}{2}\int_{D_{\ell_1}} |\mathbf{u}_\ell - \mathbf{u}_{\ell+r}|^2 dx + \frac{1}{2}\int_{D_{\ell_1}} |\nabla\beta|^2 dx.$$

So by (8.13), we obtain for a constant C depending only on ω'

$$\int_{\Omega_{\ell_1}} |\nabla(\mathbf{u}_\ell - \mathbf{u}_{\ell+r})|^2 dx \leq \int_{D_{\ell_1}} |\nabla(\mathbf{u}_\ell - \mathbf{u}_{\ell+r})|^2 dx$$

$$+ C\int_{D_{\ell_1}} |\mathbf{u}_\ell - \mathbf{u}_{\ell+r}|^2 dx.$$

Since $\mathbf{u}_\ell - \mathbf{u}_{\ell+r}$ vanishes on the lateral boundary of the cylinder, we have a Poincaré inequality of the type

$$\int_{D_{\ell_1}} |(\mathbf{u}_\ell - \mathbf{u}_{\ell+r})|^2 dx \leq C\int_{D_{\ell_1}} |\nabla\mathbf{u}_\ell - \mathbf{u}_{\ell+r}|^2 dx$$

where $C = C(\omega')$. Thus we are ending up for some constant depending only on ω' to

$$\int_{\Omega_{\ell_1}} |\nabla(\mathbf{u}_\ell - \mathbf{u}_{\ell+r})|^2 dx \leq C\int_{D_{\ell_1}} |\nabla(\mathbf{u}_\ell - \mathbf{u}_{\ell+r})|^2 dx$$

$$\leq C\int_{\Omega_{\ell_1+1}} |\nabla(\mathbf{u}_\ell - \mathbf{u}_{\ell+r})|^2 dx - C\int_{\Omega_{\ell_1}} |\nabla(\mathbf{u}_\ell - \mathbf{u}_{\ell+r})|^2 dx$$

i.e.

$$\int_{\Omega_{\ell_1}} |\nabla(\mathbf{u}_\ell - \mathbf{u}_{\ell+r})|^2 dx \leq a\int_{\Omega_{\ell_1+1}} |\nabla(\mathbf{u}_\ell - \mathbf{u}_{\ell+r})|^2 dx$$

with $a = \frac{C}{C+1} < 1$. Using our usual technique starting from $\ell_1 = \frac{\ell}{2}$ we arrive easily to

$$\int_{\Omega_{\frac{\ell}{2}}} |\nabla(\mathbf{u}_\ell - \mathbf{u}_{\ell+r})|^2 dx \leq \frac{1}{a} e^{-\frac{\ell}{2} \ln \frac{1}{a}} \int_{\Omega_\ell} |\nabla(\mathbf{u}_\ell - \mathbf{u}_{\ell+r})|^2 dx$$

i.e.

$$||\nabla(\mathbf{u}_\ell - \mathbf{u}_{\ell+r})||_{2,\Omega_{\ell/2}} \leq \frac{1}{\sqrt{a}} e^{-\frac{\ell}{4} \ln \frac{1}{a}} \{||\nabla \mathbf{u}_\ell||_{2,\Omega_\ell} + ||\nabla \mathbf{u}_{\ell+r}||_{2,\Omega_\ell}\}. \quad (8.14)$$

The estimate of the last quantity is easy. Taking $\mathbf{v} = \mathbf{u}_\ell$ in (8.3), we obtain

$$\mu \int_{\Omega_\ell} \nabla \mathbf{u}_\ell \cdot \nabla \mathbf{u}_\ell dx = \int_{\Omega_\ell} \mathbf{f} \cdot \mathbf{u}_\ell \leq ||\mathbf{f}||_{2,\Omega_\ell} ||\mathbf{u}_\ell||_{2,\Omega_\ell}$$

$$\leq C ||\mathbf{f}||_{2,\Omega_\ell} ||\nabla \mathbf{u}_\ell||_{2,\Omega_\ell}$$

by the Poincaré inequality for some constant $C = C(\omega')$. This gives thanks to (8.8)

$$||\nabla \mathbf{u}_\ell||_{2,\Omega_\ell} \leq C ||\mathbf{f}||_{2,\Omega_\ell} \leq C \ell^\gamma. \quad (8.15)$$

Going back to (8.14) we obtain, recalling that $r \in [0,1]$ and assuming $\ell \geq 1$

$$||\nabla(\mathbf{u}_\ell - \mathbf{u}_{\ell+r})||_{2,\Omega_{\ell/2}} \leq C\{\ell^\gamma + (\ell+r)^\gamma\} e^{-\frac{\ell}{4} \ln \frac{1}{a}}$$

$$\leq C\ell^\gamma \{1 + (1 + \frac{r}{\ell})^\gamma\} e^{-\frac{\ell}{4} \ln \frac{1}{a}}$$

$$\leq C\ell^\gamma \{1 + 2^\gamma\} e^{-\frac{\ell}{4} \ln \frac{1}{a}}$$

$$\leq C\ell^\gamma e^{-\frac{\ell}{4} \ln \frac{1}{a}} \leq C e^{-\alpha \ell}$$

for any $\alpha < \frac{1}{4} \ln \frac{1}{a}$. Using the triangle inequality (see (2.11)), we obtain for any t

$$||\nabla(\mathbf{u}_\ell - \mathbf{u}_{\ell+t})||_{2,\Omega_{\ell/2}} \leq C e^{-\alpha \ell} \quad (8.16)$$

for possibly some other constant C independent of ℓ. Then $\mathbf{u}_\ell$ is a Cauchy "sequence" in $\mathbb{H}_0^1(\Omega_{\ell_0})$ for any $\ell_0 < \ell/2$ and converges when $\ell \to \infty$ toward some $\mathbf{u}_\infty \in \mathbb{H}_0^1(\Omega_{\ell_0})$. Of course this convergence implies that

$$\mathbf{u}_\infty = 0 \text{ on } \mathbb{R} \times \partial \omega', \quad \text{div}\, \mathbf{u}_\infty = 0 \text{ in } \Omega_\infty, \quad \int_{\omega'} u_\infty^1 dx' = 0 \text{ for a.e. } x_1 \in \mathbb{R}.$$

From (8.3) one has

$$\mu \int_{\Omega_{\ell_0}} \nabla \mathbf{u}_\ell \cdot \nabla \mathbf{v} dx = \int_{\Omega_{\ell_0}} \mathbf{f} \cdot \mathbf{v} dx \quad \forall \mathbf{v} \in \hat{\mathbb{H}}_0^1(\Omega_{\ell_0}).$$

Letting $\ell \to \infty$, we derive for any ℓ_0

$$\mu \int_{\Omega_{\ell_0}} \nabla \mathbf{u}_\infty \cdot \nabla \mathbf{v} dx = \int_{\Omega_{\ell_0}} \mathbf{f} \cdot \mathbf{v} dx \quad \forall \mathbf{v} \in \hat{\mathbb{H}}_0^1(\Omega_{\ell_0}).$$

Finally passing to the limit in (8.16) provides

$$\left\|\nabla(\mathbf{u}_\ell - \mathbf{u}_\infty)\right\|_{2,\Omega_{\ell/2}} \le Ce^{-\alpha\ell}$$

i.e.

$$\left\|\nabla(\mathbf{u}_{2\ell} - \mathbf{u}_\infty)\right\|_{2,\Omega_\ell} \le Ce^{-2\alpha\ell}.$$

It follows that

$$\left\|\nabla\mathbf{u}_\infty\right\|_{2,\Omega_\ell} \le Ce^{-\alpha 2\ell} + \left\|\nabla\mathbf{u}_{2\ell}\right\|_{2,\Omega_\ell}$$

and by (8.15)

$$\left\|\nabla\mathbf{u}_\infty\right\|_{2,\Omega_\ell} \le C\ell^\gamma.$$

This proves the last property mentioned in (8.9). To prove uniqueness of $\mathbf{u}_\infty$, notice that if $\mathbf{u}'_\infty$ denotes another solution, we would have

$$\mu \int_{\Omega_\ell} \nabla(\mathbf{u}_\infty - \mathbf{u}'_\infty) \cdot \nabla \mathbf{v} dx = 0 \quad \forall \mathbf{v} \in \hat{\mathbb{H}}_0^1(\Omega_\ell).$$

Then arguing like we did when exploiting (8.11), we arrive to an inequality similar to (8.14), namely

$$\left\|\nabla(\mathbf{u}_\infty - \mathbf{u}'_\infty)\right\|_{2,\Omega_{\ell/2}} \le Ce^{-\alpha'\ell}\{\left\|\nabla\mathbf{u}_\infty\right\|_{2,\Omega_\ell} + \left\|\nabla\mathbf{u}'_\infty\right\|_{2,\Omega_\ell}\}$$

$$\le C\ell^\gamma e^{-\alpha'\ell} \to 0$$

when $\ell \to \infty$. This implies that $\nabla\mathbf{u}_\infty = \nabla\mathbf{u}'_\infty$ in Ω_∞ and thus $\mathbf{u}_\infty = \mathbf{u}'_\infty$.

Remark 8.2. Note that the uniqueness of $\mathbf{u}_\infty$ remains true if we take in (8.9) for the growth of $\left\|\nabla\mathbf{u}_\infty\right\|_{2,\Omega_\ell}$ any power of ℓ, i.e. if we replace the last condition of (8.9) by

$$\left\|\nabla\mathbf{u}_\infty\right\|_{2,\Omega_\ell} = O(\ell^\delta), \quad \delta \ge 0.$$

We claim now that there exists $p_\infty \in \hat{L}_{loc}^2(\overline{\Omega}_\infty)$ such that

$$-\mu\Delta\mathbf{u}_\infty + \nabla p_\infty = \mathbf{f} \quad \text{in } \mathcal{D}'(\Omega_\infty).$$

One has

$$\langle \mu\Delta\mathbf{u}_\infty + \mathbf{f}, \mathbf{v} \rangle = 0 \quad \forall \mathbf{v} \in \hat{\mathbb{H}}_0^1(\Omega_\ell).$$

($\langle \, , \, \rangle$ denotes the duality between $\mathbb{H}^{-1}(\Omega_\ell)$ and $\mathbb{H}_0^1(\Omega_\ell)$). Then –see [51], [73], [1]– there exists $p_{\infty,\ell} \in L^2(\Omega_\ell)$ such that

$$\mu \Delta \mathbf{u}_\infty + \mathbf{f} = \nabla p_{\infty,\ell} \quad \text{in } \mathbb{H}^{-1}(\Omega_\ell). \tag{8.17}$$

This is true for every ℓ. Suppose now that we consider $\ell \in \mathbb{N}^*$. We are going to construct p_∞ inductively. First we set

$$p_\infty = p_{\infty,1} \quad \text{on } \Omega_1.$$

Assuming p_∞ constructed on $\Omega_{\ell-1}$, there exists $p_{\infty,\ell} \in L^2(\Omega_\ell)$ satisfying (8.17). Then clearly on $\Omega_{\ell-1}$ one has

$$\nabla(p_\infty - p_{\infty,\ell}) = 0$$

and since $\Omega_{\ell-1}$ is connected, one has for some constant C_ℓ

$$p_\infty - p_{\infty,\ell} = C_\ell \quad \text{in } \Omega_{\ell-1}.$$

One sets then

$$p_\infty = p_{\infty,\ell} + C_\ell \quad \text{in } \Omega_\ell.$$

This makes p_∞ coincide with our previous definition on $\Omega_{\ell-1}$ and completes the proof of existence and uniqueness of a solution to (8.9). We claim now that there exists a constant C independent of ℓ large enough satisfying

$$|p_\ell - p_\infty|_{2,\Omega_{\ell/2}} \leq C\ell ||\nabla(\mathbf{u}_\ell - \mathbf{u}_\infty)||_{2,\Omega_{\ell/2}}. \tag{8.18}$$

By subtraction of (8.1) and (8.9) we obtain in $\mathbb{H}^{-1}(\Omega_\ell)$

$$-\nabla(p_\ell - p_\infty) = -\mu \Delta(\mathbf{u}_\ell - \mathbf{u}_\infty)$$

which is equivalent to

$$\int_{\Omega_\ell} (p_\ell - p_\infty)\mathrm{div}\,\mathbf{v}dx = \mu \int_{\Omega_\ell} \nabla(\mathbf{u}_\ell - \mathbf{u}_\infty) \cdot \nabla\mathbf{v}dx \quad \forall \mathbf{v} \in \mathbb{H}_0^1(\Omega_\ell). \tag{8.19}$$

We suppose the representative of $(p_\ell - p_\infty)$ chosen such that

$$\int_{\Omega_{\ell/2}} (p_\ell - p_\infty)dx = 0.$$

Then, by Lemma 8.2, there exists $\mathbf{v} \in \mathbb{H}_0^1(\Omega_{\ell/2})$ such that

$$\begin{cases} \mathrm{div}\,\mathbf{v} = p_\ell - p_\infty & \text{in } \Omega_{\ell/2}, \\ ||\nabla\mathbf{v}||_{2,\Omega_{\ell/2}} \leq C\ell|p_\ell - p_\infty|_{2,\Omega_{\ell/2}}, \end{cases}$$

for some constant C independent of ℓ. Extending $\mathbf{v}$ by 0 outside of $\Omega_{\ell/2}$ and using it in (8.19), it comes

$$\int_{\Omega_{\ell/2}} (p_\ell - p_\infty)^2 dx = \mu \int_{\Omega_{\ell/2}} \nabla(\mathbf{u}_\ell - \mathbf{u}_\infty) \cdot \nabla v dx$$

$$\leq \mu ||\nabla(\mathbf{u}_\ell - \mathbf{u}_\infty)||_{2,\Omega_{\ell/2}} ||\nabla v||_{2,\Omega_{\ell/2}}$$

$$\leq C\ell ||\nabla(\mathbf{u}_\ell - \mathbf{u}_\infty)||_{2,\Omega_{\ell/2}} |p_\ell - p_\infty|_{2,\Omega_{\ell/2}}$$

which is precisely (8.18). Letting $t \to \infty$ in (8.16) we get

$$||\nabla(\mathbf{u}_\ell - \mathbf{u}_\infty)||_{2,\Omega_{\ell/2}} \leq Ce^{-\alpha\ell}$$

for any $\alpha < \frac{1}{4} \ln \frac{1}{a}$. From (8.18) we obtain similarly

$$|p_\ell - p_\infty|_{2,\Omega_{\ell/2}} \leq Ce^{-\alpha\ell}$$

for any $\alpha < \frac{1}{4} \ln \frac{1}{a}$. This completes the proof of the theorem. $\qquad\square$

In the case where $\mathbf{f} = \mathbf{f}(x')$ depends only on the section of the cylinder, it is interesting to see if at the limit, i.e. when $\ell \to \infty$, $\mathbf{u}_\ell$ depends only on the section of the cylinder. The answer will be given in the following result. We will need the space

$$\mathbb{V}(\omega') = \{\mathbf{v} = (v^1, \mathbf{v}') \in H_0^1(\omega') \times \hat{\mathbb{H}}_0^1(\omega') \mid \int_{\omega'} v^1 dx' = 0\}.$$

Then we have:

Theorem 8.2. *Suppose that* $\mathbf{f} = \mathbf{f}(x') \in \mathbb{L}^2(\omega') = (L^2(\omega'))^n$. *Let* $(\mathbf{u}_\ell, p_\ell)$ *be the solution to* (8.2) *and* $\mathbf{u}_\infty$ *the solution to*

$$\begin{cases} \mathbf{u}_\infty \in \mathbb{V}(\omega'), \\ \mu \int_{\omega'} \nabla' \mathbf{u}_\infty \cdot \nabla' v dx' = \int_{\omega'} \mathbf{f} \cdot \mathbf{v} dx' \quad \forall \mathbf{v} \in \mathbb{V}(\omega'). \end{cases} \qquad (8.20)$$

Moreover let $p_\infty \in \hat{L}^2(\Omega_\infty)$ *be such that*

$$-\mu \Delta \mathbf{u}_\infty + \nabla p_\infty = \mathbf{f} \quad in \; \mathcal{D}'(\Omega_\infty). \qquad (8.21)$$

Then there are positive constants C *and* α *such that*

$$||\nabla(\mathbf{u}_\ell - \mathbf{u}_\infty)||_{2,\Omega_{\ell/2}} + |p_\ell - p_\infty|_{2,\Omega_{\ell/2}} \leq Ce^{-\alpha\ell}. \qquad (8.22)$$

Proof. First by the Lax-Milgram theorem it is easy to see that the problem (8.20) possesses a unique solution. To complete the proof it is enough to show that $\mathbf{u}_\infty \in \mathbb{V}(\omega')$ satisfies (8.9). It is a trivial matter to verify all the conditions of (8.9). Note that since $\mathbf{u}_\infty$ is independent of x_1, one has

$$||\nabla \mathbf{u}_\infty||_{2,\Omega_\ell} = O(\ell).$$

By Remark 8.2, this completes the proof of the theorem. $\qquad\square$

Considering in (8.20) a test function of the type $(0, \mathbf{v}')$ with

$$\operatorname{div}'\mathbf{v}' = \sum_{i=2}^{n} \partial_{x_i} v^i = 0$$

one sees that $\mathbf{u}'_\infty$ is solution to

$$\begin{cases} \mathbf{u}'_\infty \in \hat{\mathbb{H}}^1_0(\omega'), \\ \mu \int_{\omega'} \nabla'\mathbf{u}'_\infty \cdot \nabla'\mathbf{v}dx' = \int_{\omega'} \mathbf{f}' \cdot \mathbf{v}dx' \quad \forall \mathbf{v} \in \hat{\mathbb{H}}^1_0(\omega'). \end{cases} \tag{8.23}$$

In the system above $\hat{\mathbb{H}}^1_0(\omega')$ denotes the set of solenoidal vector fields –i.e. divergence free in ω'– with $n-1$ components in $H^1_0(\omega')$. This shows that $\mathbf{u}'_\infty$ is the solution of the Stokes problem set in ω' and corresponding to a force density $\mathbf{f}'$. Note that when $n = 2$, $\mathbf{u}'_\infty = 0$.

Taking now as test function in (8.20) $(v^1, 0)$ with v^1 satisfying

$$\int_{\omega'} v^1 dx' = 0$$

one sees that u^1_∞ is the solution to

$$\begin{cases} u^1_\infty \in H^1_0(\omega') \cap \hat{L}^2(\omega'), \\ \mu \int_{\omega'} \nabla'u^1_\infty \cdot \nabla'vdx' = \int_{\omega'} f^1 vdx' \quad \forall v \in H^1_0(\omega') \cap \hat{L}^2(\omega'). \end{cases} \tag{8.24}$$

Note that the problem above is equivalent to the problem

$$\begin{cases} (u^1_\infty, k) \in (H^1_0(\omega') \cap \hat{L}^2(\omega')) \times \mathbb{R}, \\ \mu\Delta'u^1_\infty + f^1 = k \quad \text{in } H^{-1}(\omega'). \end{cases} \tag{8.25}$$

(Δ' denotes the Laplace operator in $\mathbb{R}^{(n-1)}$, i.e. in the last $n-1$ coordinates). Clearly taking as test function $v \in H^1_0(\omega') \cap \hat{L}^2(\omega')$ in (8.25) shows that u^1_∞ satisfies (8.24). Conversely suppose that u^1_∞ satisfies (8.24). Let θ be a function of $\mathcal{D}(\omega')$ such that

$$\int_{\omega'} \theta dx' = 1.$$

For $\varphi \in H^1_0(\omega')$ it is clear that

$$\varphi - \left(\int_{\omega'} \varphi dx' \right)\theta \in H^1_0(\omega') \cap \hat{L}^2(\omega')$$

and thus from (8.24) one derives

$$\mu \int_{\omega'} \nabla'u^1_\infty \cdot \nabla'\varphi dx' = \int_{\omega'} f^1 \varphi dx'$$

$$+ \left(\int_{\omega'} \mu\nabla'u^1_\infty \cdot \nabla'\theta - f^1\theta dx' \right) \int_{\omega'} \varphi dx',$$

i.e. in $H^{-1}(\omega')$

$$\mu\Delta'u_\infty^1 + f^1 = k = -\Big(\int_{\omega'}\mu\nabla'u_\infty^1 \cdot \nabla'\theta - f^1\theta dx'\Big).$$

One can compute the constant k more explicitly. Indeed let us denote by $u(f)$ the weak solution to

$$-\mu\Delta'u(f) = f \quad \text{in } \omega', \quad u(f) = 0 \quad \text{on } \partial\omega'.$$

One has by linearity of Δ'

$$\mu\Delta'\{u(f) - ku(1)\} = \mu\Delta'u(f) - k\mu\Delta'u(1) = -f + k.$$

Thus by uniqueness of the solution to (8.25) one has

$$u_\infty^1 = u(f^1) - ku(1).$$

Using the constraint

$$\int_{\omega'} u_\infty^1 dx' = 0$$

one finds immediately that

$$k = \int_{\omega'} u(f^1)dx' / \int_{\omega'} u(1)dx'.$$

Note that by the maximum principle the denominator of this fraction never vanishes.

Chapter 9

Variational inequalities

9.1 A simple result

As an heuristic introduction to the result below let us notice that the solution to

$$\begin{cases} -u'' + u = 0 \text{ in } (-\ell, \ell), \\ u(\pm \ell) = g, \end{cases}$$

where g is a constant is given by

$$u = u_\ell = g \frac{e^{-\ell} - e^{\ell}}{e^{-2\ell} - e^{2\ell}} \{e^{-x} + e^{x}\}$$

and converges toward 0 exponentially quickly when $\ell \to \infty$.

Let us now precise our notation. We denote by Ω_ℓ the open subset of $\mathbb{R}^n$ defined as

$$\Omega_\ell = (-\ell, \ell)^p \times \omega_2.$$

ℓ is a positive number, $p \geq 1$, ω_2 is a bounded open subset of $\mathbb{R}^{n-p}$. The points in $\mathbb{R}^n$ are denoted by

$$x = (X_1, X_2),$$

with

$$X_1 = (x_1, \ldots, x_p), \qquad X_2 = (x_{p+1}, \ldots, x_n).$$

Let K be a closed convex subset of $\mathbb{R}^n$ such that for some $a > 0$

$$\{(X_1, 0) \mid |X_1| \leq a\} \subset K. \tag{9.1}$$

($| \ |$ denotes the euclidean norm). Let us denote by g a function of X_2 such that

$$(0, \nabla_{X_2} g) \in K, \ g \in H_0^1(\omega_2), \ |\nabla_{X_2} g| \leq A \text{ a.e. on } \omega_2 \tag{9.2}$$

for some positive constant A. Set also

$$K_\ell = \{v \in H^1(\Omega_\ell) \mid \nabla v(x) \in K \text{ a.e. } x \in \Omega_\ell, v = g \text{ on } \partial\Omega_\ell\}. \qquad (9.3)$$

Let us denote by u_ℓ the solution to

$$\begin{cases} u_\ell \in K_\ell, \\ \int_{\Omega_\ell} \nabla u_\ell \cdot \nabla(v - u_\ell)dx \geq 0 \quad \forall v \in K_\ell. \end{cases} \qquad (9.4)$$

It is easy to show that K_ℓ is not empty ($g \in K_\ell$) and is a closed convex subset of

$$V_\ell = \{v \in H^1(\Omega_\ell) \mid v = 0 \text{ on } (-\ell, \ell)^p \times \partial\omega_2\}.$$

Moreover V_ℓ is a Hilbert space for the usual Dirichlet scalar product. Thus, (see the Appendix), there exists a unique solution to (9.4). Then we have

Theorem 9.1. u_ℓ *converges toward 0 with an exponential rate of convergence.*

Proof. First notice that due to (9.2) and the boundedness of ω_2, g is uniformly bounded by some constant g_∞. Then one has also

$$|u_\ell| \leq g_\infty. \qquad (9.5)$$

Indeed the functions $v = u_\ell - (u_\ell - g_\infty)^+$, $v = u_\ell + (-u_\ell - g_\infty)^+$ are both in K_ℓ. Using them in (9.4) one gets

$$\int_{\Omega_\ell} \nabla u_\ell \cdot \nabla(u_\ell - g_\infty)^+ dx \leq 0, \qquad \int_{\Omega_\ell} \nabla(-u_\ell) \cdot \nabla(-u_\ell - g_\infty)^+ dx \leq 0.$$

This can also be written

$$\int_{\Omega_\ell} \nabla(u_\ell - g_\infty) \cdot \nabla(u_\ell - g_\infty)^+ dx \leq 0, \qquad \int_{\Omega_\ell} \nabla(-u_\ell - g_\infty) \cdot \nabla(-u_\ell - g_\infty)^+ dx \leq 0$$

which implies that $(u_\ell - g_\infty)^+$ and $(-u_\ell - g_\infty)^+$ vanish which gives (9.5). Let us denote by C the constant given by

$$C = \frac{g_\infty + 1}{a}.$$

We claim that for $\ell_1 + C \leq \ell$

$$v = \text{sign}(u_\ell)\left\{|u_\ell| \wedge \left(\frac{|u_\ell|}{|u_\ell| + 1}a(|X_1|_\infty - \ell_1)^+\right)\right\} \in K_\ell \qquad (9.6)$$

($\text{sign}(z)$ denotes the sign of z, $|X_1|_\infty$ is the maximum norm defined as $|X_1|_\infty = \max_{i=1}^p |x_i|$ for $X_1 = (x_1, \ldots, x_p)$, $\wedge$ denotes the minimum of two numbers). Indeed this function vanishes when u_ℓ does. Moreover for $|X_1|_\infty \geq \ell_1 + C$, one has

$$a(|X_1|_\infty - \ell_1)^+ \geq aC = g_\infty + 1 \geq |u_\ell| + 1$$

and $v = u_\ell$. In particular $v = g$ for $|X_1|_\infty = \ell$ and $\nabla v \in K$. If $|X_1|_\infty \leq \ell_1$ one has $v = 0$ and $\nabla v \in K$. For $a(|X_1|_\infty - \ell_1)^+ \geq |u_\ell| + 1$, $v = u_\ell$ and again $\nabla v \in K$. Finally for $0 < a(|X_1|_\infty - \ell_1)^+ \leq |u_\ell| + 1$ one has

$$v = \frac{u_\ell}{|u_\ell| + 1}\{a|X_1|_\infty - \ell_1\}$$

and

$$\begin{aligned}
\nabla v &= \frac{u_\ell}{|u_\ell| + 1}a\nabla|X_1|_\infty + \frac{1}{(|u_\ell| + 1)^2}a(|X_1|_\infty - \ell_1)\nabla u_\ell \\
&= \frac{|u_\ell|}{|u_\ell| + 1}\operatorname{sign}(u_\ell)a\nabla|X_1|_\infty + \frac{1}{|u_\ell| + 1}\frac{a(|X_1|_\infty - \ell_1)}{|u_\ell| + 1}\nabla u_\ell \in K
\end{aligned} \tag{9.7}$$

since

$$\pm a\nabla|X_1|_\infty \in K \quad \text{and} \quad \frac{a(|X_1|_\infty - \ell_1)}{|u_\ell| + 1}\nabla u_\ell \in K.$$

Moreover one has always

$$|\nabla v| \leq a|u_\ell| + |\nabla u_\ell|. \tag{9.8}$$

Using the function v in (9.4) one gets

$$\int_{\Omega_{\ell_1+C}} |\nabla u_\ell|^2 dx \leq \int_{\Omega_{\ell_1+C}\backslash\Omega_{\ell_1}} \nabla u_\ell \cdot \nabla v \, dx \leq \int_{\Omega_{\ell_1+C}\backslash\Omega_{\ell_1}} a|u_\ell||\nabla u_\ell| + |\nabla u_\ell|^2 dx.$$

Using the Poincaré inequality (Cf. the Appendix) it follows for some constant γ

$$\int_{\Omega_{\ell_1}} |\nabla u_\ell|^2 dx \leq \gamma \int_{\Omega_{\ell_1+C}\backslash\Omega_{\ell_1}} |\nabla u_\ell|^2 dx$$

i.e.

$$\int_{\Omega_{\ell_1}} |\nabla u_\ell|^2 dx \leq \delta \int_{\Omega_{\ell_1+C}} |\nabla u_\ell|^2 dx, \quad \delta = \frac{\gamma}{\gamma + 1} < 1.$$

Iterating this formula starting from $\frac{\ell}{2}$ with a step C, i.e. making $[\frac{\ell}{2C}]$ iterations with $[\ ]$ denoting the integer part of a number, we arrive to

$$\int_{\Omega_{\frac{\ell}{2}}} |\nabla u_\ell|^2 dx \leq \delta^{[\frac{\ell}{2C}]} \int_{\Omega_{\frac{\ell}{2}+C[\frac{\ell}{2C}]}} |\nabla u_\ell|^2 dx.$$

Since $\frac{\ell}{2C} - 1 \leq [\frac{\ell}{2C}] \leq \frac{\ell}{2C}$ this leads to

$$\int_{\Omega_{\frac{\ell}{2}}} |\nabla u_\ell|^2 dx \leq \frac{1}{\delta} \delta^{\frac{\ell}{2C}} \int_{\Omega_\ell} |\nabla u_\ell|^2 dx. \qquad (9.9)$$

We have to estimate this last integral. For this one considers v given by (9.6) where we have replaced u_ℓ by g and ℓ_1 by $\ell - C$. Arguing as above, it is easy to see that $v \in K$. Using (9.4) one gets

$$\int_{\Omega_\ell} |\nabla u_\ell|^2 dx \leq \int_{\Omega_\ell \setminus \Omega_{\ell-C}} \nabla u_\ell \cdot \nabla v \, dx$$

$$\leq \{ \int_{\Omega_\ell} |\nabla u_\ell|^2 dx \}^{\frac{1}{2}} \{ \int_{\Omega_\ell \setminus \Omega_{\ell-C}} |\nabla v|^2 dx \}^{\frac{1}{2}}.$$

Thus

$$\int_{\Omega_\ell} |\nabla u_\ell|^2 dx \leq \int_{\Omega_\ell \setminus \Omega_{\ell-C}} |\nabla v|^2 dx$$

$$\leq \int_{\Omega_\ell \setminus \Omega_{\ell-C}} (a|g| + |\nabla_{X_2} g|)^2 dx$$

$$\leq (ag_\infty + A)^2 \{ (2\ell)^p - (2(\ell - C))^p \} |\omega_2|$$

$$\leq C\ell^{p-1}.$$

Going back to (9.9) we deduce for some constants C

$$\int_{\Omega_{\frac{\ell}{2}}} |\nabla u_\ell|^2 dx \leq Ce^{-\frac{\ell}{2C} \ln \frac{1}{\delta}} \ell^{p-1} \leq Ce^{-\alpha\ell}$$

provided we choose $\alpha < \frac{1}{2C} \ln \frac{1}{\delta}$.   □

Remark 9.1. The above result holds for a general elliptic operator replacing the Laplacian (possibly nonlinear), $K = K(x)$...

One cannot remove the assumption (9.1). Suppose for instance that $\Omega_\ell = (-\ell, \ell) \times (-1, 1)$ and K is given by the figure below (Fig. 9.1).

If $g = g(x_2)$, $u \in K_\ell$ implies that $\partial_{x_1} u \geq 0$ and thus $u \equiv g$, i.e. K_ℓ is reduced to g and the solution does not go to 0.

9.2   The obstacle problem in unbounded domains

We consider a domain Ω such that

$$\Omega \subset \mathbb{R}^p \times \omega_2$$

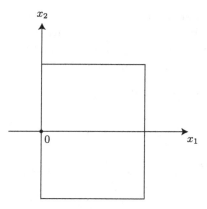

Fig. 9.1 A non suitable convex subset.

where ω_2 is a bounded open subset of $\mathbb{R}^{n-p}$. For $\ell > 0$, if ω_1 is a bounded convex open subset of $\mathbb{R}^p$, we denote by Ω_ℓ the set

$$\Omega_\ell = (\ell\omega_1 \times \omega_2) \cap \Omega.$$

We denote by $A = A(x)$ a $n \times n$ matrix satisfying for some positive constants λ and Λ

$$A(x)\xi \cdot \xi \geq \lambda|\xi|^2 \quad \text{a.e. } x \in \Omega, \forall \xi \in \mathbb{R}^n, \tag{9.10}$$

$$|A(x)\xi| \leq \Lambda|\xi| \quad \text{a.e. } x \in \Omega, \forall \xi \in \mathbb{R}^n. \tag{9.11}$$

We define V_ℓ by

$$V_\ell = \{v \in H^1(\Omega_\ell) \mid v = 0 \text{ on } \partial\Omega_\ell \cap \partial\Omega\}$$

i.e. a function v in V_ℓ vanishes only on the part of $\partial\Omega_\ell$ which is also in $\partial\Omega$. We assume that this Hilbert space is normed by

$$|v|_{V_\ell} = \|\nabla v\|_{2,\Omega_\ell} = \left(\int_{\Omega_\ell} |\nabla v|^2 dx\right)^{\frac{1}{2}}.$$

Let

$$\varphi \in H^1_{loc}(\overline{\Omega}), \ \varphi \leq 0 \text{ on } \partial\Omega. \tag{9.12}$$

For $f \in V_\ell^*$ we denote by u_ℓ the solution to

$$\begin{cases} u_\ell \in \mathcal{K}_\ell = \{v \in V_\ell \mid v(x) \geq \varphi(x) \text{ a.e. in } \Omega_\ell\}, \\ \displaystyle\int_{\Omega_\ell} A(x)\nabla u_\ell \cdot \nabla(v - u_\ell)dx \geq \langle f, v - u_\ell \rangle \quad \forall v \in \mathcal{K}_\ell, \end{cases} \tag{9.13}$$

($\langle \, , \, \rangle$ denotes the V_ℓ^*, V_ℓ duality). It is clear that $\mathcal{K}_\ell$ is a closed subset of V_ℓ which is not empty since φ^+, the positive part of φ, belongs to $\mathcal{K}_\ell$. Then the existence and uniqueness of a solution to (9.13) follows from Theorem A.3 of the Appendix. Then we can show:

Theorem 9.2. *We suppose that*

$$|\varphi^+|_{V_\ell}, \, |f|_{V_\ell^*} = O(e^{\delta\ell}) \tag{9.14}$$

when $\ell \to \infty$. Set

$$\mathcal{K}_\infty = \{v \in H^1_{loc}(\overline{\Omega}) \mid v = 0 \text{ on } \partial\Omega, \, v(x) \geq \varphi(x) \text{ a.e. } x \in \Omega\}.$$

Then there exists δ_0 such that for $\delta < \delta_0$ the problem

$$\begin{cases} u_\infty \in \mathcal{K}_\infty, \\ \displaystyle\int_{\Omega_\ell} A(x)\nabla u_\infty \cdot \nabla((v - u_\infty)\rho)dx \geq \langle f, (v - u_\infty)\rho \rangle, \\ \qquad\qquad \forall v \in \mathcal{K}_\infty, \, \forall \ell > 0, \, \forall \rho \in W_0^{1,\infty}(\ell\omega_1), \, \rho \geq 0, \\ \|\nabla u_\infty\|_{2,\Omega_\ell} = O(e^{\delta\ell}), \end{cases} \tag{9.15}$$

possesses a unique solution. Moreover there exists $\beta > 0$ such that

$$\|\nabla(u_\ell - u_\infty)\|_{2,\Omega_{\frac{\ell}{2}}} = O(e^{-\beta\ell}) \tag{9.16}$$

when $\ell \to \infty$. ($W_0^{1,\infty}(\ell\omega_1)$ denotes the space of uniformly Lipschitz continuous functions vanishing on $\partial(\ell\omega_1)$).

Proof. We start by estimating $u_\ell - u_{\ell+r}$ for $r \in [0,1]$. Denote by $\rho = \rho_{\ell_1}$, $\ell_1 \leq \ell - 1$ the function defined in (2.7). Then it is clear that

$$u_\ell - (u_\ell - u_{\ell+r})\rho \in \mathcal{K}_\ell$$

and by (9.13) noting that ρ is vanishing out of Ω_{ℓ_1+1}, we get

$$\int_{\Omega_{\ell_1+1}} A(x)\nabla u_\ell \cdot \nabla(-(u_\ell - u_{\ell+r})\rho)dx \geq \langle f, -(u_\ell - u_{\ell+r})\rho \rangle. \tag{9.17}$$

Similarly

$$u_{\ell+r} + (u_\ell - u_{\ell+r})\rho \in \mathcal{K}_{\ell+r}$$

and since ρ vanishes out of Ω_{ℓ_1+1}, we get from (9.13) where ℓ is replaced by $\ell + r$

$$\int_{\Omega_{\ell_1+1}} A(x)\nabla u_{\ell+r} \cdot \nabla((u_\ell - u_{\ell+r})\rho)dx \geq \langle f, (u_\ell - u_{\ell+r})\rho \rangle. \tag{9.18}$$

Adding (9.17) and (9.18) leads to

$$\int_{\Omega_{\ell_1+1}} A(x)\nabla(u_\ell - u_{\ell+r}) \cdot \nabla((u_\ell - u_{\ell+r})\rho)dx \leq 0$$

which can be written as

$$\int_{\Omega_{\ell_1+1}} A(x)\nabla(u_\ell - u_{\ell+r}) \cdot \nabla(u_\ell - u_{\ell+r})\rho dx$$

$$\leq -\int_{\Omega_{\ell_1+1}} A(x)\nabla(u_\ell - u_{\ell+r}) \cdot \nabla\rho(X_1)(u_\ell - u_{\ell+r})dx.$$

Using (9.10), (9.11) we get since $\rho = 1$ on Ω_{ℓ_1}

$$\lambda \int_{\Omega_{\ell_1}} |\nabla(u_\ell - u_{\ell+r})|^2 dx \leq \Lambda C \int_{\Omega_{\ell_1+1}\backslash\Omega_{\ell_1}} |\nabla(u_\ell - u_{\ell+r})||u_\ell - u_{\ell+r}|dx$$

where C is the constant appearing in (2.7). Using the Young inequality we get, setting $D_{\ell_1} = \Omega_{\ell_1+1}\backslash\Omega_{\ell_1}$

$$\int_{\Omega_{\ell_1}} |\nabla(u_\ell - u_{\ell+r})|^2 dx \leq \frac{\Lambda C}{2\lambda} \int_{D_{\ell_1}} |\nabla(u_\ell - u_{\ell+r})|^2 + |u_\ell - u_{\ell+r}|^2 dx. \quad (9.19)$$

By the Poincaré inequality (Cf. the Appendix) one has for a.e. $X_1 \in (\ell_1 + 1)\omega_1\backslash\ell_1\omega_1$

$$\int_{\omega_2} (u_\ell - u_{\ell+r})^2(X_1, \cdot)dX_2 \leq C(\omega_2) \int_{\omega_2} |\nabla_{X_2}(u_\ell - u_{\ell+r})|^2(X_1, \cdot)dX_2.$$

Integrating in X_1, this leads to

$$\int_{D_{\ell_1}} (u_\ell - u_{\ell+r})^2 dx \leq C(\omega_2) \int_{D_{\ell_1}} |\nabla(u_\ell - u_{\ell+r})|^2 dx.$$

Going back to (9.19), we obtain

$$\int_{\Omega_{\ell_1}} |\nabla(u_\ell - u_{\ell+r})|^2 dx \leq \frac{\Lambda(C(\omega_2) \vee 1)C}{2\lambda} \int_{D_{\ell_1}} |\nabla(u_\ell - u_{\ell+r})|^2 dx$$

where $(C(\omega_2) \vee 1)$ denotes the maximum of $C(\omega_2)$ and 1. Thus it comes easily

$$\int_{\Omega_{\ell_1}} |\nabla(u_\ell - u_{\ell+r})|^2 dx \leq a \int_{\Omega_{\ell_1+1}} |\nabla(u_\ell - u_{\ell+r})|^2 dx,$$

with

$$a = \frac{\Lambda(C(\omega_2) \vee 1)C}{2\lambda + \Lambda(C(\omega_2) \vee 1)C} < 1.$$

Using our usual iteration process we obtain

$$\int_{\Omega_{\frac{\ell}{2}}} |\nabla(u_\ell - u_{\ell+r})|^2 dx \le \frac{1}{a} e^{-\frac{\ell}{2}\ln\frac{1}{a}} \int_{\Omega_\ell} |\nabla(u_\ell - u_{\ell+r})|^2 dx. \qquad (9.20)$$

It remains to estimate this last integral, i.e. to estimate u_ℓ.

We notice that $\varphi^+ \in \mathcal{K}_\ell$ and by (9.13) we have

$$\int_{\Omega_\ell} A(x)\nabla u_\ell \cdot \nabla(\varphi^+ - u_\ell)dx \ge \langle f, \varphi^+ - u_\ell\rangle.$$

It follows then using the Young inequality

$$\lambda \int_{\Omega_\ell} |\nabla u_\ell|^2 dx \le \int_{\Omega_\ell} A(x)\nabla u_\ell \cdot \nabla u_\ell dx$$

$$\le \Lambda ||\nabla u_\ell||_{2,\Omega_\ell} ||\nabla \varphi^+||_{2,\Omega_\ell} + |f|_{V_\ell^*}\{||\nabla u_\ell||_{2,\Omega_\ell} + ||\nabla \varphi^+||_{2,\Omega_\ell}\}$$

$$\le \frac{\Lambda\epsilon}{2}||\nabla u_\ell||_{2,\Omega_\ell}^2 + \frac{\Lambda}{2\epsilon}||\nabla \varphi^+||_{2,\Omega_\ell}^2 + \frac{1}{2\epsilon}|f|_{V_\ell^*}^2$$

$$+ \frac{\epsilon}{2}\{||\nabla u_\ell||_{2,\Omega_\ell} + ||\nabla \varphi^+||_{2,\Omega_\ell}\}^2$$

$$\le \frac{\Lambda\epsilon}{2}||\nabla u_\ell||_{2,\Omega_\ell}^2 + \frac{\Lambda}{2\epsilon}||\nabla \varphi^+||_{2,\Omega_\ell}^2 + \frac{1}{2\epsilon}|f|_{V_\ell^*}^2$$

$$+ \epsilon\{||\nabla u_\ell||_{2,\Omega_\ell}^2 + ||\nabla \varphi^+||_{2,\Omega_\ell}^2\}$$

$$= (\epsilon + \frac{\Lambda\epsilon}{2})||\nabla u_\ell||_{2,\Omega_\ell}^2 + (\epsilon + \frac{\Lambda}{2\epsilon})||\nabla \varphi^+||_{2,\Omega_\ell}^2 + \frac{1}{2\epsilon}|f|_{V_\ell^*}^2$$

and thus for small enough ϵ by (9.14), we obtain

$$||\nabla u_\ell||_{2,\Omega_\ell}^2 \le C\{||\nabla \varphi^+||_{2,\Omega_\ell}^2 + |f|_{V_\ell^*}^2\} \le C e^{2\delta\ell}. \qquad (9.21)$$

From (9.20) we then deduce for some constant C

$$\int_{\Omega_{\frac{\ell}{2}}} |\nabla(u_\ell - u_{\ell+r})|^2 dx \le C e^{-2\alpha\ell}\{e^{2\delta\ell} + e^{2\delta(\ell+r)}\}$$

where $\alpha = \frac{1}{4}\ln\frac{1}{a}$. This can be written as

$$\int_{\Omega_{\frac{\ell}{2}}} |\nabla(u_\ell - u_{\ell+r})|^2 dx \le C e^{-2(\alpha-\delta)\ell}\{1 + e^{2\delta r}\}$$

$$\le C e^{-2(\alpha-\delta)\ell}$$

since $r \in [0,1]$. Setting $\beta = \alpha - \delta$ with

$$\delta < \delta_0 < \alpha = \frac{1}{4}\ln\frac{1}{a} \qquad (9.22)$$

we get

$$|u_\ell - u_{\ell+r}|_{V_{\frac{\ell}{2}}} \leq Ce^{-\beta\ell}, \quad r \in [0,1].$$

If now $t > 0$ is arbitrary the triangular inequality gives us

$$|u_\ell - u_{\ell+t}|_{V_{\frac{\ell}{2}}} \leq |u_\ell - u_{\ell+1}|_{V_{\frac{\ell}{2}}} + |u_{\ell+1} - u_{\ell+2}|_{V_{\frac{\ell}{2}}} + \cdots + |u_{\ell+[t]} - u_{\ell+t}|_{V_{\frac{\ell}{2}}}$$

$$\leq |u_\ell - u_{\ell+1}|_{V_{\frac{\ell}{2}}} + |u_{\ell+1} - u_{\ell+2}|_{V_{\frac{\ell+1}{2}}} + \cdots + |u_{\ell+[t]} - u_{\ell+t}|_{V_{\frac{\ell+[t]}{2}}}$$

$$\leq Ce^{-\beta\ell} + Ce^{-\beta(\ell+1)} + \cdots + Ce^{-\beta(\ell+[t])}$$

$$\leq Ce^{-\beta\ell}\{1 + e^{-\beta\ell} + e^{-2\beta\ell} + \cdots + Ce^{-\beta[t])}\}$$

$$\leq Ce^{-\beta\ell}\frac{1}{1 - e^{-\beta}}$$

i.e.

$$|u_\ell - u_{\ell+t}|_{V_{\frac{\ell}{2}}} \leq Ce^{-\beta\ell}. \tag{9.23}$$

Let us now choose $\ell_0 \leq \frac{\ell}{2}$. From the inequality above it is clear that u_ℓ is a Cauchy sequence in V_{ℓ_0}. Thus there exists $u_\infty \in V_{\ell_0}$ such that

$$u_\ell \to u_\infty \text{ in } H^1(\Omega_{\ell_0}).$$

In particular since u_ℓ vanishes on $\partial\Omega_\ell \cap \partial\Omega$, u_∞ vanishes on $\partial\Omega_{\ell_0} \cap \partial\Omega$. Let us now choose a function $\rho \in W_0^{1,\infty}(\ell_0\omega_1)$, $0 \leq \rho \leq 1$. Then for any $v \in \mathcal{K}_\infty$ one has

$$u_\ell + \rho(v - u_\ell) \in \mathcal{K}_\ell$$

thus from (9.13) we deduce

$$\int_{\Omega_{\ell_0}} A(x)\nabla u_\ell \cdot \nabla(v - u_\ell)\rho dx \geq \langle f, (v - u_\ell)\rho\rangle \quad \forall v \in \mathcal{K}_\infty.$$

Passing to the limit when $\ell \to \infty$, we derive

$$\int_{\Omega_{\ell_0}} A(x)\nabla u_\infty \cdot \nabla(v - u_\infty)\rho dx \geq \langle f, (v - u_\infty)\rho\rangle \quad \forall v \in \mathcal{K}_\infty.$$

Note that in (9.15) we did not assume $0 \leq \rho \leq 1$, but replacing possibly ρ by $\mu\rho$ for $\mu > 0$ small enough we can suppose ρ arbitrary. Now since

$$u_\ell \geq \varphi \text{ on } \Omega_{\ell_0}$$

one deduces that

$$u_\infty \geq \varphi \text{ on } \Omega_{\ell_0}$$

i.e. $u_\infty \in \mathcal{K}_\infty$ and thus the first part of (9.15) is proved. In order to show that u_∞ is solution to (9.15) we yet need to estimate $||\nabla u_\infty||_{2,\Omega_\ell}$. From (9.23) passing to the limit in $t \to \infty$ we obtain

$$|u_\ell - u_\infty|_{V_{\frac{\ell}{2}}} \le Ce^{-\beta\ell}.$$

This implies

$$|u_\infty|_{V_{\frac{\ell}{2}}} \le Ce^{-\beta\ell} + |u_\ell|_{V_{\frac{\ell}{2}}}.$$

Changing $\frac{\ell}{2}$ into ℓ we get

$$|u_\infty|_{V_\ell} \le Ce^{-2\beta\ell} + |u_{2\ell}|_{V_\ell} \le Ce^{2\delta\ell} \tag{9.24}$$

by (9.21). This is not yet the last row of (9.15) and we need to improve the estimate. For that we take in (9.15) $v = \varphi^+$ and ρ the function of X_1 defined by $\rho = \rho_\ell(X_1) = 1 \wedge \text{dist}(X_1, (\partial(\ell+1)\omega_1))$ where $\wedge$ denotes the minimum of two numbers. Note that one has

$$0 \le \rho \le 1, \ \rho = 1 \text{ on } \ell\omega_1, \ \rho = 0 \text{ outside } (\ell+1)\omega_1, \tag{9.25}$$

$$|\nabla_{X_1}\rho(X_1)| \le 1. \tag{9.26}$$

We obtain

$$\int_{\Omega_{\ell+1}} A(x)\nabla u_\infty \cdot \nabla((\varphi^+ - u_\infty)\rho)dx \ge \langle f, (\varphi^+ - u_\infty)\rho\rangle.$$

This can be written as

$$\int_{\Omega_{\ell+1}} A(x)\nabla u_\infty \cdot \nabla u_\infty \rho dx \le -\int_{\Omega_{\ell+1}} A(x)\nabla u_\infty \cdot \nabla \rho u_\infty dx$$

$$+ \int_{\Omega_{\ell+1}} A(x)\nabla u_\infty \cdot \nabla(\rho\varphi^+)dx - \langle f, (\varphi^+ - u_\infty)\rho\rangle.$$

We denote by D_ℓ the set $D_\ell = \Omega_{\ell+1}\backslash\Omega_\ell$. We then derive from above (note (9.10), (9.11), (9.25), (9.26))

$$\lambda\int_{\Omega_{\ell+1}} |\nabla u_\infty|^2\rho dx \le \Lambda\int_{D_\ell} |\nabla u_\infty||u_\infty|dx + \Lambda\int_{D_\ell} |\nabla u_\infty||\varphi^+|dx$$

$$+ \Lambda\int_{\Omega_{\ell+1}} |\nabla u_\infty||\nabla\varphi^+|\rho dx + |\langle f, (\varphi^+ - u_\infty)\rho\rangle|. \tag{9.27}$$

We then estimate each term on the right hand side of this inequality. By the Young and Poincaré inequalities we get

$$\Lambda\int_{D_\ell} |\nabla u_\infty||u_\infty|dx \le \frac{\Lambda}{2}\int_{D_\ell} |\nabla u_\infty|^2 dx + \frac{\Lambda}{2}\int_{D_\ell} |u_\infty|^2 dx$$

$$\le \frac{\Lambda}{2}(1 + C(\omega_2))\int_{D_\ell} |\nabla u_\infty|^2 dx. \tag{9.28}$$

Again by the Young and Poincaré inequalities

$$\Lambda \int_{D_\ell} |\nabla u_\infty||\varphi^+|dx \le \frac{\Lambda}{2} \int_{D_\ell} |\nabla u_\infty|^2 dx + \frac{\Lambda}{2} \int_{D_\ell} (\varphi^+)^2 dx$$

$$\le \frac{\Lambda}{2} \int_{D_\ell} |\nabla u_\infty|^2 dx + \frac{\Lambda}{2} C(\omega_2) \int_{D_\ell} |\nabla \varphi^+|^2 dx$$

$$\le \frac{\Lambda}{2} \int_{D_\ell} |\nabla u_\infty|^2 dx + \frac{\Lambda}{2} C(\omega_2) \int_{\Omega_{\ell+1}} |\nabla \varphi^+|^2 dx.$$

For the third term by the Young inequality it comes

$$\Lambda \int_{\Omega_{\ell+1}} |\nabla u_\infty||\nabla \varphi^+|\rho dx \le \epsilon \int_{\Omega_{\ell+1}} |\nabla u_\infty|^2 \rho dx + \frac{\Lambda^2}{4\epsilon} \int_{\Omega_{\ell+1}} |\nabla \varphi^+|^2 dx.$$

Finally for the last term one has

$$|\langle f, (\varphi^+ - u_\infty)\rho \rangle| \le |f|_{V_{\ell+1}^*} ||\nabla((\varphi^+ - u_\infty)\rho)||_{2,\Omega_{\ell+1}}$$

$$\le \frac{1}{2\epsilon} |f|_{V_{\ell+1}^*}^2 + \frac{\epsilon}{2} ||\nabla((\varphi^+ - u_\infty)\rho)||_{2,\Omega_{\ell+1}}^2 \qquad (9.29)$$

and

$$\int_{\Omega_{\ell+1}} |\nabla((\varphi^+ - u_\infty)\rho)|^2 dx$$

$$\le 4 \int_{\Omega_{\ell+1}} \{|\rho \nabla \varphi^+|^2 + |\varphi^+ \nabla \rho|^2 + |\rho \nabla u_\infty|^2 + |u_\infty \nabla \rho|^2\} dx$$

$$= 4 \int_{\Omega_{\ell+1}} \{|\rho \nabla \varphi^+|^2 + |\rho \nabla u_\infty|^2\} dx + 4 \int_{D_\ell} \{|\varphi^+ \nabla \rho|^2 + |u_\infty \nabla \rho|^2\} dx$$

$$\le 4 \int_{\Omega_{\ell+1}} \{|\nabla \varphi^+|^2 + |\nabla u_\infty|^2 \rho\} dx + 4 \int_{D_\ell} \{|\varphi^+|^2 + |u_\infty|^2\} dx$$

$$\le 4 \int_{\Omega_{\ell+1}} \{|\nabla \varphi^+|^2 + |\nabla u_\infty|^2 \rho\} dx + 4C(\omega_2) \int_{D_\ell} \{|\nabla \varphi^+|^2 + |\nabla u_\infty|^2\} dx$$

$$\le 4 \int_{\Omega_{\ell+1}} |\nabla u_\infty|^2 \rho dx + 4C(\omega_2) \int_{D_\ell} |\nabla u_\infty|^2 dx + 4(1+C(\omega_2)) \int_{\Omega_{\ell+1}} |\nabla \varphi^+|^2 dx$$

and by (9.29) we get

$$|\langle f, (\varphi^+ - u_\infty)\rho \rangle| \le \frac{1}{2\epsilon} |f|_{V_{\ell+1}^*}^2 + 2\epsilon \int_{\Omega_{\ell+1}} |\nabla u_\infty|^2 \rho dx$$

$$+ 2\epsilon C(\omega_2) \int_{D_\ell} |\nabla u_\infty|^2 dx + 2\epsilon(1 + C(\omega_2)) \int_{\Omega_{\ell+1}} |\nabla \varphi^+|^2 dx. \qquad (9.30)$$

Collecting the estimates above it comes

$$\lambda \int_{\Omega_{\ell+1}} |\nabla u_\infty|^2 \rho dx \leq \{\frac{\Lambda}{2}(1 + C(\omega_2)) + \frac{\Lambda}{2} + 2\epsilon C(\omega_2)\} \int_{D_\ell} |\nabla u_\infty|^2 dx$$

$$+ 3\epsilon \int_{\Omega_{\ell+1}} |\nabla u_\infty|^2 \rho dx$$

$$+ \{\frac{\Lambda}{2}C(\omega_2) + \frac{\Lambda^2}{4\epsilon} + 2\epsilon(1 + C(\omega_2))\} \int_{\Omega_{\ell+1}} |\nabla \varphi^+|^2 dx$$

$$+ \frac{1}{2\epsilon} |f|^2_{V^*_{\ell+1}}$$

hence by (9.14)

$$\lambda \int_{\Omega_{\ell+1}} |\nabla u_\infty|^2 \rho dx \leq 3\epsilon \int_{\Omega_{\ell+1}} |\nabla u_\infty|^2 \rho dx$$

$$+ C(\epsilon) \int_{D_\ell} |\nabla u_\infty|^2 dx + C(\epsilon) e^{2\delta(\ell+1)}.$$

Choosing $3\epsilon = \frac{\lambda}{2}$ we arrive to

$$\int_{\Omega_{\ell+1}} |\nabla u_\infty|^2 \rho dx \leq C_1 e^{2\delta(\ell+1)} + C_2 \int_{D_\ell} |\nabla u_\infty|^2 dx$$

for some constants C_1 and C_2. From this we derive

$$\int_{\Omega_\ell} |\nabla u_\infty|^2 dx \leq C_1 e^{2\delta(\ell+1)} + C_2 \int_{\Omega_{\ell+1}} |\nabla u_\infty|^2 dx - C_2 \int_{\Omega_\ell} |\nabla u_\infty|^2 dx$$

i.e.

$$\int_{\Omega_\ell} |\nabla u_\infty|^2 dx \leq C_0 e^{2\delta(\ell+1)} + \gamma \int_{\Omega_{\ell+1}} |\nabla u_\infty|^2 dx \qquad (9.31)$$

where $C_0 = \frac{C_1}{1+C_2}$ and $\gamma = \frac{C_2}{1+C_2} < 1$. Note that C_0 and γ are independent of ℓ and δ. We iterate the inequality (9.31) $[\ell]$ times to get

$$\int_{\Omega_\ell} |\nabla u_\infty|^2 dx \leq C_0 e^{2\delta(\ell+1)} + \gamma \{C_0 e^{2\delta(\ell+2)} + \gamma \int_{\Omega_{\ell+2}} |\nabla u_\infty|^2 dx\}$$

$$\leq C_0 e^{2\delta(\ell+1)} \{1 + \gamma e^{2\delta} + (\gamma e^{2\delta})^2 + \cdots + (\gamma e^{2\delta})^{[\ell]-1}\} + \gamma^{[\ell]} \int_{\Omega_{\ell+[\ell]}} |\nabla u_\infty|^2 dx.$$

If we choose

$$\gamma e^{2\delta} < 1 \iff \delta < \frac{1}{2}\ln(\frac{1}{\gamma}) \qquad (9.32)$$

then

$$1 + \gamma e^{2\delta} + (\gamma e^{2\delta})^2 + \cdots + (\gamma e^{2\delta})^{[\ell]-1} \leq \frac{1}{1 - \gamma e^{2\delta}}$$

so since $\ell - 1 < [\ell] \leq \ell$ we have

$$\int_{\Omega_\ell} |\nabla u_\infty|^2 dx \leq C_0 e^{2\delta(\ell+1)} \frac{1}{1 - \gamma e^{2\delta}} + \frac{1}{\gamma} \gamma^\ell \int_{\Omega_{2\ell}} |\nabla u_\infty|^2 dx.$$

This implies by (9.24)

$$\int_{\Omega_\ell} |\nabla u_\infty|^2 dx \leq \frac{C_0 e^{2\delta}}{1 - \gamma e^{2\delta}} e^{2\delta\ell} + C e^{-\ell \ln(\frac{1}{\gamma})} e^{8\delta\ell}.$$

Now if we also choose

$$8\delta - \ln(\frac{1}{\gamma}) < 2\delta \iff \delta < \frac{1}{6} \ln(\frac{1}{\gamma}) \tag{9.33}$$

we obtain

$$\int_{\Omega_\ell} |\nabla u_\infty|^2 dx \leq C e^{2\delta\ell}$$

i.e. the last line of (9.15). Thus by (9.22), (9.32), (9.33) choosing

$$\delta < \delta_0 = (\frac{1}{4} \ln(\frac{1}{a})) \wedge (\frac{1}{6} \ln(\frac{1}{\gamma}))$$

we obtain (9.15), (9.16).

It remains to show the uniqueness of the solution to (9.15).

Let $\rho = \rho_{\ell_1}$ be the function defined by (2.7) and u_∞, u'_∞ two solutions to (9.15). Taking $v = u'_\infty$ in (9.15) and $v = u_\infty$ in (9.15) written for u'_∞ one gets

$$\int_{\Omega_\ell} A(x) \nabla u_\infty \cdot \nabla((u'_\infty - u_\infty)\rho) dx \geq \langle f, (u'_\infty - u_\infty)\rho \rangle$$

and

$$\int_{\Omega_\ell} A(x) \nabla u'_\infty \cdot \nabla((u_\infty - u'_\infty)\rho) dx \geq \langle f, (u_\infty - u'_\infty)\rho \rangle .$$

Adding these two inequalities it comes

$$\int_{\Omega_\ell} A(x) \nabla(u_\infty - u'_\infty) \cdot \nabla((u_\infty - u'_\infty)\rho) dx \leq 0$$

and then arguing as after (9.18) with u_ℓ, $u_{\ell+r}$ replaced respectively by u_∞, u'_∞, we arrive to (see (9.20))

$$\int_{\Omega_{\frac{\ell}{2}}} |\nabla(u_\infty - u'_\infty)|^2 dx \leq \frac{1}{a} e^{-\alpha\ell} \int_{\Omega_\ell} |\nabla(u_\infty - u'_\infty)|^2 dx$$

for $\alpha = \frac{1}{2} \ln \frac{1}{a}$. It follows that

$$\int_{\Omega_{\frac{\ell}{2}}} |\nabla(u_\infty - u'_\infty)|^2 dx \leq \frac{2}{a} e^{-\alpha\ell} \left\{ \int_{\Omega_\ell} |\nabla u_\infty|^2 dx + \int_{\Omega_\ell} |\nabla u'_\infty|^2 dx \right\}$$

$$\leq C e^{-\alpha\ell} e^{2\delta\ell}$$

and the uniqueness follows since we have chosen $\delta < \frac{\alpha}{2}$ and the right hand side above converges towards 0. This completes the proof of the theorem.

$\square$

Remark 9.2. Suppose that

$$\Omega = \mathbb{R}^p \times \omega_2, \quad \Omega_\ell = \ell\omega_1 \times \omega_2,$$

$$f \in L^2(\omega_2), \quad \varphi \in H^1(\omega_2), \quad \varphi \leq 0 \text{ on } \partial\omega_2.$$

In addition suppose that A is an elliptic matrix (i.e. satisfying (9.10) and (9.11)) of the type

$$A(x) = \begin{pmatrix} A_{11}(x) & A_{12}(X_2) \\ A_{21}(x) & A_{22}(X_2) \end{pmatrix}$$

where A_{11}, A_{22} are respectively $p \times p$ and $(n-p) \times (n-p)$-matrices. Then there exists a unique $\tilde{u}_\infty$ solution to

$$\begin{cases} \tilde{u}_\infty \in K_\infty = \{v \in H_0^1(\omega_2) \mid v(X_2) \geq \varphi(X_2) \text{ a.e. } X_2 \in \omega_2\}, \\ \int_{\omega_2} A_{22}(X_2)\nabla_{X_2}\tilde{u}_\infty \cdot \nabla_{X_2}(v - \tilde{u}_\infty)dX_2 \\ \qquad\qquad \geq \int_{\omega_2} f(X_2)(v - \tilde{u}_\infty)dX_2 \quad \forall v \in K_\infty. \end{cases} \tag{9.34}$$

It is clear that f and φ satisfy (9.14). If $v \in \mathcal{K}_\infty$ one has for a.e. $X_1 \in \ell\omega_1$

$$v(X_1, \cdot) \in K_\infty.$$

Thus from (9.34) one deduces for a.e. $X_1 \in \ell\omega_1$

$$\int_{\omega_2} A_{22}(X_2)\nabla_{X_2}\tilde{u}_\infty \cdot \nabla_{X_2}(v - \tilde{u}_\infty)dX_2 \geq \int_{\omega_2} f(v - \tilde{u}_\infty)dX_2.$$

Now if $\rho = \rho(X_1) \geq 0$ is in $W_0^{1,\infty}(\ell\omega_2)$, due to its independence of X_1 the above inequality implies

$$\int_{\omega_2} A_{22}(X_2)\nabla_{X_2}\tilde{u}_\infty \cdot \nabla_{X_2}(v - \tilde{u}_\infty)\rho dX_2 \geq \int_{\omega_2} f(v - \tilde{u}_\infty)\rho dX_2.$$

Integrating this inequality on $\ell\omega_1$ we get

$$\int_{\Omega_\ell} A_{22}(X_2)\nabla_{X_2}\tilde{u}_\infty \cdot \nabla_{X_2}(v - \tilde{u}_\infty)\rho dx$$

$$\geq \int_{\Omega_\ell} f(X_2)(v - \tilde{u}_\infty)\rho dx \quad \forall v \in \mathcal{K}_\infty. \tag{9.35}$$

Since $\tilde{u}_\infty$ is independent of X_1 it is easy to see that

$$A(x)\nabla\tilde{u}_\infty = \begin{pmatrix} A_{11}(x) & A_{12}(X_2) \\ A_{21}(x) & A_{22}(X_2) \end{pmatrix} \begin{pmatrix} 0 \\ \nabla_{X_2}\tilde{u}_\infty \end{pmatrix} = \begin{pmatrix} A_{12}(X_2)\nabla_{X_2}\tilde{u}_\infty \\ A_{22}(X_2)\nabla_{X_2}\tilde{u}_\infty \end{pmatrix}.$$

Thus one has

$$\int_{\Omega_\ell} A(x)\nabla\tilde{u}_\infty \cdot \nabla((v-\tilde{u}_\infty)\rho)dx$$

$$= \int_{\Omega_\ell} A_{12}(X_2)\nabla_{X_2}\tilde{u}_\infty \cdot \nabla_{X_1}((v-\tilde{u}_\infty)\rho)dx$$

$$+ \int_{\Omega_\ell} A_{22}(X_2)\nabla_{X_2}\tilde{u}_\infty \cdot \nabla_{X_2}((v-\tilde{u}_\infty)\rho)dx$$

$$= \int_{\Omega_\ell} A_{22}(X_2)\nabla_{X_2}\tilde{u}_\infty \cdot \nabla_{X_2}((v-\tilde{u}_\infty)\rho)dx.$$

Indeed, since $A_{12}(X_2)\nabla_{X_2}\tilde{u}_\infty$ is independent of X_1, one has

$$\int_{\Omega_\ell} A_{12}(X_2)\nabla_{X_2}\tilde{u}_\infty \cdot \nabla_{X_1}((v-\tilde{u}_\infty)\rho)dx$$

$$= \int_{\Omega_\ell} \nabla_{X_1} \cdot \{\rho(v-\tilde{u}_\infty)A_{12}(X_2)\nabla_{X_2}\tilde{u}_\infty\}dx$$

$$= \int_{\omega_2} \int_{\ell\omega_1} \nabla_{X_1} \cdot \{\rho(v-\tilde{u}_\infty)A_{12}(X_2)\nabla_{X_2}\tilde{u}_\infty\}dX_1dX_2 = 0$$

since $(v-\tilde{u}_\infty)\rho$ vanishes on $\partial\ell\omega_1$. Thus from (9.35) we derive

$$\int_{\Omega_\ell} A(x)\nabla\tilde{u}_\infty \cdot \nabla((v-\tilde{u}_\infty)\rho)dx \geq \int_{\Omega_\ell} f(v-\tilde{u}_\infty)\rho dX_2 \quad \forall v \in \mathcal{K}_\infty.$$

Since it is easy to see that

$$||\nabla\tilde{u}_\infty||_{2,\Omega_\ell} = O(\ell^{\frac{p}{2}})$$

it is clear that $\tilde{u}_\infty$ satisfies (9.15) and is equal to u_∞. Thus in this case, the independence of the data with respect to X_1 forces u_ℓ at the limit to be independent of X_1 and the rate of convergence of u_ℓ toward u_∞ is exponential.

9.3   A general framework

We suppose until the end of the chapter that

$$\Omega_\ell = \ell\omega_1 \times \omega_2,$$

where ω_1, ω_2 are such that

ω_1 is an open bounded convex subset of $\mathbb{R}^p$ containing 0

ω_2 is a bounded open subset of $\mathbb{R}^{n-p}$.

We consider also A an elliptic matrix (i.e. satisfying (9.10), (9.11) for $\Omega = \mathbb{R}^p \times \omega_2$) of the type

$$A(x) = \begin{pmatrix} A_{11}(x) & A_{12}(X_2) \\ A_{21}(x) & A_{22}(X_2) \end{pmatrix}$$

where A_{11}, A_{22} are respectively $p \times p$ and $(n-p) \times (n-p)$-matrices. In addition we denote by $a = a(X_2)$ and $f = f(X_2)$ functions depending only on X_2 and such that

$$f \in L^2(\omega_2),$$

$$a \in L^\infty(\omega_2), \ 0 < \lambda \le a(X_2) \le \Lambda \ \text{ a.e. } X_2 \in \omega_2. \tag{9.36}$$

Let K_ℓ and K_∞ be two subsets of $H^1(\Omega_\ell)$, $H^1(\omega_2)$ respectively and u_ℓ, u_∞ two functions satisfying

$$\begin{cases} u_\ell \in K_\ell, \\ \int_{\Omega_\ell} A(x) \nabla u_\ell \cdot \nabla (v - u_\ell) + a(X_2) u_\ell (v - u_\ell) dx \\ \qquad\qquad \ge \int_{\Omega_\ell} f(X_2)(v - u_\ell) dx \ \ \forall v \in K_\ell, \end{cases} \tag{9.37}$$

$$\begin{cases} u_\infty \in K_\infty, \\ \int_{\omega_2} A_{22}(X_2) \nabla_{X_2} u_\infty \cdot \nabla_{X_2}(v - u_\infty) + a(X_2) u_\infty (v - u_\infty) dX_2 \\ \qquad\qquad \ge \int_{\omega_2} f(X_2)(v - u_\ell) dX_2 \ \ \forall v \in K_\infty. \end{cases} \tag{9.38}$$

Note that K_ℓ, K_∞ are not necessarily closed convex sets, i.e. we enlarge slightly the scope of our results not restricting ourselves to variational inequalities in the strict sense.

We denote by $\rho = \rho_{\ell_1}$, $\ell_1 \le \ell - 1$ the function defined by (2.7). Then we have:

Theorem 9.3. *Suppose that there exists v_0 independent of ℓ and X_1 such that*

$$v_0 \in K_\ell, \ \forall \ell \text{ large enough}, \tag{9.39}$$

and that for any $\ell_1 \le \ell - 1$

$$u_\ell - \rho_{\ell_1}(u_\ell - u_\infty) \in K_\ell, \tag{9.40}$$

$$u_\infty + \rho_{\ell_1}(X_1)(u_\ell(X_1, \cdot) - u_\infty) \in K_\infty \ \text{ a.e. } X_1 \in \ell_1 \omega_1. \tag{9.41}$$

Then there exist positive constants C, α independent of ℓ such that

$$|u_\ell - u_\infty|_{H^1(\Omega_{\frac{\ell}{2}})} \le C e^{-\alpha \ell} \tag{9.42}$$

$(|v|_{H^1(O)} = (\int_O |\nabla v|^2 + v^2 dx)^{\frac{1}{2}}).$

Proof. By (9.41) taking $v = u_\infty + \rho_{\ell_1}(X_1)(u_\ell(X_1, \cdot) - u_\infty)$ in (9.38) one gets for a.e. $X_1 \in \ell\omega_1$

$$\int_{\omega_2} A_{22}(X_2)\nabla_{X_2}u_\infty \cdot \nabla_{X_2}(\rho_{\ell_1}(X_1)(u_\ell(X_1, \cdot) - u_\infty))$$

$$+ a(X_2)u_\infty \rho_{\ell_1}(X_1)(u_\ell(X_1, \cdot) - u_\infty)dX_2$$

$$\geq \int_{\omega_2} f\rho_{\ell_1}(X_1)(u_\ell(X_1, \cdot) - u_\infty)dX_2.$$

Integrating in X_1 we obtain

$$\int_{\Omega_\ell} A_{22}(X_2)\nabla_{X_2}u_\infty \cdot \nabla_{X_2}(\rho_{\ell_1}(u_\ell - u_\infty))$$

$$+ a(X_2)u_\infty \rho_{\ell_1}(u_\ell - u_\infty)dx \qquad (9.43)$$

$$\geq \int_{\Omega_\ell} f(X_2)\rho_{\ell_1}(u_\ell - u_\infty)dx.$$

Since u_∞ is independent of X_1 one has

$$A(x)\nabla u_\infty = \begin{pmatrix} A_{11}(x) & A_{12}(X_2) \\ A_{21}(x) & A_{22}(X_2) \end{pmatrix}\begin{pmatrix} 0 \\ \nabla_{X_2}u_\infty \end{pmatrix} = \begin{pmatrix} A_{12}(X_2)\nabla_{X_2}u_\infty \\ A_{22}(X_2)\nabla_{X_2}u_\infty \end{pmatrix}$$

and for $v \in H^1(\Omega_\ell)$

$$A(x)\nabla u_\infty \cdot \nabla v = A_{12}(X_2)\nabla_{X_2}u_\infty \cdot \nabla_{X_1}v + A_{22}(X_2)\nabla_{X_2}u_\infty \cdot \nabla_{X_2}v.$$

Since ρ_{ℓ_1} vanishes on the boundary of $\ell\omega_1$ and since $A_{12}\nabla_{X_2}u_\infty$ is independent of X_1 one has

$$\int_{\Omega_\ell} A_{12}(X_2)\nabla_{X_2}u_\infty \cdot \nabla_{X_1}(\rho_{\ell_1}(u_\ell - u_\infty))dx$$

$$= \int_{\Omega_\ell} \nabla_{X_1} \cdot \{\rho_{\ell_1}(u_\ell - u_\infty)A_{12}(X_2)\nabla_{X_2}u_\infty\}dx$$

$$= \int_{\omega_2}\int_{\ell\omega_1} \nabla_{X_1} \cdot \{\rho_{\ell_1}(u_\ell - u_\infty)A_{12}(X_2)\nabla_{X_2}u_\infty\}dX_1dX_2 = 0.$$

Thus (9.43) can be written

$$\int_{\Omega_\ell} A(x)\nabla u_\infty \cdot \nabla\{\rho_{\ell_1}(u_\ell - u_\infty)\} + a(X_2)u_\infty\{\rho_{\ell_1}(u_\ell - u_\infty)\}dx$$

$$\qquad (9.44)$$

$$\geq \int_{\Omega_\ell} f\{\rho_{\ell_1}(u_\ell - u_\infty)\}dx.$$

Using now $v = u_\ell - \rho_{\ell_1}(u_\ell - u_\infty)$ in (9.37) we get

$$\int_{\Omega_\ell} A(x)\nabla u_\ell \cdot \nabla\{-\rho_{\ell_1}(u_\ell - u_\infty)\} + a(X_2)u_\ell\{-\rho_{\ell_1}(u_\ell - u_\infty)\}dx$$

$$\geq \int_{\Omega_\ell} f\{-\rho_{\ell_1}(u_\ell - u_\infty)\}dx.$$

Adding this inequality to (9.44) we obtain

$$\int_{\Omega_\ell} A(x)\nabla(u_\ell - u_\infty) \cdot \nabla\{\rho_{\ell_1}(u_\ell - u_\infty)\} + a(u_\ell - u_\infty)^2 \rho_{\ell_1}\, dx \leq 0.$$

Since ρ_{ℓ_1} vanishes outside Ω_{ℓ_1+1} this leads to

$$\int_{\Omega_{\ell_1+1}} \{A(x)\nabla(u_\ell - u_\infty) \cdot \nabla(u_\ell - u_\infty) + a(u_\ell - u_\infty)^2\}\rho_{\ell_1}\, dx$$

$$\leq \int_{\Omega_{\ell_1+1}} A(x)\nabla_{X_1}(u_\ell - u_\infty) \cdot \nabla_{X_1}\rho_{\ell_1}(u_\ell - u_\infty)\, dx.$$

Since $\rho_{\ell_1} = 1$ on Ω_{ℓ_1}, using (2.7), (9.10), (9.11) and (9.36), we derive easily

$$\lambda\int_{\Omega_{\ell_1}} |\nabla(u_\ell - u_\infty)|^2 + (u_\ell - u_\infty)^2 dx$$

$$\leq \int_{D_{\ell_1}} \Lambda|\nabla_{X_1}(u_\ell - u_\infty)||\nabla_{X_1}\rho_{\ell_1}||u_\ell - u_\infty|dx$$

$$\leq C\Lambda\int_{D_{\ell_1}} |\nabla_{X_1}(u_\ell - u_\infty)||u_\ell - u_\infty|dx$$

where C is the constant in (2.7), $D_{\ell_1} = \Omega_{\ell_1+1}\backslash\Omega_{\ell_1}$. Then using the Young inequality $ab \leq \frac{1}{2}(a^2 + b^2)$, we derive

$$\int_{\Omega_{\ell_1}} |\nabla(u_\ell - u_\infty)|^2 + (u_\ell - u_\infty)^2 dx$$

$$\leq \frac{C\Lambda}{2\lambda}\int_{D_{\ell_1}} |\nabla_{X_1}(u_\ell - u_\infty)|^2 + (u_\ell - u_\infty)^2 dx \quad (9.45)$$

$$\leq \frac{C\Lambda}{2\lambda}\int_{D_{\ell_1}} |\nabla(u_\ell - u_\infty)|^2 + (u_\ell - u_\infty)^2 dx$$

and thus

$$|u_\ell - u_\infty|^2_{H^1(\Omega_{\ell_1})} \leq \frac{C\Lambda}{2\lambda}\{|u_\ell - u_\infty|^2_{H^1(\Omega_{\ell_1+1})} - |u_\ell - u_\infty|^2_{H^1(\Omega_{\ell_1})}\}$$

which leads to

$$|u_\ell - u_\infty|_{H^1(\Omega_{\ell_1})} \leq \gamma|u_\ell - u_\infty|_{H^1(\Omega_{\ell_1+1})} \quad (9.46)$$

for $\gamma = (\frac{C\Lambda}{2\lambda+C\Lambda})^{\frac{1}{2}} < 1$.

Remark 9.3. Under the assumptions of the theorem and if K_ℓ, K_∞ are subsets of $H_0^1(\Omega_\ell)$, $H_0^1(\omega_2)$ respectively, one can assume instead of (9.36)

$$0 \leq a(X_2) \leq \Lambda \text{ a.e. } X_2 \in \omega_2. \quad (9.47)$$

Then instead of (9.45) one arrives to

$$\int_{\Omega_{\ell_1}} |\nabla(u_\ell - u_\infty)|^2 dx \leq \frac{C\Lambda}{2\lambda} \int_{D_{\ell_1}} |\nabla_{X_1}(u_\ell - u_\infty)|^2 + (u_\ell - u_\infty)^2 dx$$

$$\leq C' \int_{D_{\ell_1}} |\nabla(u_\ell - u_\infty)|^2 dx$$

thanks to the Poincaré inequality. Then (9.46) follows since - again by the Poincaré inequality - the Dirichlet norm is equivalent to the $H^1(\Omega_\ell)$ norm with equivalence constants independent of ℓ.

Iterating now (9.46) from $\ell_1 = \frac{\ell}{2}$, $[\frac{\ell}{2}]$ times we obtain

$$|u_\ell - u_\infty|_{H^1(\Omega_{\frac{\ell}{2}})} \leq \gamma^{[\frac{\ell}{2}]} |u_\ell - u_\infty|_{H^1(\Omega_{\frac{\ell}{2} + [\frac{\ell}{2}]})}$$

$$\leq \frac{1}{\gamma} \gamma^{\frac{\ell}{2}} |u_\ell - u_\infty|_{H^1(\Omega_\ell)} \tag{9.48}$$

$$= \frac{1}{\gamma} e^{-\frac{\ell}{2} \ln \frac{1}{\gamma}} |u_\ell - u_\infty|_{H^1(\Omega_\ell)}$$

since $\frac{\ell}{2} - 1 < [\frac{\ell}{2}] \leq \frac{\ell}{2}$. It remains to estimate the right hand side of the above inequality. Taking $v = v_0$ in (9.37) we get

$$\int_{\Omega_\ell} A(x)\nabla u_\ell \cdot \nabla(v_0 - u_\ell) + a(X_2)u_\ell(v_0 - u_\ell)dx \geq \int_{\Omega_\ell} f(v_0 - u_\ell)dx.$$

This can be written as

$$\int_{\Omega_\ell} A(x)\nabla u_\ell \cdot \nabla u_\ell + au_\ell^2 dx$$

$$\leq \int_{\Omega_\ell} A(x)\nabla u_\ell \cdot \nabla v_0 + au_\ell v_0 dx + \int_{\Omega_\ell} f(u_\ell - v_0)dx \tag{9.49}$$

$$\leq \Lambda \int_{\Omega_\ell} |\nabla u_\ell||\nabla v_0| + |u_\ell||v_0|dx + \int_{\Omega_\ell} |f|(|u_\ell| + |v_0|)dx.$$

By (9.10), (9.11) we derive

$$\int_{\Omega_\ell} |\nabla u_\ell|^2 + u_\ell^2 dx$$

$$\leq \frac{\Lambda}{\lambda} \int_{\Omega_\ell} |\nabla u_\ell||\nabla v_0| + |u_\ell||v_0|dx + \frac{1}{\lambda} \int_{\Omega_\ell} |f||u_\ell| + |f||v_0|dx.$$

Using the Young inequality we obtain for some constant $C = C(\lambda, \Lambda)$

$$\int_{\Omega_\ell} |\nabla u_\ell|^2 + u_\ell^2 dx$$

$$\leq \frac{1}{2} \int_{\Omega_\ell} |\nabla u_\ell|^2 + u_\ell^2 dx + C \int_{\Omega_\ell} |\nabla v_0|^2 + |v_0|^2 + |f|^2 dx$$

and thus since v_0, f are independent of X_1

$$\int_{\Omega_\ell} |\nabla u_\ell|^2 + u_\ell^2 dx \leq 2C \int_{\Omega_\ell} |\nabla v_0|^2 + |v_0|^2 + |f|^2 dx$$

$$= 2C \int_{\ell\omega_1} \int_{\omega_2} |\nabla v_0|^2 + |v_0|^2 + |f|^2 dX_2 dX_1$$

$$= 2C|\ell\omega_1|\{|v_0|_{H^1(\omega_2)}^2 + |f|_{2,\omega_2}^2\}$$

$$= 2C|\omega_1|\{|v_0|_{H^1(\omega_2)}^2 + |f|_{2,\omega_2}^2\}\ell^p$$

i.e. for some constant C independent of ℓ

$$|u_\ell|_{H^1(\Omega_\ell)} \leq C\ell^{\frac{p}{2}}. \tag{9.50}$$

Similarly since u_∞ is independent of X_1 we have

$$|u_\infty|_{H^1(\Omega_\ell)}^2 = \int_{\ell\omega_1} \int_{\omega_2} |\nabla u_\infty|^2 + |u_\infty|^2 dX_2 dX_1$$

$$= |\ell\omega_1||u_\infty|_{H^1(\omega_2)}^2 \leq C\ell^p.$$

Going back to (9.48) we get for some constant C

$$|u_\ell - u_\infty|_{H^1(\Omega_{\frac{\ell}{2}})} \leq Ce^{-\frac{\ell}{2}\ln\frac{1}{\gamma}}\ell^{\frac{p}{2}} \leq Ce^{-\alpha\ell}$$

provided $\alpha < \frac{1}{2}\ln\frac{1}{\gamma}$. This completes the proof of the theorem. $\qquad\square$

Remark 9.4. In the case where K_ℓ, K_∞ are subsets of $H_0^1(\Omega_\ell)$, $H_0^1(\omega_2)$ respectively and a satisfies (9.47) one obtains the estimate (9.50) by noting that (9.49) would imply

$$\int_{\Omega_\ell} |\nabla u_\ell|^2 dx$$

$$\leq \frac{\Lambda}{\lambda} \int_{\Omega_\ell} |\nabla u_\ell||\nabla v_0| + |u_\ell||v_0|dx + \frac{1}{\lambda}\int_{\Omega_\ell} |f||u_\ell| + |f||v_0|dx$$

$$\leq \epsilon \int_{\Omega_\ell} |\nabla u_\ell|^2 + |u_\ell|^2 dx + C_\epsilon \int_{\Omega_\ell} |\nabla v_0|^2 + |v_0|^2 + |f|^2 dx$$

$$\leq C\epsilon \int_{\Omega_\ell} |\nabla u_\ell|^2 dx + C_\epsilon \int_{\Omega_\ell} |\nabla v_0|^2 + |v_0|^2 + |f|^2 dx$$

by the Young and Poincaré inequalities. This would lead to

$$\int_{\Omega_\ell} |\nabla u_\ell|^2 dx \leq C\ell^p$$

and to (9.50) by the Poincaré inequality again, i.e. the equivalence of the Dirichlet norm and the $H^1(\Omega_\ell)$-norm.

9.4 Some applications

For almost every $X_2 \in \omega_2$ suppose that

$K(X_2)$ is a closed convex subset of $\mathbb{R}^{n-p+1}$ containing 0.

Then

$$K_\ell = \{v \in H^1(\Omega_\ell) \mid (v, \nabla_{X_2} v)(x) \in K(X_2) \text{ a.e. } x = (X_1, X_2) \in \Omega_\ell\},$$

$$K_\infty = \{v \in H^1(\omega_2) \mid (v, \nabla_{X_2} v) \in K(X_2) \text{ a.e. } X_2 \in \omega_2\}$$

are closed convex subsets of $H^1(\Omega_\ell)$, $H^1(\omega_2)$ respectively. They are both non empty containing 0. Thus there exist u_ℓ, u_∞ unique solutions to (9.37), (9.38) respectively. Moreover one has

$$v_\ell = u_\ell - \rho_{\ell_1}(X_1)(u_\ell - u_\infty) = (1 - \rho_{\ell_1}(X_1))u_\ell + \rho_{\ell_1}(X_1)u_\infty$$

and thus

$$(v_\ell, \nabla_{X_2} v_\ell) = ((1 - \rho_{\ell_1})u_\ell + \rho_{\ell_1} u_\infty, (1 - \rho_{\ell_1})\nabla_{X_2} u_\ell + \rho_{\ell_1} \nabla_{X_2} u_\infty)$$

$$= (1 - \rho_{\ell_1})(u_\ell, \nabla_{X_2} u_\ell)$$

$$+ \rho_{\ell_1}(u_\infty, \nabla_{X_2} u_\infty) \in K(X_2) \text{ a.e. } x \in \Omega_\ell.$$

Similarly if

$$w_\ell = u_\infty + \rho_{\ell_1}(X_1)(u_\ell(X_1, \cdot) - u_\infty) = (1 - \rho_{\ell_1}(X_1))u_\infty(X_1, \cdot) + \rho_{\ell_1}(X_1)u_\ell$$

one has $(w_\ell, \nabla_{X_2} w_\ell) \in K(X_2)$ a.e. $X_2 \in \omega_2$. Thus from Theorem 9.3 we deduce

$$|u_\ell - u_\infty|_{H^1(\Omega_{\frac{\ell}{2}})} \leq C e^{-\alpha \ell}$$

i.e. $u_\ell \to u_\infty$ exponentially quickly.

In the case where

$$K(X_2) = K_1(X_2) \times \mathbb{R}^{n-p}$$

with $K_1(X_2)$ a closed convex subset of $\mathbb{R}$ containing 0, i.e.

$$K_1(X_2) = [\varphi(X_2), \psi(X_2)] \tag{9.51}$$

for some functions φ, ψ one has

$$K_\ell = \{v \in H^1(\Omega_\ell) \mid \varphi(X_2) \leq v(x) \leq \psi(X_2) \text{ a.e. } x = (X_1, X_2) \in \Omega_\ell\},$$

$$K_\infty = \{v \in H^1(\omega_2) \mid \varphi(X_2) \leq v(X_2) \leq \psi(X_2) \text{ a.e. } X_2 \in \omega_2\}$$

i.e. a double obstacle problem. Tolerating $\psi = +\infty$ in (9.51) we recover the obstacle problem.

Remark 9.5. Taking into account the Remarks 9.3 and 9.4, similar results hold for

$$K_\ell = \{v \in H^1_0(\Omega_\ell) \mid (v, \nabla_{X_2} v)(x) \in K(X_2) \text{ a.e. } x = (X_1, X_2) \in \Omega_\ell\},$$

$$K_\infty = \{v \in H^1_0(\omega_2) \mid (v, \nabla_{X_2} v) \in K(X_2) \text{ a.e. } X_2 \in \omega_2\}$$

under the relaxed assumption (9.47) on a.

As another example, suppose now that ω_2' is a measurable subset of ω_2 and consider

$$K_\ell = \{v \in H^1(\Omega_\ell) \mid \int_{\omega_2'} |\nabla_{X_2} v(X_1, X_2)|^2 dX_2 \leq 1 \text{ a.e. } X_1 \in \ell\omega_1\},$$

$$K_\infty = \{v \in H^1(\omega_2) \mid \int_{\omega_2'} |\nabla_{X_2} v(X_2)|^2 dX_2 \leq 1\}.$$

Then K_ℓ, K_∞ are both closed convex subsets of $H^1(\Omega_\ell)$, $H^1(\omega_2)$ respectively. They both contain 0. Then there exist u_ℓ, u_∞ unique solutions to (9.10), (9.11) and it is easy to see that (9.40), (9.41) hold in such a way that Theorem 9.3 applies.

Remark 9.6. In the case where

$$K_\ell = \{v \in H^1(\Omega_\ell) \mid |\nabla v(x)| \leq 1 \text{ a.e. } x \in \Omega_\ell\}$$

i.e. in the case of the elastic plastic torsion problem, then (9.40) does not hold. Thus some other techniques are needed (see [40], [41]).

Chapter 10

Calculus of variations

10.1 Monotonicity properties

We collect in this part some results which are perhaps known but for which we have been unable to find proper references.

Let $F : \Omega \times \mathbb{R} \times \mathbb{R}^n \to \mathbb{R}$ be a function such that

$$\mathcal{F}(v) = F(x, v(x), \nabla v(x)) \in L^1(\Omega) \quad \forall v \in W_0^{1,q}(\Omega). \tag{10.1}$$

For $f \in W^{-1,q'}(\Omega)$ the dual of $W_0^{1,q}(\Omega)$ we set

$$E_\Omega(v) = E_{\Omega,f}(v) = \int_\Omega \mathcal{F}(v) \, dx - \langle f, v \rangle \quad \forall v \in W_0^{1,q}(\Omega) \tag{10.2}$$

where $\langle \, , \, \rangle$ denotes the duality bracket between $W^{-1,q'}(\Omega)$ and $W_0^{1,q}(\Omega)$. We consider the minimization problem of E_Ω on $W_0^{1,q}(\Omega)$, i.e. we denote by $u = u_f$ a function in $W_0^{1,q}(\Omega)$ such that

$$E_\Omega(u) \leq E_\Omega(v) \quad \forall v \in W_0^{1,q}(\Omega). \tag{10.3}$$

Then we have:

Theorem 10.1. *Let $u_1 = u_{f_1}$ and $u_2 = u_{f_2}$ be two minimizers of E_Ω on $W_0^{1,q}(\Omega)$ corresponding to f_1 and f_2 respectively. One has*

$$\langle f_1 - f_2, (u_1 - u_2)^+ \rangle \geq 0. \tag{10.4}$$

$(\ )^+$ *denotes the positive part of a function.*

Proof. Set

$$A = \{x \in \Omega \mid u_2(x) > u_1(x)\}. \tag{10.5}$$

Consider

$$v = u_1 \wedge u_2 = u_1 - (u_1 - u_2)^+.$$

213

One has - if A^C denotes the set $\Omega \backslash A$

$$E_{\Omega,f_1}(v) = \int_\Omega \mathcal{F}(v) \, dx - \langle f_1, v \rangle = \int_A \mathcal{F}(u_1) \, dx + \int_{A^C} \mathcal{F}(u_2) \, dx$$
$$- \langle f_1, u_1 \rangle + \langle f_1, (u_1 - u_2)^+ \rangle.$$

We consider next

$$w = u_1 \vee u_2 = u_2 + (u_1 - u_2)^+.$$

One has

$$E_{\Omega,f_2}(w) = \int_\Omega \mathcal{F}(w) \, dx - \langle f_2, w \rangle = \int_A \mathcal{F}(u_2) \, dx + \int_{A^C} \mathcal{F}(u_1) \, dx$$
$$- \langle f_2, u_2 \rangle - \langle f_2, (u_1 - u_2)^+ \rangle.$$

Adding up the last two equalities we get

$$E_{\Omega,f_1}(v) + E_{\Omega,f_2}(w)$$
$$= E_{\Omega,f_1}(u_1) + E_{\Omega,f_2}(u_2) + \langle f_1 - f_2, (u_1 - u_2)^+ \rangle. \qquad (10.6)$$

Thus (10.4) follows since u_1, u_2 are minimizers of E_{Ω,f_1}, E_{Ω,f_2} respectively. This completes the proof of the theorem. $\qquad \square$

Remark 10.1. Exchanging the role of f_1 and f_2 we have

$$\langle f_2 - f_1, (u_2 - u_1)^+ \rangle = \langle f_2 - f_1, (u_1 - u_2)^- \rangle$$
$$= \langle f_1 - f_2, -(u_1 - u_2)^- \rangle \geq 0$$

and thus adding to (10.4)

$$\langle f_1 - f_2, u_1 - u_2 \rangle \geq 0 \qquad (10.7)$$

which means that the operator $f \to u_f$ is monotone.

Definition 10.1. We say that $f_1 \geq f_2$ when this inequality holds in the distributional sense, i.e. when we have

$$\langle f_1, \varphi \rangle \geq \langle f_2, \varphi \rangle \quad \forall \varphi \in \mathcal{D}(\Omega), \quad \varphi \geq 0,$$

which by density of $\mathcal{D}(\Omega)$ in $W_0^{1,q}(\Omega)$ means

$$\langle f_1, v \rangle \geq \langle f_2, v \rangle \quad \forall v \in W_0^{1,q}(\Omega), \quad v \geq 0. \qquad (10.8)$$

Then we have:

Corollary 10.1. *Let $u_1 = u_{f_1}$ and $u_2 = u_{f_2}$ be two minimizers of E_Ω on $W_0^{1,q}(\Omega)$ corresponding to f_1 and f_2 respectively. If*

$$f_1 \leq f_2 \qquad (10.9)$$

one has

$$\langle f_1 - f_2, (u_1 - u_2)^+ \rangle = 0. \qquad (10.10)$$

Proof. This follows immediately from the definition above and (10.4). □

Remark 10.2. In the case where f_1 and f_2 are functions satisfying (10.9), one has $u_1 \leq u_2$ on the set where $f_1 < f_2$.

Corollary 10.2. *Let* $u_1 = u_{f_1}$ *and* $u_2 = u_{f_2}$ *be two minimizers of* E_Ω *on* $W_0^{1,q}(\Omega)$ *corresponding to* f_1 *and* f_2 *respectively. If* (10.9) *holds and either* u_1 *or* u_2 *is unique then*

$$u_1 \leq u_2. \tag{10.11}$$

Proof. Indeed in this case from (10.6) one derives

$$E_{\Omega,f_1}(v) - E_{\Omega,f_1}(u_1) = E_{\Omega,f_2}(u_2) - E_{\Omega,f_2}(w) = 0.$$

Then either $v = u_1$ or $w = u_2$ which in both cases means $u_1 \leq u_2$. □

As an immediate consequence one has

Corollary 10.3. *Suppose that* 0 *is solution to* (10.3) *for* $f = 0$ *and that* (10.3) *has a unique solution* u_f *for* $f \geq 0$. *Then one has* $u_f \geq 0$.

As a remarkable property we have also

Proposition 10.1. *The following assertions are equivalent:*

(i) The minimization problem (10.3) *admits a unique solution.*

(ii) If $u = u_f$ *denotes a minimizer of* E_Ω, *the mapping* $f \to u_f$ *is monotone in the sense that*

$$f_1 \leq f_2 \Longrightarrow u_{f_1} \leq u_{f_2}.$$

Proof.

• $(ii) \Longrightarrow (i)$.

Indeed if u_1 and u_2 are two solutions corresponding to the same f, the monotonicity property implies $u_1 \geq u_2$ and $u_2 \geq u_1$ i.e. $u_1 = u_2$.

• $(i) \Longrightarrow (ii)$ follows directly from Corollary 10.2.

□

Remark 10.3. The same results hold if the minimization on $W_0^{1,q}(\Omega)$ is replaced by a minimization on a subset K of $W_0^{1,q}(\Omega)$ having the property that

$$u_1, u_2 \in K \implies u_1 \wedge u_2, \ u_1 \vee u_2 \in K.$$

Note also that our results do not involve any convexity on F or K.

One has also the following additional monotonicity property with respect to the domain Ω.

Theorem 10.2. *Suppose that $f \geq 0$ in the distributional sense and that (10.3) admits a unique solution for two open subsets Ω and Ω' such that*

$$\Omega \subset \Omega'. \tag{10.12}$$

If u, u' denote the solutions to (10.3) corresponding to Ω, Ω' respectively and if 0 is solution to (10.3) for $f = 0$ for Ω, Ω', one has

$$u' \geq u \geq 0. \tag{10.13}$$

Proof. $u, u' \geq 0$ is a consequence of Corollary 10.3. Suppose that u is extended by 0 outside Ω. Consider then $v = u \wedge u' = u' - (u' - u)^+$ and

$$A = \{x \in \Omega' \mid u(x) > u'(x)\}.$$

Note that A is included in Ω. One has

$$\int_\Omega \mathcal{F}(v) \, dx - \langle f, v \rangle = \int_{A^C} \mathcal{F}(u) \, dx$$
$$+ \int_A \mathcal{F}(u') \, dx - \langle f, v \rangle.$$

If

$$\int_A \mathcal{F}(u') \, dx < \int_A \mathcal{F}(u) \, dx - \langle f, u \rangle + \langle f, v \rangle$$

one gets a contradiction with the definition of u which is not a minimizer. Thus it holds

$$\int_A \mathcal{F}(u) \, dx \leq \int_A \mathcal{F}(u') \, dx + \langle f, u \rangle - \langle f, v \rangle.$$

If now one considers $w = u \vee u' = u + (u' - u)^+$ it comes

$$\int_{\Omega'} \mathcal{F}(w) \, dx - \langle f, w \rangle = \int_{A^C} \mathcal{F}(u') \, dx$$
$$+ \int_A \mathcal{F}(u) \, dx - \langle f, w \rangle$$

and thus

$$\int_{\Omega'} \mathcal{F}(w) \, dx - \langle f, w \rangle \leq \int_{\Omega'} \mathcal{F}(u') \, dx$$
$$+ \langle f, u \rangle - \langle f, v \rangle - \langle f, w \rangle$$
$$= \int_{\Omega'} \mathcal{F}(u') \, dx - \langle f, u' \rangle.$$

By uniqueness of the solution one has $w = u'$ which implies that $u' \geq u$. This completes the proof of the theorem. $\qquad\square$

10.2 A convergence result

We denote by Ω_ℓ the open subset of $\mathbb{R}^n$ defined as

$$\Omega_\ell = \ell\omega_1 \times \omega_2$$

where

ω_1 is a bounded convex open subset of $\mathbb{R}^p$ containing 0, $\qquad$ (10.14)

ω_2 being a bounded open subset of $\mathbb{R}^{(n-p)}$.

Thus for every $\ell_0 \leq \ell - 1$ there exists a function $\rho = \rho_{\ell_0}(X_1)$ such that

$$0 \leq \rho \leq 1, \quad \rho = 1 \text{ on } \ell_0\omega_1, \quad \rho = 0 \text{ outside } (\ell_0 + 1)\omega_1, \quad |\nabla\rho| \leq C \quad (10.15)$$

where C is a constant independent of ℓ_0. We consider then a function $F : \mathbb{R}^n \to \mathbb{R}$ such that

$$F \text{ is convex}, \qquad (10.16)$$

and such that there exist positive constants $\lambda, \Lambda, \Lambda'$ such that

$$\lambda|\xi|^q - \Lambda' \leq F(\xi) \leq \Lambda|\xi|^q + \Lambda' \quad \forall \xi \in \mathbb{R}^n, \quad F(0) = -\Lambda' \qquad (10.17)$$

(q is a fixed number such that $q > 1$).

For such a function we have (see [48]):

$$|F(P) - F(Q)| \leq C_q\Lambda\{1 + |P|^{q-1} + |Q|^{q-1}\}|P - Q| \quad \forall P, Q \in \mathbb{R}^n. \quad (10.18)$$

Then for

$$f = f(X_2) \in L^{q'}(\omega_2), \quad \frac{1}{q} + \frac{1}{q'} = 1, \qquad (10.19)$$

we consider the minimization of the functional

$$E_{\Omega_\ell}(v) = \int_{\Omega_\ell} F(\nabla v) - fv \, dx \qquad (10.20)$$

on $W_0^{1,q}(\Omega_\ell)$. We have:

Theorem 10.3. *There exists u_ℓ solution to*

$$u_\ell \in W_0^{1,q}(\Omega_\ell), \quad E_{\Omega_\ell}(u_\ell) \leq E_{\Omega_\ell}(v) \quad \forall v \in W_0^{1,q}(\Omega_\ell). \qquad (10.21)$$

Moreover if F is strictly convex, the solution of the problem above is unique.

Proof. First note that by (10.17) and (10.19), $E_{\Omega_\ell}(v)$ is well defined for any $v \in W_0^{1,q}(\Omega_\ell)$. Moreover by the Hölder and Poincaré inequalities, one has

$$\int_{\Omega_\ell} fv \, dx \leq |f|_{q',\Omega_\ell} |v|_{q,\Omega_\ell} \leq C|f|_{q',\Omega_\ell} ||\nabla v||_{q,\Omega_\ell} \tag{10.22}$$

and thus -see (10.17)-

$$E_{\Omega_\ell}(v) \geq \lambda ||\nabla v||_{q,\Omega_\ell}^q - C|f|_{q',\Omega_\ell} ||\nabla v||_{q,\Omega_\ell} - \Lambda'|\Omega_\ell|.$$

This implies that E_{Ω_ℓ} is coercive or that every minimizing sequence is bounded in $W_0^{1,q}(\Omega_\ell)$. To conclude it is enough to show that E_{Ω_ℓ} is lower semicontinuous for the weak topology of $W_0^{1,q}(\Omega_\ell)$. E_{Ω_ℓ} being convex it is enough to show the lower semicontinuity of E_{Ω_ℓ} for the strong topology of $W_0^{1,q}(\Omega_\ell)$. If

$$v_n \to v \text{ in } W_0^{1,q}(\Omega_\ell)$$

up to a subsequence, one has

$$v_n \to v \text{ in } L^q(\Omega_\ell), \quad \nabla v_n \to \nabla v \text{ a.e. in } \Omega_\ell.$$

By Fatou's lemma one deduces

$$\liminf_{n\to\infty} E_{\Omega_\ell}(v_n) = \liminf_{n\to\infty} \int_{\Omega_\ell} F(\nabla v_n) \, dx - \int_{\Omega_\ell} fv \, dx$$

$$\geq \int_{\Omega_\ell} \liminf_{n\to\infty} F(\nabla v_n) \, dx - \int_{\Omega_\ell} fv \, dx = E_{\Omega_\ell}(v).$$

This completes the proof of the theorem.   □

If 0 denotes the 0 vector in $\mathbb{R}^p$ and ∇_{X_2} the gradient in X_2, i.e.

$$\nabla_{X_2} = (\partial_{x_{p+1}}, \cdots, \partial_{x_n})$$

one sets

$$E_{\omega_2}(v) = \int_{\omega_2} F(0, \nabla_{X_2} v) - fv \, dX_2 \quad \forall v \in W_0^{1,q}(\omega_2). \tag{10.23}$$

Clearly $E_{\omega_2}(v)$ is well defined and with the same proof as above, one has

Theorem 10.4. *There exists u_∞ solution to*

$$u_\infty \in W_0^{1,q}(\omega_2), \quad E_{\omega_2}(u_\infty) \leq E_{\omega_2}(v) \quad \forall v \in W_0^{1,q}(\omega_2). \tag{10.24}$$

Moreover if $F(0, \cdot)$ is strictly convex, the solution of the problem above is unique.

Our goal is now to show:

Theorem 10.5. *Suppose that $f \geq 0$ and F strictly convex then one has*

$$. \ u_\ell \to u_\infty \tag{10.25}$$

when $\ell \to +\infty$.

Remark 10.4. For the time being we are not more precise on the way the convergence above occurs. This result completes the results of [28]. See also [19].

We will need several lemmas.

One denotes by $W_{lat}^{1,q}(\Omega_\ell)$ the space

$$W_{lat}^{1,q}(\Omega_\ell) = \{v \in W^{1,q}(\Omega_\ell) \mid v = 0 \text{ on } \ell\omega_1 \times \partial\omega_2\} \tag{10.26}$$

and by $V^{1,q}(\Omega_\ell)$ the space

$$V^{1,q}(\Omega_\ell) = \{v \in W_{lat}^{1,q}(\Omega_\ell) \mid \fint_{\ell\omega_1} \nabla_{X_1} v \, dX_1 = 0 \text{ a.e. } X_2 \in \omega_2\}. \tag{10.27}$$

($\fint_{\ell\omega_1} = \frac{1}{|\ell\omega_1|} \int_{\ell\omega_1}$ where $|\ell\omega_1|$ denotes the measure of $\ell\omega_1$).

Lemma 10.1. u_∞ *is the unique solution to*

$$u_\infty \in V^{1,q}(\Omega_\ell), \quad E_{\Omega_\ell}(u_\infty) \leq E_{\Omega_\ell}(v) \quad \forall v \in V^{1,q}(\Omega_\ell). \tag{10.28}$$

Proof. First notice that $u_\infty \in V^{1,q}(\Omega_\ell)$. Indeed one has since u_∞ is independent of X_1

$$\fint_{\ell\omega_1} \nabla_{X_1} u_\infty \, dX_1 = 0.$$

Next let $u \in V^{1,q}(\Omega_\ell)$. Set

$$v(X_2) = \fint_{\ell\omega_1} u(., X_2) \, dX_1.$$

It is easy to see that $v \in W_0^{1,q}(\omega_2)$. Thus

$$E_{\omega_2}(u_\infty) \leq E_{\omega_2}(v) = \int_{\omega_2} F(0, \nabla_{X_2} v) - fv \, dX_2.$$

Now using the fact that $u \in V^{1,q}(\Omega_\ell)$ one has

$$0 = \fint_{\ell\omega_1} \nabla_{X_1} u \, dX_1.$$

Also by differentiation under the integral, we get

$$\nabla_{X_2} v = \fint_{\ell\omega_1} \nabla_{X_2} u \, dX_1.$$

Therefore we have by Jensen's inequality

$$E_{\omega_2}(u_\infty) \leq \int_{\omega_2} \left\{ F\left(\frac{1}{|\ell\omega_1|} \int_{\ell\omega_1} \nabla_{X_1} u \, dX_1, \frac{1}{|\ell\omega_1|} \int_{\ell\omega_1} \nabla_{X_2} u \, dX_1 \right) \right\} dX_2$$

$$- \int_{\omega_2} f \left\{ \frac{1}{|\ell\omega_1|} \int_{\ell\omega_1} u \, dX_1 \right\} dX_2$$

$$\leq \int_{\omega_2} \frac{1}{|\ell\omega_1|} \int_{\ell\omega_1} \{ F(\nabla u) - f u \} \, dX_1 dX_2 = \frac{E_{\Omega_\ell}(u)}{|\ell\omega_1|}.$$

Clearly equality holds in the above inequality if $u = u_\infty$. Then the claim follows from the uniqueness of the minimizer. $\qquad\square$

One can then show:

Lemma 10.2. *One has for $\ell' > \ell$*

$$0 \leq u_\ell \leq u_{\ell'} \leq u_\infty. \tag{10.29}$$

Proof. Recall that we are assuming $f \geq 0$. The nonnegativity of all the functions above results from Corollary 10.3. Indeed for $f = 0$ one has for instance

$$E_{\Omega_\ell}(v) = \int_{\Omega_\ell} F(\nabla v) dx \geq \int_{\Omega_\ell} -\Lambda' dx$$

$$= \int_{\Omega_\ell} F(0) dx = E_{\Omega_\ell}(0) \quad \forall v \in W_0^{1,q}(\Omega_\ell)$$

i.e. 0 is a minimizer. The monotonicity property $\ell \to u_\ell$ follows from Theorem 10.2. To prove that u_ℓ (or u'_ℓ) is less than u_∞ one proceeds as follows. One considers

$$A_\ell = \{ x \in \Omega_\ell \mid u_\ell(x) > u_\infty(x) \}.$$

One has

$$E_{A_\ell}(u_\ell) \leq E_{A_\ell}(u_\infty) \qquad \left(E_{A_\ell}(\cdot) = \int_{A_\ell} F(\nabla \cdot) - f \cdot \, dx \right). \tag{10.30}$$

Indeed, if not, setting $v_\ell = u_\ell \wedge u_\infty$ one has $v_\ell \in W_0^{1,q}(\Omega_\ell)$ and

$$E_{\Omega_\ell}(v_\ell) = E_{A_\ell}(v_\ell) + E_{A_\ell^C}(v_\ell) = E_{A_\ell}(u_\infty) + E_{A_\ell^C}(u_\ell) < E_{\Omega_\ell}(u_\ell)$$

and a contradiction with the definition of u_ℓ (A_ℓ^C denotes the set $\Omega \backslash A_\ell$). Assuming then (10.30) and considering $w_\ell = u_\ell \vee u_\infty$ one has $w_\ell \in V^{1,q}(\Omega_\ell)$. Indeed if ν denotes the normal to $\partial(\ell\omega_1)$ this follows from

$$\fint_{\ell\omega_1} \nabla_{X_1} w_\ell(., X_2) \, dX_1 = \frac{1}{|\ell\omega_1|} \int_{\partial(\ell\omega_1)} w_\ell(., X_2)\nu \, d\sigma(X_1)$$

$$= \frac{1}{|\ell\omega_1|} \int_{\partial(\ell\omega_1)} u_\infty(., X_2)\nu \, d\sigma(X_1)$$

$$= \fint_{\ell\omega_1} \nabla_{X_1} u_\infty(., X_2) \, dX_1 = 0.$$

Then

$$E_{\Omega_\ell}(w_\ell) = E_{A_\ell}(w_\ell) + E_{A_\ell^C}(w_\ell) = E_{A_\ell}(u_\ell) + E_{A_\ell^C}(u_\infty) \leq E_{\Omega_\ell}(u_\infty)$$

and thus $w_\ell = u_\infty$, i.e. $u_\ell \leq u_\infty$. $\qquad\square$

Clearly it follows from Lemma 10.2 that there exists $\tilde{u}_\infty$ such that

$$u_\ell \leq \tilde{u}_\infty \leq u_\infty \quad \forall \ell, \qquad u_\ell \to \tilde{u}_\infty \text{ a.e. in } \mathbb{R}^p \times \omega_2. \tag{10.31}$$

Lemma 10.3. *The function $\tilde{u}_\infty$ is independent of X_1.*

Proof. For $i = 1, \cdots, p$ we set

$$\tau_h^i v(x) = v(x - he_i), \quad h > 0$$

where e_i denotes the i-th vector of the canonical basis of $\mathbb{R}^p$. We claim that

$$u_{\ell+h} \geq \tau_{h'}^i u_\ell \quad \text{for } 0 < h' \leq \lambda h \tag{10.32}$$

$\lambda \leq 1$ being so small that

$$\lambda e_i \in \omega_1. \tag{10.33}$$

Indeed if (10.33) holds we have for $X_1 - h'e_i \in \ell\omega_1$ and some $Y_1 \in \omega_1$

$$X_1 = \ell Y_1 + h'e_i = (\ell + h)\{\frac{\ell}{\ell + h}Y_1 + \frac{h}{\ell + h}\frac{h'}{h}e_i\} \in (\ell + h)\omega_1$$

(since $Y_1, \frac{h'}{h}e_i \in \omega_1$ and ω_1 is convex containing 0). Thus the support of $\tau_{h'}^i u_\ell$ is contained in $\Omega_{\ell+h}$. One defines

$$\mathcal{A} = \{x \in \Omega_{\ell+h} \mid \tau_{h'}^i u_\ell(x) > u_{\ell+h}(x)\}.$$

One has $E_{\mathcal{A}}(u_{\ell+h}) \leq E_{\mathcal{A}}(\tau_{h'}^i u_\ell)$. Indeed if not

$$w_\ell = u_{\ell+h} \vee \tau_{h'}^i u_\ell = u_{\ell+h} + (\tau_{h'}^i u_\ell - u_{\ell+h})^+ \in W_0^{1,q}(\Omega_{\ell+h})$$

and

$$E_{\Omega_{\ell+h}}(w_\ell) = E_{\mathcal{A}}(\tau_{h'}^i u_\ell) + E_{\mathcal{A}^c}(u_{\ell+h})$$
$$< E_{\mathcal{A}}(u_{\ell+h}) + E_{\mathcal{A}^c}(u_{\ell+h}) = E_{\Omega_{\ell+h}}(u_{\ell+h})$$

and a contradiction with the definition of $u_{\ell+h}$. Let us set

$$\mathcal{A}' = \{x \in \Omega_\ell \mid \tau_{-h'}^i(u_{\ell+h})(x) < u_\ell(x)\}.$$
$$x \in \mathcal{A}' \Leftrightarrow x + h'e_i \in \Omega_{\ell+h} \text{ and } u_{\ell+h}(x + h'e_i) < u_\ell(x)$$
$$\Leftrightarrow y = x + h'e_i \in \Omega_{\ell+h} \text{ and } u_{\ell+h}(y) < \tau_{h'}^i u_\ell(y)$$
$$\Leftrightarrow x + h'e_i \in \mathcal{A},$$

i.e. one has

$$\mathcal{A}' = \mathcal{A} - h'e_i. \tag{10.34}$$

Consider next

$$v_\ell = u_\ell \wedge \tau_{-h'}^i(u_{\ell+h}) = u_\ell - (u_\ell - \tau_{-h'}^i(u_{\ell+h}))^+.$$

Clearly $v_\ell \in W_0^{1,q}(\Omega_\ell)$ (recall that $\tau_{-h'}^i(u_{\ell+h}) \geq 0$). Then

$$E_{\Omega_\ell}(v_\ell) = E_{\mathcal{A}'}(\tau_{-h'}^i(u_{\ell+h})) + E_{\mathcal{A}'^c}(u_\ell).$$

Using (10.34) one has

$$E_{\mathcal{A}'}(\tau_{-h'}^i(u_{\ell+h})) = \int_{\mathcal{A}'} \{F(\nabla u_{\ell+h}(x + h'e_i)) - f(X_2)u_{\ell+h}(x + h'e_i)\}dx$$
$$= \int_{\mathcal{A}} \{F(\nabla u_{\ell+h}(y)) - f(X_2)u_{\ell+h}(y)\}dy = E_{\mathcal{A}}(u_{\ell+h})$$

and

$$E_{\mathcal{A}}(\tau_{h'}^i(u_\ell)) = \int_{\mathcal{A}'} \{F(\nabla u_\ell(x - h'e_i)) - f(X_2)u_\ell(x - h'e_i)\} \, dx$$
$$= \int_{\mathcal{A}'} \{F(\nabla u_\ell(y)) - f(X_2)u_\ell(y)\}dy = E_{\mathcal{A}'}(u_\ell).$$

Thus we obtain from above – recall that $E_{\mathcal{A}}(u_{\ell+h}) \leq E_{\mathcal{A}}(\tau_{h'}^i u_\ell)$

$$E_{\Omega_\ell}(v_\ell) = E_{\mathcal{A}}(u_{\ell+h}) + E_{\mathcal{A}'^c}(u_\ell) \leq E_{\mathcal{A}}(\tau_{h'}^i u_\ell) + E_{\mathcal{A}'^c}(u_\ell)$$
$$= E_{\mathcal{A}'}(u_\ell) + E_{\mathcal{A}'^c}(u_\ell) = E_{\Omega_\ell}(u_\ell).$$

This is only possible if $v_\ell = u_\ell$, i.e. $(u_\ell - \tau_{-h'}^i(u_{\ell+h}))^+ = 0$ which is also

$$\tau_{h'}^i(u_\ell) \leq u_{\ell+h}.$$

Similary one would get

$$\tau_{-h'}^i(u_\ell) \leq u_{\ell+h}.$$

Thus, passing to the limit in ℓ in the inequalities above, one derives

$$\tilde{u}_\infty(x - h'e_i) \leq \tilde{u}_\infty(x), \quad \tilde{u}_\infty(x + h'e_i) \leq \tilde{u}_\infty(x) \quad \text{a.e. } x,$$

which implies

$$\tilde{u}_\infty(x) \leq \tilde{u}_\infty(x - h'e_i) \leq \tilde{u}_\infty(x) \quad \text{a.e. } x, \, \forall i = 1, \cdots, p, \, \forall h' \text{ small.}$$

This completes the proof of the lemma. $\qquad\square$

Lemma 10.4. *The function $\tilde{u}_\infty$ belongs to $W_0^{1,q}(\omega_2)$.*

Proof. Let $\rho = \rho_{\ell_0}$ be the function defined in (10.15). Since $(1-\rho)u_\ell \in W_0^{1,q}(\Omega_\ell)$, one derives from (10.21)

$$\int_{\Omega_\ell} \{F(\nabla u_\ell) - fu_\ell\}dx \leq \int_{\Omega_\ell} \{F(\nabla((1-\rho)u_\ell)) - f(1-\rho)u_\ell\}dx.$$

Since $(1-\rho) = 1$ outside of Ω_{ℓ_0+1} and 0 on Ω_{ℓ_0} we get

$$\int_{\Omega_{\ell_0+1}} \{F(\nabla u_\ell) - fu_\ell\}dx$$

$$\leq \int_{\Omega_{\ell_0}} F(0)dx + \int_{D_{\ell_0}} \{F(\nabla((1-\rho)u_\ell)) - f(1-\rho)u_\ell\}\,dx$$

$$\leq \int_{D_{\ell_0}} \{F(\nabla((1-\rho)u_\ell)) - f(1-\rho)u_\ell\}dx$$

(with $D_{\ell_0} = \Omega_{\ell_0+1}\backslash\Omega_{\ell_0}$) and thus

$$\int_{\Omega_{\ell_0+1}} F(\nabla u_\ell)dx \leq \int_{D_{\ell_0}} F(\nabla((1-\rho)u_\ell))dx + \int_{\Omega_{\ell_0+1}} |f||u_\ell|dx$$

$$\leq \int_{D_{\ell_0}} F(\nabla((1-\rho)u_\ell))dx + |f|_{q',\Omega_{\ell_0+1}}|u_\ell|_{q,\Omega_{\ell_0+1}}$$

$$\leq \int_{D_{\ell_0}} F(\nabla((1-\rho)u_\ell))dx + C|f|_{q',\Omega_{\ell_0+1}}||\nabla u_\ell||_{q,\Omega_{\ell_0+1}},$$

by the Poincaré inequality. Using the assumptions on F and Young's inequality it comes

$$\lambda\int_{\Omega_{\ell_0+1}} |\nabla u_\ell|^q \,dx - \Lambda'|\Omega_{\ell_0+1}| \leq \Lambda\int_{D_{\ell_0}} |\nabla((1-\rho)u_\ell)|^q \,dx$$

$$+ \epsilon||\nabla u_\ell||^q_{q,\Omega_{\ell_0+1}} + C(\epsilon)|f|^{q'}_{q',\Omega_{\ell_0+1}} + \Lambda'|\Omega_{\ell_0+1}|.$$

Choosing $\epsilon = \frac{\lambda}{2}$ we get

$$\frac{\lambda}{2}\int_{\Omega_{\ell_0+1}} |\nabla u_\ell|^q \,dx$$

$$\leq \Lambda\int_{D_{\ell_0}} |-u_\ell\nabla\rho + (1-\rho)\nabla u_\ell|^q \,dx + C(\lambda)|f|^{q'}_{q',\Omega_{\ell_0+1}} + 2\Lambda'|\Omega_{\ell_0+1}|$$

$$\leq C\int_{D_{\ell_0}} |\nabla u_\ell|^q \,dx + C(\lambda)\int_{(\ell_0+1)\omega_1}\int_{\omega_2} |f|^{q'} \,dX_2dX_1 + 2\Lambda'|\Omega_{\ell_0+1}|$$

(we have used here the Poincaré inequality again). Thus we derive

$$\frac{\lambda}{2} \int_{\Omega_{\ell_0+1}} |\nabla u_\ell|^q \, dx$$

$$\leq C \int_{D_{\ell_0}} |\nabla u_\ell|^q \, dx + (\ell_0 + 1)^p |\omega_1| \{ C(\lambda)|f|_{q',\omega_2}^{q'} + 2\Lambda'|\omega_2| \}.$$

If we still denote by C the maximum $\frac{2C}{\lambda} \vee \{ \frac{2}{\lambda}|\omega_1|(C(\lambda)|f|_{q',\omega_2}^{q'} + 2\Lambda'|\omega_2|) \}$ we obtain

$$\int_{\Omega_{\ell_0+1}} |\nabla u_\ell|^q \, dx \leq C \int_{D_{\ell_0}} |\nabla u_\ell|^q \, dx + C(\ell_0 + 1)^p$$

where C is independent of ℓ_0 and ℓ. Thus it comes

$$\int_{\Omega_{\ell_0}} |\nabla u_\ell|^q \, dx \leq C \int_{\Omega_{\ell_0+1}} |\nabla u_\ell|^q \, dx - C \int_{\Omega_{\ell_0}} |\nabla u_\ell|^q \, dx + C(\ell_0 + 1)^p$$

and then

$$\int_{\Omega_{\ell_0}} |\nabla u_\ell|^q \, dx \leq \gamma \int_{\Omega_{\ell_0+1}} |\nabla u_\ell|^q \, dx + \gamma(\ell_0 + 1)^p$$

with $\gamma = \frac{C}{C+1} < 1$. Iterating this inequality $[\ell - \ell_0]$ times where $[\ ]$ denotes the integer part of a number leads to

$$\int_{\Omega_{\ell_0}} |\nabla u_\ell|^q \, dx \leq \gamma \{ \gamma \int_{\Omega_{\ell_0+2}} |\nabla u_\ell|^q \, dx + \gamma(\ell_0 + 2)^p \} + \gamma(\ell_0 + 1)^p$$

$$\leq \gamma^2 \int_{\Omega_{\ell_0+2}} |\nabla u_\ell|^q \, dx + \gamma^2(\ell_0 + 2)^p + \gamma(\ell_0 + 1)^p$$

$$\leq \gamma^{[\ell-\ell_0]} \int_{\Omega_{\ell_0+[\ell-\ell_0]}} |\nabla u_\ell|^q \, dx + \sum_{k=1}^{[\ell-\ell_0]} \gamma^k(\ell_0 + k)^p.$$

Using the fact that

$$\ell - \ell_0 - 1 < [\ell - \ell_0] \leq \ell - \ell_0$$

we derive

$$\int_{\Omega_{\ell_0}} |\nabla u_\ell|^q \, dx \leq \frac{1}{\gamma^{\ell_0+1}} \gamma^\ell \int_{\Omega_\ell} |\nabla u_\ell|^q \, dx + \sum_{k=1}^{\infty} \gamma^k(\ell_0 + k)^p. \qquad (10.35)$$

Since

$$\lim_{k \to +\infty} \frac{\gamma^{k+1}(\ell_0 + k + 1)^p}{\gamma^k(\ell_0 + k)^p} = \gamma < 1,$$

the series above converges and has for sum some real number that we will denote by $\Sigma(\ell_0)$. We know on the other hand taking $v = 0$ in (10.21) that

$$\int_{\Omega_\ell} F(\nabla u_\ell) \, dx - \int_{\Omega_\ell} f u_\ell \, dx \leq \int_{\Omega_\ell} F(0) \, dx$$

i.e.

$$\int_{\Omega_\ell} F(\nabla u_\ell) \, dx - \int_{\Omega_\ell} F(0) \, dx \leq \int_{\Omega_\ell} f u_\ell dx$$
$$\leq |f|_{q',\Omega_\ell} |u_\ell|_{q,\Omega_\ell} \leq C|f|_{q',\Omega_\ell} ||\nabla u_\ell||_{q,\Omega_\ell}$$

by Hölder's inequality. Thus

$$\lambda \int_{\Omega_\ell} |\nabla u_\ell|^q \, dx \leq C|f|_{q',\Omega_\ell} ||\nabla u_\ell||_{q,\Omega_\ell}.$$

This leads to

$$\int_{\Omega_\ell} |\nabla u_\ell|^q \, dx \leq C \int_{\ell\omega_1} \int_{\omega_2} |f|^{q'} \, dX_2 dX_1 \leq C\ell^p.$$

Thus, going back to (10.35) we obtain

$$\int_{\Omega_{\ell_0}} |\nabla u_\ell|^q \, dx \leq \frac{1}{\gamma^{\ell_0+1}} C\gamma^\ell \ell^p + \Sigma(\ell_0) \leq K(\ell_0) \quad \forall \ell \tag{10.36}$$

where $K(\ell_0)$ is a constant independent of ℓ. It follows that u_ℓ is bounded in $W_{lat}^{1,q}(\Omega_{\ell_0})$ and thus up to a subsequence one has that there exists $\hat{u}_\infty \in W_{lat}^{1,q}(\Omega_{\ell_0})$ such that

$$u_\ell \rightharpoonup \hat{u}_\infty \text{ in } W_{lat}^{1,q}(\Omega_{\ell_0}), \quad u_\ell \to \hat{u}_\infty \text{ in } L^q(\Omega_{\ell_0}).$$

Since by the dominated convergence theorem one has

$$u_\ell \to \tilde{u}_\infty \text{ in } L^q(\Omega_{\ell_0})$$

one obtains

$$\hat{u}_\infty = \tilde{u}_\infty \in W_{lat}^{1,q}(\Omega_{\ell_0})$$

i.e. $\tilde{u}_\infty \in W_0^{1,q}(\omega_2)$. This completes the proof of the lemma. $\qquad \square$

Lemma 10.5. *One has*

$$\tilde{u}_\infty = u_\infty. \tag{10.37}$$

Proof. First notice that from (10.35), (10.36) one has

$$\int_{\Omega_{\ell_0}} |\nabla u_\ell|^q \, dx \le 1 + \ell_0^p \sum_{k=1}^{+\infty} \gamma^k (1+k)^p$$

$$= 1 + C\ell_0^p,$$

(10.38)

for ℓ large enough i.e. $\ell \ge \ell(\ell_0)$, $\ell_0 \ge 1$. If $\rho = \rho_{\ell_0}$ is the function defined in (10.15) one has

$$(1-\rho)u_\ell + \rho u_\infty \in W_0^{1,q}(\Omega_\ell)$$

and thus from (10.21)

$$\int_{\Omega_\ell} \{F(\nabla u_\ell) - f u_\ell\} \, dx$$

$$\le \int_{\Omega_\ell} \left\{ F(\nabla\{(1-\rho)u_\ell + \rho u_\infty\}) - f\{(1-\rho)u_\ell + \rho u_\infty\} \right\} \, dx.$$

Since $(1-\rho)u_\ell + \rho u_\infty = u_\ell$ on $\Omega_\ell \backslash \Omega_{\ell_0+1}$ and is equal to u_∞ on Ω_{ℓ_0} we get

$$\int_{\Omega_{\ell_0+1}} \{F(\nabla u_\ell) - f u_\ell\} \, dx \le \int_{\Omega_{\ell_0}} \{F(\nabla u_\infty) - f u_\infty\} \, dx$$

$$+ \int_{D_{\ell_0}} \left\{ F(\nabla\{(1-\rho)u_\ell + \rho u_\infty\}) - f\{(1-\rho)u_\ell + \rho u_\infty\} \right\} \, dx.$$

(10.39)

Let us denote by I the last integral above. One has

$$I = \int_{D_{\ell_0}} \left\{ F(\nabla\{(1-\rho)u_\ell + \rho u_\infty\}) - F((1-\rho)\nabla u_\ell + \rho\nabla u_\infty) \right.$$

$$\left. + F((1-\rho)\nabla u_\ell + \rho\nabla u_\infty) - f\{(1-\rho)u_\ell + \rho u_\infty\} \right\} \, dx$$

$$\le \int_{D_{\ell_0}} \left\{ F(\nabla\{(1-\rho)u_\ell + \rho u_\infty\}) - F((1-\rho)\nabla u_\ell + \rho\nabla u_\infty) \right\} \, dx$$

$$+ \int_{D_{\ell_0}} (1-\rho)\{F(\nabla u_\ell) - f u_\ell\} \, dx + \rho\{F(\nabla u_\infty) - f u_\infty\} \, dx$$

due to the convexity of F. It follows from (10.39) that

$$\int_{\Omega_{\ell_0+1}} \rho\{F(\nabla u_\ell) - f u_\ell\} dx \le \int_{\Omega_{\ell_0+1}} \rho\{F(\nabla u_\infty) - f u_\infty\} dx$$

$$+ \int_{D_{\ell_0}} \left\{ F(\nabla\{(1-\rho)u_\ell + \rho u_\infty\}) - F((1-\rho)\nabla u_\ell + \rho\nabla u_\infty) \right\} dx.$$

(10.40)

Let J be the last integral above. Since

$$\nabla\{(1-\rho)u_\ell + \rho u_\infty\} - \{(1-\rho)\nabla u_\ell + \rho\nabla u_\infty\} = -(u_\ell - u_\infty)\nabla\rho$$

we deduce from (10.18)

$$J \le C \int_{D_{\ell_0}} \Big(1 + |\nabla\{(1-\rho)u_\ell + \rho u_\infty\}|^{q-1}$$

$$+ |\{(1-\rho)\nabla u_\ell + \rho\nabla u_\infty\}|^{q-1}\Big)|u_\ell - u_\infty|dx$$

$$\le C \int_{D_{\ell_0}} \Big(1 + |\nabla u_\ell|^{q-1} + |u_\ell|^{q-1} + |\nabla u_\infty|^{q-1} + |u_\infty|^{q-1}\Big)|u_\ell - u_\infty|dx$$

(we used the inequality $(a+b)^{q-1} \le C(a^{q-1}+b^{q-1})$ valid for $q \ge 1$, $a, b \ge 0$). Using Hölder's inequality we derive

$$J \le C\Big(\int_{D_{\ell_0}} |u_\ell - u_\infty|^q \, dx\Big)^{\frac{1}{q}}$$

$$\Big\{ (\int_{D_{\ell_0}} 1 \, dx)^{\frac{1}{q'}} + (\int_{D_{\ell_0}} |\nabla u_\ell|^q \, dx)^{\frac{1}{q'}} + (\int_{D_{\ell_0}} |u_\ell|^q \, dx)^{\frac{1}{q'}}$$

$$+ (\int_{D_{\ell_0}} |\nabla u_\infty|^q \, dx)^{\frac{1}{q'}} + (\int_{D_{\ell_0}} |u_\infty|^q \, dx)^{\frac{1}{q'}} \Big\}.$$

Due to the Poincaré inequality it comes

$$J \le C\Big(\int_{D_{\ell_0}} |u_\ell - u_\infty|^q \, dx\Big)^{\frac{1}{q}}$$

$$\Big\{ |D_{\ell_0}|^{1-\frac{1}{q}} + (\int_{\Omega_{\ell_0+1}} |\nabla u_\ell|^q \, dx)^{1-\frac{1}{q}} + (\int_{\Omega_{\ell_0+1}} |\nabla u_\infty|^q \, dx)^{1-\frac{1}{q}} \Big\}.$$

Using (10.38) we deduce that for ℓ large enough

$$J = J_\ell \le C\Big(\int_{D_{\ell_0}} |u_\ell - u_\infty|^q \, dx\Big)^{\frac{1}{q}}(\ell_0 + 1)^{p(1-\frac{1}{q})}.$$

Thus, for large ℓ, (10.40) can be written as

$$\int_{\Omega_{\ell_0+1}} \rho\{F(\nabla u_\ell) - f u_\ell\} \, dx \le \int_{\Omega_{\ell_0+1}} \rho\{F(\nabla u_\infty) - f u_\infty\} \, dx$$

$$+ C\Big(\int_{D_{\ell_0}} |u_\ell - u_\infty|^q \, dx\Big)^{\frac{1}{q}}(\ell_0 + 1)^{p(1-\frac{1}{q})}. \tag{10.41}$$

Passing to the lim inf and using the weak lower semicontinuity of the left hand side (recall that F is convex) we get

$$\int_{\Omega_{\ell_0+1}} \rho\{F(\nabla\tilde{u}_\infty) - f\tilde{u}_\infty\} \, dx \le \int_{\Omega_{\ell_0+1}} \rho\{F(\nabla u_\infty) - f u_\infty\} \, dx$$

$$+ C\Big(\int_{D_{\ell_0}} |\tilde{u}_\infty - u_\infty|^q \, dx\Big)^{\frac{1}{q}}(\ell_0 + 1)^{p(1-\frac{1}{q})}.$$

Considering each of the terms above one has

$$\int_{\Omega_{\ell_0+1}} \rho\{F(\nabla\tilde{u}_\infty) - f\tilde{u}_\infty\}dx$$

$$= \int_{(\ell_0+1)\omega_1} \int_{\omega_2} \rho\{F(\nabla\tilde{u}_\infty) - f\tilde{u}_\infty\}dx$$

$$= E_{\omega_2}(\tilde{u}_\infty) \int_{(\ell_0+1)\omega_1} \rho dX_1.$$

Similarly

$$\int_{\Omega_{\ell_0+1}} \rho\{F(\nabla u_\infty) - f u_\infty\}\, dx = E_{\omega_2}(u_\infty) \int_{(\ell_0+1)\omega_1} \rho dX_1.$$

Finally

$$\left(\int_{D_{\ell_0}} |\tilde{u}_\infty - u_\infty|^q\, dx\right)^{\frac{1}{q}} = \left(\int_{(\ell_0+1)\omega_1\backslash\ell_0\omega_1} \int_{\omega_2} |\tilde{u}_\infty - u_\infty\}|^q\, dx\right)^{\frac{1}{q}}$$

$$= |\tilde{u}_\infty - u_\infty|_{q,\omega_2}|(\ell_0+1)\omega_1\backslash\ell_0\omega_1|^{\frac{1}{q}}$$

$$= |\tilde{u}_\infty - u_\infty|_{q,\omega_2}(|(\ell_0+1)\omega_1| - |\ell_0\omega_1|)^{\frac{1}{q}}$$

$$\leq C\ell_0^{(p-1)\frac{1}{q}}.$$

where C is independent of ℓ_0. Thus we have obtained

$$E_{\omega_2}(\tilde{u}_\infty) \leq E_{\omega_2}(u_\infty) + C\ell_0^{(p-1)\frac{1}{q}} \ell_0^{p(1-\frac{1}{q})} / \int_{\Omega_{\ell_0+1}} \rho\, dX_1$$

$$\leq E_{\omega_2}(u_\infty) + C\frac{\ell_0^{(p-1)\frac{1}{q}} \ell_0^{p(1-\frac{1}{q})}}{\ell_0^p}$$

for some constant C, i.e.

$$E_{\omega_2}(\tilde{u}_\infty) \leq E_{\omega_2}(u_\infty) + C\ell_0^{-\frac{1}{q}}.$$

Letting $\ell_0 \to +\infty$ we obtain $E_{\omega_2}(\tilde{u}_\infty) \leq E_{\omega_2}(u_\infty)$. By definition of u_∞ this implies that $\tilde{u}_\infty = u_\infty$ and the result follows. $\square$

Remark 10.5. We have now proved the theorem 10.5. Note that the convergence (10.25) takes place pointwise and also in $L^q(\Omega_{\ell_0})$-strong and $W^{1,q}_{lat}(\Omega_{\ell_0})$-weak for any ℓ_0.

Remark 10.6. The function

$$F(\xi) = \frac{1}{q}|\xi|^q$$

satisfies the assumptions (10.16), (10.17). The solution of the minimization problem (10.21) is also weak solution of the q-Laplace Dirichlet problem

$$\begin{cases} -\text{div}\,(|\nabla u_\ell|^{q-2}\nabla u_\ell) = f \text{ in } \Omega_\ell, \\ u_\ell = 0 \text{ on } \partial\Omega_\ell. \end{cases}$$

One recovers the results of [39], i.e. the convergence of u_ℓ toward the weak solution u_∞ to

$$\begin{cases} -\text{div}\,_{X_2}(|\nabla_{X_2} u_\infty|^{q-2}\nabla_{X_2} u_\infty) = f \text{ in } \omega_2, \\ u_\infty = 0 \text{ on } \partial\omega_2 \end{cases}$$

with a clear meaning for div $_{X_2}$.

To conclude this section we would like to consider the asymptotic behavior of

$$\frac{E_{\Omega_\ell}(u_\ell)}{|\ell\omega_1|}$$

when $\ell \to \infty$.

In particular we would like to prove (Cf. [28]):

Theorem 10.6. (Convergence of the energy) *One has for some constant $C > 0$ and sufficiently large ℓ,*

$$E_{\omega_2}(u_\infty) \leq \frac{E_{\Omega_\ell}(u_\ell)}{|\ell\omega_1|} \leq E_{\omega_2}(u_\infty) + \frac{C}{\ell}$$

where u_ℓ, u_∞ are the solutions to (10.21) and (10.24) respectively.

Proof. Since $u_\ell \in W_0^{1,q}(\Omega_\ell) \subset V^{1,q}(\Omega_\ell)$ the left hand side inequality is an immediate consequence of (10.28) which can be written for $v = u_\ell$

$$|\ell\omega_1|E_{\omega_2}(u_\infty) = E_{\Omega_\ell}(u_\infty) \leq E_{\Omega_\ell}(u_\ell).$$

To prove the right hand side inequality, one considers $\rho = \rho_{\ell-1}(X_1)$, as defined in (10.15). Since $\rho u_\infty \in W_0^{1,q}(\Omega_\ell)$ from (10.21), we have

$$E_{\Omega_\ell}(u_\ell) \leq E_{\Omega_\ell}(\rho u_\infty).$$

Thus from (10.15), (10.17) we deduce for C as in (10.15)

$$E_{\Omega_\ell}(\rho u_\infty) = E_{\Omega_{\ell-1}}(u_\infty) + \int_{\Omega_\ell \setminus \Omega_{\ell-1}} F(\nabla(\rho u_\infty)) - f u_\infty \rho dx$$

$$\leq E_{\Omega_\ell}(u_\infty) + \int_{\Omega_\ell \setminus \Omega_{\ell-1}} F(\nabla(\rho u_\infty)) - F(\nabla u_\infty) - f u_\infty(\rho - 1) dx$$

$$\leq |\ell\omega_1| E_{\omega_2}(u_\infty) + \int_{\Omega_\ell \setminus \Omega_{\ell-1}} 2\Lambda' + \Lambda|\nabla(\rho u_\infty)|^q + \Lambda|\nabla u_\infty|^q + |f||u_\infty| dx$$

$$\leq |\ell\omega_1| E_{\omega_2}(u_\infty) + \int_{\Omega_\ell \setminus \Omega_{\ell-1}} 2\Lambda' + 2\Lambda|\nabla u_\infty|^q + \Lambda C|u_\infty|^q + |f||u_\infty| dx$$

$$\leq |\ell\omega_1| E_{\omega_2}(u_\infty)$$
$$+ C'\{|\ell\omega_1| - |(\ell-1)\omega_1|\} \int_{\omega_2} 1 + |\nabla u_\infty|^q + |u_\infty|^q + |f||u_\infty| dx.$$

Dividing by $|\ell\omega_1|$ the result follows, i.e. one has for some constant C

$$\frac{E_{\Omega_\ell}(u_\ell)}{|\ell\omega_1|} \leq E_{\omega_2}(u_\infty) + \frac{C}{\ell}.$$

This completes the proof of the theorem. $\qquad\qquad\qquad\qquad\square$

10.3   The case of the q-Laplace operator

As we have seen in Remark 10.6 above, for $f = f(X_2) \geq 0$ the solution to

$$\begin{cases} u_\ell \in W_0^{1,q}(\Omega_\ell), \\ \int_{\Omega_\ell} |\nabla u_\ell|^{q-2} \nabla u_\ell \cdot \nabla v dx = \int_{\Omega_\ell} f v dx \ \forall v \in W_0^{1,q}(\Omega_\ell), \end{cases} \tag{10.42}$$

converges toward the solution to the q-Laplace equation on the section ω_2 of the cylinder namely the solution to

$$\begin{cases} u_\infty \in W_0^{1,q}(\omega_2), \\ \int_{\omega_2} |\nabla_{X_2} u_\infty|^{q-2} \nabla_{X_2} u_\infty \cdot \nabla_{X_2} v dX_2 \\ \qquad\qquad = \int_{\omega_2} f v dX_2 \ \ \forall v \in W_0^{1,q}(\omega_2). \end{cases} \tag{10.43}$$

The method that we developed above does not provide any rate of convergence. This is what we would like to consider here allowing changes of sign for $f \in L^{q'}(\omega_2)$. We follow here [76]. First let us recall the following lemma.

Lemma 10.6. *There exist constant c_q, C_q such that*

$$||\xi|^{q-2}\xi - |\eta|^{q-2}\eta| \leq C_q|\xi - \eta|(|\xi| + |\eta|)^{q-2} \ \forall \xi, \eta \in \mathbb{R}^n, \tag{10.44}$$

$$(|\xi|^{q-2}\xi - |\eta|^{q-2}\eta) \cdot (\xi - \eta) \geq c_q|\xi - \eta|^2(|\xi| + |\eta|)^{q-2} \ \forall \xi, \eta \in \mathbb{R}^n. \tag{10.45}$$

We refer the reader to [16] for a proof.

Then we have

Theorem 10.7. *Let u_ℓ, u_∞ be the solutions to* (10.42), (10.43). *There exist constants C_1, C_2 independent of ℓ such that*

$$||\nabla(u_\ell - u_\infty)||^q_{q,\Omega_{\ell/2}} \leq C_1 \ell^{-(\frac{2q}{q-2}-p)} \quad \text{for } q > 2, \tag{10.46}$$

$$||\nabla(u_\ell - u_\infty)||^q_{q,\Omega_{\ell/2}} \leq C_2 \ell^{-(\frac{q^2}{2-q}-p)} \quad \text{for } q < 2. \tag{10.47}$$

Proof. Note that if $v \in W_0^{1,q}(\Omega_\ell)$ then $v(X_1, \cdot) \in W_0^{1,q}(\omega_2)$ for almost every $X_1 \in \ell\omega_1$. Thus from (10.43) one deduces that for almost every $X_1 \in \ell\omega_1$

$$\int_{\omega_2} |\nabla_{X_2} u_\infty|^{q-2} \nabla_{X_2} u_\infty \cdot \nabla_{X_2} v(X_1, \cdot) dX_2 = \int_{\omega_2} fv(X_1, \cdot) dX_2.$$

Integrating in X_1 on $\ell\omega_1$ and using the fact that u_∞ is independent of X_1 we get

$$\int_{\Omega_\ell} |\nabla u_\infty|^{q-2} \nabla u_\infty \cdot \nabla v dx = \int_{\Omega_\ell} fv dx.$$

Subtracting this equation from (10.42) we obtain

$$\int_{\Omega_\ell} (|\nabla u_\ell|^{q-2} \nabla u_\ell - |\nabla u_\infty|^{q-2} \nabla u_\infty) \cdot \nabla v dx = 0 \quad \forall v \in W_0^{1,q}(\Omega_\ell). \tag{10.48}$$

Denote by $\rho_1 = \rho_1(X_1)$ a smooth function such that

$$0 \leq \rho_1 \leq 1, \ \rho_1 = 1 \text{ on } \frac{1}{2}\omega_1, \ \rho_1 = 0 \text{ on } \partial\omega_1, \ |\nabla_{X_1}\rho_1| \leq C \tag{10.49}$$

and set

$$\rho = \rho_\ell = \rho_1\left(\frac{X_1}{\ell}\right).$$

It is clear that for any $\alpha \geq 1$ one has

$$(u_\ell - u_\infty)\rho^\alpha \in W_0^{1,q}(\Omega_\ell).$$

Thus we derive from (10.48) that

$$\int_{\Omega_\ell} (|\nabla u_\ell|^{q-2} \nabla u_\ell - |\nabla u_\infty|^{q-2} \nabla u_\infty) \cdot \nabla\{(u_\ell - u_\infty)\rho^\alpha\} dx = 0.$$

This can also be written

$$\int_{\Omega_\ell} (|\nabla u_\ell|^{q-2} \nabla u_\ell - |\nabla u_\infty|^{q-2} \nabla u_\infty) \cdot \nabla(u_\ell - u_\infty)\rho^\alpha dx$$

$$= -\frac{\alpha}{\ell} \int_{\Omega_\ell} |\nabla u_\ell|^{q-2} \nabla_{X_1} u_\ell \cdot \nabla_{X_1} \rho_1 \rho^{\alpha-1} (u_\ell - u_\infty) dx.$$

Using (10.45), (10.49) we obtain

$$
\begin{aligned}
c_q \int_{\Omega_\ell} (|\nabla u_\ell| + |\nabla u_\infty|)^{q-2} |\nabla(u_\ell - u_\infty)|^2 \rho^\alpha dx \\
\leq \frac{\alpha C}{\ell} \int_{\Omega_\ell} |\nabla u_\ell|^{q-2} |\nabla_{X_1} u_\ell| |u_\ell - u_\infty| \rho^{\alpha-1} dx.
\end{aligned}
\tag{10.50}
$$

The case $q > 2$.

We suppose here that $\alpha \geq \frac{2q}{q-2}$. The inequality (10.50) can be written for some constant C independent of ℓ

$$
\int_{\Omega_\ell} (|\nabla u_\ell| + |\nabla u_\infty|)^{q-2} |\nabla(u_\ell - u_\infty)|^2 \rho^\alpha dx
$$

$$
\leq \frac{C}{\ell} \int_{\Omega_\ell} |\nabla u_\ell|^{\frac{q-2}{2}} |\nabla_{X_1}(u_\ell - u_\infty)| \rho^{\frac{\alpha}{2}} |\nabla u_\ell|^{\frac{q-2}{2}} |u_\ell - u_\infty| \rho^{\frac{\alpha}{2}-1} dx
$$

$$
\leq \frac{C}{\ell} \{ \int_{\Omega_\ell} |\nabla u_\ell|^{q-2} |\nabla_{X_1}(u_\ell - u_\infty)|^2 \rho^\alpha dx \}^{\frac{1}{2}}
$$

$$
\{ \int_{\Omega_\ell} |\nabla u_\ell|^{q-2} |u_\ell - u_\infty|^2 \rho^{(\alpha-2)} dx \}^{\frac{1}{2}}.
$$

(We used the Cauchy-Schwarz inequality in the second integral). Thus we get

$$
\int_{\Omega_\ell} (|\nabla u_\ell| + |\nabla u_\infty|)^{q-2} |\nabla(u_\ell - u_\infty)|^2 \rho^\alpha dx
$$

$$
\leq \frac{C^2}{\ell^2} \int_{\Omega_\ell} |\nabla u_\ell|^{q-2} |u_\ell - u_\infty|^2 \rho^{(\alpha-2)} dx.
$$

Since $|\nabla u_\ell| + |\nabla u_\infty| \geq |\nabla u_\ell - \nabla u_\infty|$ we derive using Hölder's inequality in the integral of the right hand side

$$
\int_{\Omega_\ell} |\nabla(u_\ell - u_\infty)|^q \rho^\alpha dx
$$

$$
\leq \frac{C^2}{\ell^2} \{ \int_{\Omega_\ell} |u_\ell - u_\infty|^q \rho^{(\alpha-2)\frac{q}{2}} dx \}^{\frac{2}{q}} \{ \int_{\Omega_\ell} |\nabla u_\ell|^q dx \}^{1-\frac{2}{q}}.
$$

Notice that

$$
\alpha \geq \frac{2q}{q-2} \quad \Leftrightarrow \quad (\alpha - 2)\frac{q}{2} \geq \alpha.
$$

Thus since $0 \leq \rho \leq 1$ we obtain

$$
\int_{\Omega_\ell} |\nabla(u_\ell - u_\infty)|^q \rho^\alpha dx
$$

$$
\leq \frac{C^2}{\ell^2} \{ \int_{\Omega_\ell} |u_\ell - u_\infty|^q \rho^\alpha dx \}^{\frac{2}{q}} \{ \int_{\Omega_\ell} |\nabla u_\ell|^q dx \}^{1-\frac{2}{q}}
$$

which leads, after applying the Poincaré inequality in the first integral of the right hand side, to

$$\int_{\Omega_\ell} |\nabla(u_\ell - u_\infty)|^q \rho^\alpha dx \leq (\frac{C}{\ell})^{\frac{2q}{q-2}} \int_{\Omega_\ell} |\nabla u_\ell|^q dx. \qquad (10.51)$$

It remains to estimate this last integral. Taking $v = u_\ell$ in (10.42) we obtain

$$\int_{\Omega_\ell} |\nabla u_\ell|^q dx = \int_{\Omega_\ell} f u_\ell dx \leq |f|_{q',\Omega_\ell} |u_\ell|_{q,\Omega_\ell} \leq C(q,\omega_2)|f|_{q',\Omega_\ell} ||\nabla u_\ell||_{q,\Omega_\ell}$$

by the Poincaré inequality. Thus for some constant C independent of ℓ

$$\begin{aligned}
\int_{\Omega_\ell} |\nabla u_\ell|^q dx &\leq C \int_{\Omega_\ell} |f|^{q'} dx = C \int_{\ell\omega_1} \int_{\omega_2} |f|^{q'} dX_2 dX_1 \\
&= C|\ell\omega_1| \int_{\omega_2} |f|^{q'} dX_2 \leq C\ell^p.
\end{aligned} \qquad (10.52)$$

($|\ell\omega_1|$ denotes the Lebesgue measure of $\ell\omega_1$). Going back to (10.51) we derive

$$||\nabla(u_\ell - u_\infty)||^q_{q,\Omega_{\ell/2}} \leq C_1 \ell^{-(\frac{2q}{q-2}-p)},$$

which completes the proof of (10.46).

The case $q < 2$.

We suppose here that $\alpha \geq \frac{2q}{2-q}$. From (10.50) we derive

$$\{\int_{\Omega_\ell} (|\nabla u_\ell| + |\nabla u_\infty|)^{q-2} |\nabla(u_\ell - u_\infty)|^2 \rho^\alpha dx\}^{\frac{1}{q}}$$

$$\{\int_{\Omega_\ell} (|\nabla u_\ell| + |\nabla u_\infty|)^{q-2} |\nabla(u_\ell - u_\infty)|^2 \rho^\alpha dx\}^{\frac{1}{q'}} \qquad (10.53)$$

$$\leq \frac{C}{\ell} \int_{\Omega_\ell} ||\nabla u_\ell|^{q-2} \nabla_{X_1} u_\ell - |\nabla u_\infty|^{q-2} \nabla_{X_1} u_\infty||u_\ell - u_\infty|\rho^{\alpha-1} dx$$

(recall that u_∞ is independent of X_1). Since $q' = \frac{q}{q-1} > 2$ one has

$$\begin{aligned}
(|\nabla u_\ell| + |\nabla u_\infty|)^{q-2}|\nabla(u_\ell - u_\infty)|^2 \\
= (|\nabla u_\ell| + |\nabla u_\infty|)^{q-q'} (|\nabla u_\ell| + |\nabla u_\infty|)^{q'-2}|\nabla(u_\ell - u_\infty)|^2 \\
\geq (|\nabla u_\ell| + |\nabla u_\infty|)^{q-q'} |\nabla u_\ell - \nabla u_\infty|^{q'-2}|\nabla(u_\ell - u_\infty)|^2 \\
= (|\nabla u_\ell| + |\nabla u_\infty|)^{q-q'} |\nabla(u_\ell - u_\infty)|^{q'}.
\end{aligned}$$

Taking into account (10.53), (10.44) we derive

$$\{\int_{\Omega_\ell} (|\nabla u_\ell| + |\nabla u_\infty|)^{q-2}|\nabla(u_\ell - u_\infty)|^2 \rho^\alpha dx\}^{\frac{1}{q}}$$

$$\{\int_{\Omega_\ell} (|\nabla u_\ell| + |\nabla u_\infty|)^{q-q'}|\nabla(u_\ell - u_\infty)|^{q'} \rho^\alpha dx\}^{\frac{1}{q'}}$$

$$\leq \frac{C}{\ell} \int_{\Omega_\ell} (|\nabla u_\ell| + |\nabla u_\infty|)^{q-2}|\nabla(u_\ell - u_\infty)||u_\ell - u_\infty|\rho^{\alpha-1} dx$$

$$= \frac{C}{\ell} \int_{\Omega_\ell} (|\nabla u_\ell| + |\nabla u_\infty|)^{q-2}|\nabla(u_\ell - u_\infty)|\rho^{\frac{\alpha}{q'}}|u_\ell - u_\infty|\rho^{\frac{\alpha}{q}-1} dx$$

$$\leq \frac{C}{\ell}\{\int_{\Omega_\ell} |u_\ell - u_\infty|^q \rho^{\alpha-q} dx\}^{\frac{1}{q}}$$

$$\{\int_{\Omega_\ell} (|\nabla u_\ell| + |\nabla u_\infty|)^{(q-2)q'}|\nabla(u_\ell - u_\infty)|^{q'} \rho^\alpha dx\}^{\frac{1}{q'}}.$$

Noting that

$$q - q' = q - \frac{q}{q-1} = q\frac{q-2}{q-1} = q'(q-2)$$

we derive for some constant C

$$\int_{\Omega_\ell} (|\nabla u_\ell| + |\nabla u_\infty|)^{q-2}|\nabla(u_\ell - u_\infty)|^2 \rho^\alpha dx$$

$$\leq \frac{C}{\ell^q} \int_{\Omega_\ell} |u_\ell - u_\infty|^q \rho^{\alpha-q} dx \qquad (10.54)$$

$$\leq \frac{C}{\ell^q} \int_{\Omega_\ell} |\nabla(u_\ell - u_\infty)|^q \rho^{\alpha-q} dx$$

by the Poincaré Inequality. Using now Hölder's inequality we have

$$\int_{\Omega_\ell} |\nabla(u_\ell - u_\infty)|^q \rho^{\alpha-q} dx$$

$$= \int_{\Omega_\ell} (|\nabla u_\ell| + |\nabla u_\infty|)^{\frac{(q-2)q}{2}}|\nabla(u_\ell - u_\infty)|^q \rho^{\alpha-q}(|\nabla u_\ell| + |\nabla u_\infty|)^{\frac{(2-q)q}{2}} dx$$

$$\leq \{\int_{\Omega_\ell} (|\nabla u_\ell| + |\nabla u_\infty|)^{q-2}|\nabla(u_\ell - u_\infty)|^2 \rho^{(\alpha-q)\frac{2}{q}} dx\}^{\frac{q}{2}}$$

$$\{\int_{\Omega_\ell} (|\nabla u_\ell| + |\nabla u_\infty|)^q dx\}^{1-\frac{q}{2}}.$$

We then notice that

$$\alpha \geq \frac{2q}{2-q} \quad \Leftrightarrow \quad (\alpha-q)\frac{2}{q} \geq \alpha$$

and thus

$$\int_{\Omega_\ell} |\nabla(u_\ell - u_\infty)|^q \rho^{\alpha-q} dx$$

$$\leq \{\int_{\Omega_\ell} (|\nabla u_\ell| + |\nabla u_\infty|)^{q-2} |\nabla(u_\ell - u_\infty)|^2 \rho^\alpha dx\}^{\frac{q}{2}}$$

$$\{\int_{\Omega_\ell} (|\nabla u_\ell| + |\nabla u_\infty|)^q dx\}^{1-\frac{q}{2}}.$$

Using (10.54) we get

$$\int_{\Omega_\ell} |\nabla(u_\ell - u_\infty)|^q \rho^{\alpha-q} dx$$

$$\leq \{\frac{C}{\ell^q} \int_{\Omega_\ell} |\nabla(u_\ell - u_\infty)|^q \rho^{\alpha-q} dx\}^{\frac{q}{2}}$$

$$\{\int_{\Omega_\ell} (|\nabla u_\ell| + |\nabla u_\infty|)^q dx\}^{1-\frac{q}{2}}$$

and thus

$$\int_{\Omega_\ell} |\nabla(u_\ell - u_\infty)|^q \rho^{\alpha-q} dx \leq \frac{C}{(\ell^{\frac{q}{2}})^{\frac{2}{2-q}}} \int_{\Omega_\ell} (|\nabla u_\ell| + |\nabla u_\infty|)^q dx$$

$$\leq C\ell^{-(\frac{q^2}{2-q}-p)}$$

(see (10.52)). Since $\alpha > q$ we get

$$\int_{\Omega_{\ell/2}} |\nabla(u_\ell - u_\infty)|^q dx \leq C\ell^{-(\frac{q^2}{2-q}-p)}$$

i.e. (10.47). This completes the proof of the theorem. $\square$

Remark 10.7. In the case $q > 2$ we have obtained the convergence of u_ℓ toward u_∞ provided $\frac{2q}{q-2} > p$. Note that when $p \leq 2$ this inequality is always true. Else for p fixed the inequality fails for $q \geq \frac{2p}{p-2}$.

In the case $q < 2$ we have obtained the convergence of u_ℓ toward u_∞ provided $\frac{q^2}{2-q} > p$. Note that when $p = 1$ this inequality is always true. Else for p fixed the inequality fails for $q \leq \frac{4}{\sqrt{1+\frac{8}{p}}+1}$.

To summarize, we have obtained convergence and a rate of convergence of u_ℓ toward u_∞ for every $q \in (\frac{4}{\sqrt{1+\frac{8}{p}}+1}, \frac{2p}{p-2})$.

Some concluding remarks

We presented here only a few problems which were addressed in the recent years. Many other developments occur. The most important achievement along the years was to reach, with relatively simple techniques, an exponential rate of convergence for the solution of problems set in a domain of the type Ω_ℓ toward a solution set in Ω_∞. This behavior carries often the name of Saint-Venant and this gives credit to the numerical analysts' pioneering vision. They indeed did not wait for the theoretical results to reduce in their computations the dimension of the problem.

We have seen that what can be done for the Dirichlet problem can be extended to the stokes problem in the case of forces independent of one direction. It would be interesting to extend this to the two dimensional time dependent Navier-Stokes problem in the spirit of our Section 1.7. Note that we did not develop too much the parabolic version of our elliptic results. Some results are given in [13]. An updated version should be considered since one can easily obtain an exponential rate of convergence there. One can address also quasilinear variants and periodic parabolic problems (see [55], [57]).

The extension in the case of the stationary Navier-Stokes problem is more delicate since uniqueness might fail - the same issue occurs for the higher dimensional evolution Navier-Stokes problem. Some higher order problems are investigated in [9], [56]. Hyperbolic issues are considered in [10], [53], [58], [59]. Regarding singular perturbations a self contained study is available in [24] - see also [17], [18], [20], [54], [56], [58]. Some interesting applications can be found in [22] in the parabolic case and in [23] in the elliptic case.

As we saw in the case of the q-Laplace operator, exponential rates of convergence are more complicated to reach for nonlinear problems (Cf.

also [65]). It can go through for obstacle problems but remains open in many situation in the framework of variational inequalities (see [41], [43], [78]). Some quasilinear problems are investigated in [13] and also various different norms are considered (i.e. not only the norm of the gradient). To this respect many results can be extended to reach an exponential rate of convergence.

Since everything extends to the periodic case in the spirit of Chapter 4, many issues remain open there (see [2], [36], [37]).

As we might have convinced the reader, any problem set in a cylinder or in a generalized cylinder of the type $\ell\omega_1 \times \omega_2$ can be analyze the way we did and lead to interesting issues. In the seventies, one of the first application of the obstacle problem and of variational inequalities was the so called Dam Problem. It was set and studied widely in two dimensions and the invoked reason was that one was assuming that the dam at stake was of cylindrical shape and long enough so that the flow in this porous medium was the same in every cross section of the dam. A rigorous proof of this assertion is still pending and challenging. (See [3], [7]) for notions on the problem.

Appendix A

The goal of this appendix is to quote some classical results of analysis that we used in the text and allow the reader to have a direct access to them. With this idea to ease the reading, we list also below the notation which occurs very often along the book.

A.1 Dictionary of the main notation

ℓ parameter going to $+\infty$.

$[\,]$ integer part of a real number.

$\Omega_\ell = \ell\omega_1 \times \omega_2$ open subset of $\mathbb{R}^n$ (so called "cylinder").

ω_1 convex bounded open subset of $\mathbb{R}^p$ containing 0.

ω_2 bounded open subset of $\mathbb{R}^{(n-p)}$.

$x = (X_1, X_2)$ point in Ω_ℓ.

$X_1 = x_1, \cdots, x_p$ first p coordinates of $x = (X_1, X_2)$.

$X_2 = x_{p+1}, \cdots, x_n$ last $n - p$ coordinates of $x = (X_1, X_2)$.

$\partial_{x_i} = \frac{\partial}{\partial_{x_i}}$ derivative in the direction x_i.

$\nabla_{X_1} =$ gradient in the p first variables $= (\partial_{x_1}, \partial_{x_2}, \cdots, \partial_{x_p})$.

$\nabla_{X_2} =$ gradient in the $n - p$ last variables $= (\partial_{x_{p+1}}, \cdots, \partial_{x_n})$.

$\rho = \rho_{\ell_1} :$ a function of X_1 such that

$$0 \le \rho_{\ell_1} \le 1, \ \rho_{\ell_1} = 1 \text{ on } \ell_1\omega_1, \ \rho_{\ell_1} = 0 \text{ outside } (\ell_1 + 1)\omega_1,$$

$$|\nabla_{X_1}\rho_{\ell_1}| \le C.$$

$|\ |$ euclidean norm of vectors or Lebesgue measure of sets.

$|\ |_{p,O}$ usual $L^p(O)$-norm, $|v|_{p,O} = (\int_O |v|^p dx)^{\frac{1}{p}}$.

$\|\nabla v\|_{2,O} = (\int_O |\nabla v|^2 dx)^{\frac{1}{2}}$ the Dirichlet norm.

$\mathcal{D}(\Omega_\ell)$, $\mathcal{D}(O)$ spaces of C^∞ functions with compact support in Ω_ℓ, O.

$H^1(O) = \{v \in L^2(O) \mid \partial_{x_i} v \in L^2(O) \ \forall i = 1, \cdots, n\}$.

$H_0^1(\Omega_\ell)$, $H_0^1(O)$ closure of $\mathcal{D}(\Omega_\ell)$, $\mathcal{D}(O)$ in $H^1(\Omega_\ell)$, $H^1(O)$.

$W_0^{1,\infty}(O)$ the space of uniformly Lipschitz continuous functions vanishing on ∂O.

∂O the boundary of the set O.

$V_{loc}(\overline{O}) = \{v \mid v \in V(U) \ \forall \ U \subset O$ open and bounded$\}$, $V = H^1, L^2, \dots$.

$H^1_{\mathrm{per}}(T)$ the closure of $\{v \in C^\infty(\overline{T}) \mid v \ T\text{-periodic}\}$ in $H^1(T)$.

A.2　Some existence results

Let us recall some classical results. We assume now that V is a reflexive Banach space with norm $|\cdot|_V$ and dual V'. We denote by $\langle\,,\,\rangle$ the duality bracket between V' and V. Let $A : V \to V'$ be an operator from V into V'. We shall say that A is monotone iff

$$\langle Au - Av, u - v \rangle \ge 0 \quad \forall u, v \in V. \tag{A.1}$$

A is strictly monotone if the equality in (A.1) holds only for $u = v$. We say that A is coercive iff for some $v_0 \in V$

$$\langle Av - Av_0, v - v_0 \rangle / |v - v_0|_V \to +\infty \quad \text{when} \quad |v|_V \to +\infty. \tag{A.2}$$

We shall say that $A : V \to V'$ is continuous on finite dimensional subspaces of V if for every finite dimensional subspace M of V the mapping

$$A : M \to V'$$

is weakly continuous, that is to say for every $v \in V$

$$u \to \langle Au, v \rangle$$

is continuous on M.

Under the assumptions above we have:

Theorem A.1. *Let $A : V \to V'$ be monotone, coercive, continuous on finite dimensional subspaces of V. For $f \in V'$ there exists $u \in V$ solution to*

$$Au = f. \tag{A.3}$$

Moreover if A is strictly monotone the solution of (A.3) is unique.

We refer to [49], [16] for a proof.

As an application we have:

Theorem A.2. *(Lax-Milgram) Let $a = a(u, v)$ be a bilinear form on H a real Hilbert space such that*

$$a \text{ is continuous } -i.e. \ |a(u, v)| \leq \Lambda |u|_H |v|_H \ \forall u, \ v \in H, \tag{A.4}$$

$$a \text{ is coercive } -i.e. \ a(u, u) \geq \lambda |u|_H^2 \ \forall u \in H, \tag{A.5}$$

(for some positive constants $\lambda, \Lambda, \ | \ |_H$ denotes the norm in H). Then for $f \in H'$ the dual of H there exists a unique $u \in H$ such that

$$a(u, v) = \langle f, v \rangle \ \forall v \in H. \tag{A.6}$$

Proof. For $u \in H$, $v \to a(u, v)$ is a continuous linear form on H (see (A.4)) that we denote by $A(u)$. By (A.5)

$$\langle A(w) - A(u), w - u \rangle = a(w - u, w - u) \geq \lambda |w - u|^2 \ \forall w, u \in H.$$

Then all the assumptions of the previous theorem are satisfied and there exists a unique $u \in H$ such that

$$A(u) = f$$

which is equivalent to

$$a(u, v) = \langle f, v \rangle \ \forall v \in H.$$

This completes the proof of the theorem. $\qquad\square$

The theorem above extends to convex subsets of H giving raise to the theory of variational inequalities (see [64], [5], [60], [68], [74], [12]). Indeed we have

Theorem A.3. *(Variational Inequalities) Let $K \neq \emptyset$ be a closed convex subset of a real Hilbert space H. Let a be a bilinear form on H satisfying (A.4), (A.5). Then for $f \in H'$ there exists a unique u solution to*

$$\begin{cases} u \in K, \\ a(u, v - u) \geq \langle f, v - u \rangle \quad \forall v \in K. \end{cases} \tag{A.7}$$

Proof. We give the proof when a is symmetric, i.e. when

$$a(u, v) = a(v, u) \quad \forall u, v \in H.$$

In this case

$$[u, v] = a(u, v)$$

is a scalar product on H which defines the same topology as the initial scalar product (see (A.4), (A.5)). Then by the Riesz representation theorem there exists $\tilde{f} \in H$ such that

$$\langle f, v \rangle = [\tilde{f}, v] \quad \forall v \in H.$$

Then u solution to (A.7) is the only element in K satisfying

$$[u, v - u] \geq [\tilde{f}, v - u] \quad \forall v \in K$$

i.e. u is the projection of $\tilde{f}$ on the closed convex set K. This completes the proof of the theorem. $\qquad\square$

Remark A.8. When K is equal to H, taking $v = u \pm w$ in (A.7) one obtains

$$a(u.w) = \langle f, w \rangle \quad \forall w \in H$$

i.e. one recovers the Theorem A.2.

Example: The obstacle problem.

We introduce this example in such a way that the reader could bridge the gaps that might remain in Chapter 4 and 9 when for instance we did not show that the different convex sets introduced were closed.

Let Ω be an open subset of $\mathbb{R}^n$ bounded in one direction, $H = H_0^1(\Omega)$, $\varphi \in H_0^1(\Omega)$, $f \in H^{-1}(\Omega) = (H_0^1(\Omega))'$, $a(u, v) = \int_\Omega \nabla u \cdot \nabla v dx$,

$$K = \{v \in H_0^1(\Omega) \mid v(x) \geq \varphi(x) \text{ a. e. } x \in \Omega\}.$$

We claim that K is a closed convex subset of $H_0^1(\Omega)$.

- K is convex.

For u, $v \in K$ one has for $\alpha \in [0, 1]$

$$(\alpha u + (1 - \alpha)v)(x) \geq \alpha\varphi(x) + (1 - \alpha)\varphi(x) = \varphi(x) \quad \text{a. e. } x \in \Omega.$$

This shows the convexity of K.

- K is closed.

Let $u_n \in K$ such that $u_n \to u$ in $H_0^1(\Omega)$. One has then

$$u_n \to u \text{ in } L^2(\Omega)$$

and for a subsequence n_k

$$u_{n_k}(x) \to u(x) \text{ on } \Omega \backslash N \tag{A.8}$$

where N is a set of measure 0. One has also

$$u_{n_k}(x) \geq \varphi(x) \text{ on } \Omega \backslash N_k \tag{A.9}$$

where N_k is a set of measure 0. Then (A.8) and (A.9) are both satisfied on $\Omega \backslash (\cup_k N_k \cup N)$ and thus passing to the limit one gets

$$u(x) \geq \varphi(x) \quad \forall x \in \Omega \backslash (\cup_k N_k \cup N)$$

i.e.

$$u(x) \geq \varphi(x) \text{ a.e. } x \in \Omega.$$

This shows that $u \in K$ and that K is closed. Since $K \neq \emptyset$ $(\varphi \in K)$ there exists a unique solution to

$$\begin{cases} u \in K, \\ \int_\Omega \nabla u \cdot \nabla(v - u) dx \geq \langle f, v - u \rangle \quad \forall v \in K. \end{cases}$$

u is the solution of the so-called "Obstacle Problem".

A.3 Poincaré's Inequality

We used frequently the following Poincaré inequality in domains of the type Ω_ℓ. We state it in the L^2-framework but the reader will transpose it easily in the L^p-case. One has

Theorem A.4. *(Poincaré's Inequality) Let $u \in H^1(\Omega_\ell)$ such that*

$$u = 0 \text{ on } \ell\omega_1 \times \partial\omega_2. \tag{A.10}$$

There exists a constant $C = C(\omega_2)$ such that

$$\int_{\Omega_\ell} u^2 dx \leq C(\omega_2) \int_{\Omega_\ell} |\nabla_{X_2} u|^2 dx \quad \forall u.$$

Proof. If $u \in H^1(\Omega_\ell)$ u satisfies $u(X_1, \cdot) \in H_0^1(\omega_2)$ for a.e. $X_1 \in \ell\omega_1$. Since ω_2 is bounded - in fact it would be enough here to be bounded in one direction - we have by the usual Poincaré inequality

$$\int_{\omega_2} u^2(X_1, \cdot) dX_2 \leq C(\omega_2) \int_{\omega_2} |\nabla_{X_2} u(X_1, \cdot)|^2 dX_2 \quad \text{a.e. } X_1 \in \ell\omega_1.$$

Integrating in X_1 this inequality we obtain

$$\int_{\ell\omega_1} \int_{\omega_2} u^2(X_1, X_2) dX_2 dX_1 \leq C(\omega_2) \int_{\ell\omega_1} \int_{\omega_2} |\nabla_{X_2} u(X_1, X_2)|^2 dX_2 dX_1.$$

This completes the proof of the theorem. $\square$

Remark A.9. Note that as a consequence one has

$$\int_{\Omega_\ell} u^2 dx \leq C(\omega_2) \int_{\Omega_\ell} |\nabla u|^2 dx \quad \forall u.$$

Bibliography

[1] C. Amrouche, V. Girault, *Decomposition of vector spaces and application to the Stokes problem in arbitrary dimension*, Czechoslovak Math. J. 44 (1994), 109-140.

[2] S. Baillet, A. Henrot, T. Takahashi, *Convergence results for a semilinear problem and for a Stokes problem in a periodic geometry*, Asymptotic Anal. 50, (2006), 325-337.

[3] C. Baiocchi, *Su un problema di frontiera libera connesso a questioni di idraulica*, Ann. Mat. Pura Appl. (4) 92 (1972), 107-127.

[4] J. Bourgain, H. Brezis, *On the equation* div $Y = f$ *and application to control of phases*, J. Amer. Math. Soc. 16 (2003), 393-426.

[5] H. Brezis, *Problèmes unilatéraux*, J. Math. Pures Appl. 51 (1972), 1-168.

[6] H. Brezis, Functional Analysis, Sobolev Spaces and Partial Differential Equations. Springer, 2010.

[7] H. Brezis, D. Kinderlehrer and G. Stampacchia, *Sur une nouvelle formulation du problème de l'écoulement à travers une digue*, (French) C. R. Acad. Sci. Paris Sr. A-B 287 (1978), no. 9, 711-714.

[8] H. Brezis, M. Chipot and Y. Xie, *Some remarks on Liouville type theorems*, proceedings of the international conference in nonlinear analysis, Hsinchu, Taiwan (2006), World Scientific Edt, (2008) p. 43-65.

[9] B. Brighi, S. Guesmia, *On elliptic boundary value problems of order 2m in cylindrical domain of large size*, Adv. Math. Sci. Appl. 18(1), (2008), 237-250.

[10] B. Brighi, S. Guesmia, *Asymptotic behavior of solutions of hyperbolic problems on a cylindrical domain*, Disc. Cont. Dyn. Syst. Suppl. (2007), 160-169.

[11] N. Bruyère, *Personal communication*.

[12] M. Chipot, Variational inequalities and flow in porous media. Springer Verlag, New York, 1984.

[13] M. Chipot, ℓ goes to plus infinity. Birkhäuser Advanced Text, 2002.

[14] M. Chipot, *ℓ goes to plus infinity : an update*, J. KSIAM, vol 18, 2, 107–127.

[15] M. Chipot, Elements of Nonlinear Analysis. Birkhäuser Advanced Text, 2000.

[16] M. Chipot, Elliptic Equations: An Introductory Course. Birkhäuser Advanced Text, 2009.

[17] M. Chipot, *On some anisotropic singular perturbation problems,* Asymptotic Analysis, 55 (2007), 125-144.

[18] M. Chipot, *Some Remarks on Anisotropic Singular Perturbation Problems,* proceedings of IUTAM Symposium on the relations of shell, plate, beam and 3D models, Tbilisi, Georgia, April 2007, Springer Edt., (2008), 91-100.

[19] M. Chipot, *On the asymptotic behaviour of some problems of the calculus of variations,* J. Elliptic Parabol. Equ. 1, (2015), 307-323.

[20] M. Chipot, S. Guesmia, *On the asymptotic behavior of elliptic, anisotropic singular perturbations problems,* Comm. Pure Applied Anal., 8, 1, (2009), 179-194.

[21] M. Chipot, S. Guesmia, *Corrector for some asymptotic problems,* Proceedings of the Steklov Institue of Mathematics, Vol 270, (2010), 263-277.

[22] M. Chipot, S. Guesmia, *On some anisotropic, nonlocal, parabolic singular perturbations problems,* Applicable Analysis, Vol. 90, No. 12, (2011), 1775-1789.

[23] M. Chipot, S. Guesmia, *On a class of integro-differential problems,* Comm. Pure Applied Analysis, 9, 5, (2010), 1249-1262.

[24] M. Chipot, S. Guesmia and A. Sengouga, *Singular Perturbations of Some Nonlinear Problems,* Journal of Mathematical Sciences, Vol 176, 6, (2011), 828-843.

[25] M. Chipot, S. Mardare, *Asymptotic behaviour of the Stokes problem in cylinders becoming unbounded in one direction,* J. Math. Pures Appl. 90, (2008), 133-159.

[26] M. Chipot, S. Mardare, *On correctors for the Stokes problem in cylinders,* proceedings of the conference on nonlinear phenomena with energy dissipation, Chiba, November 2007, P. Colli and all Edts, Gakuto International Series, Mathematical Sciences and Applications, Vol 29, Gakkotosho, (2008), 37-52.

[27] M. Chipot, S. Mardare, *The Neumann problem in cylinders becoming unbounded in one direction,* J. Math.Pures Appl.104, (2015), 921-941.

[28] M. Chipot, A. Mojsic and P. Roy, *On some variational problems set on domains tending to infinity,* Disc. Cont. Dyn. Syst., Ser. A, 36 (7), (2016), 3603-3621.

[29] M. Chipot, A. Rougirel, *On the asymptotic behaviour of the solution of elliptic problems in cylindrical domains becoming unbounded,* Communications in Contemporary Mathematics, Vol 4, 1, (2002), 15-44.

[30] M. Chipot, A. Rougirel, *On the asymptotic behaviour of the solution of parabolic problems in cylindrical domains of large size in some directions,* Discrete an Continuous Dynamical Systems, Series B, 1, (2001), 319-338.

[31] M. Chipot, A. Rougirel, *Remarks on the asymptotic behaviour of the solution to parabolic problems in domains becoming unbounded,* Nonlinear Analysis, Vol 47, (2001), 3-11.

[32] M. Chipot, A. Rougirel, *Local stability under changes of boundary conditions at a far away location,* Proceedings of the Fourth European Conference on Elliptic and Parabolic Problems, Rolduc-Gaeta 2001, World Scientific, (2002), 52-65.

[33] M. Chipot, A. Rougirel, *On the asymptotic behaviour of the eigenmodes for elliptic problems in domains becoming unbounded,* Trans. Amer. Math. Soc. 360 (2008), 3579-3602.

[34] M. Chipot, P. Roy and I. Shafrir, *Asymptotics of eigenstates of elliptic problems with mixed boundary data on domains tending to infinity,* Asymptotic Analysis 85, (2013), 199-227.

[35] M. Chipot, Y. Xie, *On the asymptotic behaviour of the p-Laplace equation in cylinders becoming unbounded,* Proceedings of the International Conference: Nonlinear PDE's and their Applications, N. Kenmochi, M. Ôtani, S. Zheng Edts, Gakkotosho, (2004), 16-27.

[36] M. Chipot, Y. Xie, *On the asymptotic behaviour of elliptic problems with periodic data,* C. R. Acad. Sci. Paris, Ser. I 339, (2004), 477-482.

[37] M. Chipot, Y. Xie, *Elliptic problems with periodic data: an asymptotic analysis,* Journ. Math. Pures et Appl., 85 (2006), 345-370.

[38] M. Chipot, Y. Xie, *Asymptotic behaviour of nonlinear parabolic problems with periodic data,* Progress in PDE and their applications, Vol 63, Nonlinear elliptic and parabolic problems: A special tribute to the work of H. Brezis, (2005), 147-156, Birkhäuser.

[39] M. Chipot, Y. Xie, *Some issues on the p-Laplace equation in cylindrical domains,* Proceedings of the Steklov Institue of Mathematics, 261, (2008), 287-294.

[40] M. Chipot, K. Yeressian, *Exponential rates of convergence by an iteration technique,* C. R. Acad. Sci. Paris, Ser. I 346, (2008), 21-26.

[41] M. Chipot, K. Yeressian, *On the asymptotic behavior of variational inequalities set in cylinders,* Disc. Cont. Dyn. Syst., Series A, Vol 33, 11 & 12, (2013), 4875-4890.

[42] M. Chipot, K. Yeressian, *On Some Variational Inequalities in Unbounded Domains,* Boll. del UMI, (9), V, (2012), 243-262.

[43] M. Chipot, K. Yeressian, *Asymptotic behaviour of the solution to variational inequalities with joint constraints on its value and its gradient,* Contemporary Mathematics 594, (2013), 137-154.

[44] P. G. Ciarlet, Mathematical Elasticity, Vol 1: Three Dimentional Elasticity. North- Holland, 1988.

[45] P. G. Ciarlet, Mathematical Elasticity, Vol 2: Theory of Plates. North-Holland, 1997.

[46] P. G. Ciarlet, P. Destuynder, *A justification of the two-dimensional plate model*, J. Mécanique, 18, (1979), 315-344

[47] D. Cioranescu and P. Donato: An introduction to homogenization. Oxford Lecture Series in Mathematics and its Applications # 17, 1999.

[48] B. Dacorogna, Direct Methods in the Calculus of Variations. Applied Mathematical Sciences Series # 78, Second Edition, Springer, 2008.

[49] Dautray R., Lions J.-L.: Mathematical Analysis and Numerical Methods for Science and Technology. Springer-Verlag, Berlin, 1990.

[50] L.C. Evans, Partial differential equations. Graduate Studies in Mathematics 19, AMS, 1998.

[51] P. G. Galdi, An Introduction to the Mathematical Theory of the Navier-Stokes Equations, Steady-state Problems. Springer, New York, 2011.

[52] D. Gilbarg, N.S. Trudinger, Elliptic Partial Differential Equations of Second Order. Springer-Verlag, Berlin, 1983.

[53] S. Guesmia, *Some results on the asymptotic behavior for hyperbolic problems in cylindrical domains becoming unbounded,* J. Math. Anal. Appl. 341(2), (2008), 1190-1212.

[54] S. Guesmia, *Asymptotic behavior of elliptic boundary-value problems with some small coefficients,* Electron. J. Diff. Equ. (2008) (59), (2008), 1-13.

[55] S. Guesmia, *Some convergence results for quasilinear parabolic boundary value problems in cylindrical domain of large size,* Nonlinear Anal. 70(9), (2009), 3320-3331.

[56] S. Guesmia, *Résultats sur l'analyse asymptotique des équations aux dérivées partielles.* Habilitation Université de Haute Alsace, France (2013).

[57] S. Guesmia, *Large time and space size behaviour of the heat equation in non-cylindrical domains,* Arch. Math., (2013), 101 (3), 293-299.

[58] S. Guesmia, A. Sengouga, *Some singular perturbations results for semilinear hyperbolic problems,* Discrete. Cont. Dyn. Syst. Ser. S, 5(3), (2012), 567-580.

[59] S. Guesmia, A. Sengouga, *Anisotropic singular perturbations for hyperbolic problems,* Appl. Math. Comput. 217 (22), (2011), 8983-8996.

[60] D. Kinderlehrer and G. Stampacchia, An introduction to Variational Inequalities and their applications. Academic Press, San Francisco-London, 1980.

[61] O. A. Ladyzhenskaya, The Mathematical Theory of Viscous Incompressible Flow. Gordon and Breach, 1969.

[62] J.-L. Lions, Quelques méthodes de résolution des problèmes aux limites non linéaires. Dunod-Gauthier-Villars, Paris, 1969.

[63] J. L. Lions, Perturbations Singulières dans les Problèmes aux Limites et en Contrôle Optimal. Lecture Notes in Mathematics # 323, Springer-Verlag, 1973.

[64] J. L. Lions, G. Stampacchia, *Variational inequalities,* Comm. Pure Appl. Math. 20, (1967), 493–519.

[65] D. Mugnai, *A limit problem for degenerate quasilinear variational inequalities in cylinders,* Recent trends in nonlinear partial differential equations. I. Evolution problems, Contemp. Math., 594, Amer. Math. Soc., Providence, RI, (2013), 281-293.

[66] O. A. Oleinik and G. A. Yosifian, *Boundary value problems for second order elliptic equations in unbounded domains and Saint-Venants principle,* Ann. Scuola Norm. Sup. Pisa Cl. Sci. (4), 4, no. 2, (1977), 269-290.

[67] P. A. Raviart, J. M. Thomas, Introduction à l'analyse numérique des équations aux dérivées partielles. Masson, 1983.

[68] J. F. Rodriges: Obstacle problems in mathematical physics. Math. Studies 134, North Holland, Amsterdam, 1987.

[69] P. Roy, *Some results in asymptotic analysis and nonlocal problems.* Thesis University of Zürich, (2013).

[70] E. Sandier, I Shafrir, *On the symmetry of minimizing harmonic maps in n dimensions,* Differential and Integral Equations, 6, (1996), 113-122.

[71] L. Schwartz: Théorie des distributions. Hermann, Paris, 1966.

[72] G. Stampacchia: Equations elliptiques du second ordre à coefficients discontinus. Presse de l'Université de Montréal, 1966.

[73] R. Temam: Navier-Stokes equations: theory and numerical analysis (3rd edition). North-Holland, Amsterdam, 1984.

[74] G. M. Troianiello: Elliptic differential equations and obstacle problems. Plenum, New York, 1987.

[75] Y. Xie, *Some convergence results for elliptic problems with periodic data,* Recent Advances on Elliptic and Parabolic Issues, Proceedings of the 2004 Swiss-Japanese Seminar, World Scientific, (2006), 265-282.

[76] Y. Xie, *On asymptotic problems in Cylinders and other mathematical issues.* Thesis University of Zürich, (2006).

[77] K. Yeressian, *Asymptotic Behavior of Elliptic Nonlocal Equations Set in Cylinders,* Asymptot. Anal. 89 no. 1-2, (2014), 21-35.

[78] K. Yeressian, *Spatial asymptotic behaviour of elliptic equations and variational inequalities.* Thesis University of Zürich, (2010).

[79] K. Yosida: Functional Analysis. Springer Verlag, Berlin, 1978.

Index

Printed in the United States
By Bookmasters